Sport Physiology for Coaches

Brian J. Sharkey, PhD
USDA Forest Service

Steven E. Gaskill, PhD
University of Montana

Human Kinetics

Library of Congress Cataloging-in-Publication Data

Sharkey, Brian J.

Sport physiology for coaches / Brian J. Sharkey, Steven E. Gaskill.

p. cm.

Includes bibliographical references and index.

ISBN 0-7360-5172-4 (soft cover)

1. Sports--Physiological aspects. 2. Athletes--Training of. 3. Muscle strength. 4. Coaches (Athletics) I. Gaskill, Steven E., 1952- .

II. Title.

RC1235.S53 2006

613.7'11--dc22

2005026507

ISBN-10: 0-7360-5172-4
ISBN-13: 978-0-7360-5172-9

The Web addresses cited in this text were current as of November 28, 2005, unless otherwise noted.

Acquisitions Editors: Scott Parker and Amy Tocco; **Developmental Editor:** Renee Thomas Pyrtel **Assistant Editor:** Kevin Matz; **Copyeditor:** Joyce Sexton; **Proofreader:** Anne Rogers; **Indexer:** Sharon Duffy; **Permission Manager:** Dalene Reeder; **Graphic Designer:** Robert Reuther; **Graphic Artist:** Kathleen Boudreau-Fuoss; **Photo Manager:** Sarah Ritz; **Cover Designer:** Keith Blomberg; **Photographer (cover):** Dan Wendt; **Photographs (interior):** © Human Kinetics unless otherwise noted; **Art Manager:** Kelly Hendren; **Illustrator:** Mic Greenburg; **Printer:** Sheridan Books

Copies of this book are available at special discounts for bulk purchase for sales promotions, premiums, fund-raising, or educational use. Special editions or book excerpts can also be created to specifications. For details, contact the Special Sales Manager at Human Kinetics.

Printed in the United States of America 10 9 8 7 6 5 4 3 2 1

Human Kinetics
Web site: www.HumanKinetics.com

United States: Human Kinetics, P.O. Box 5076, Champaign, IL 61825-5076
800-747-4457
e-mail: humank@hkusa.com

Canada: Human Kinetics
475 Devonshire Road Unit 100, Windsor, ON N8Y 2L5
800-465-7301 (in Canada only)
e-mail: orders@hkcanada.com

Europe: Human Kinetics
107 Bradford Road, Stanningley, Leeds LS28 6AT, United Kingdom
+44 (0) 113 255 5665
e-mail: hk@hkeurope.com

Australia: Human Kinetics
57A Price Avenue, Lower Mitcham, South Australia 5062
08 8277 1555
e-mail: liaw@hkaustralia.com

New Zealand: Human Kinetics
Division of Sports Distributors NZ Ltd., P.O. Box 300 226 Albany, North Shore City, Auckland
0064 9 448 1207
e-mail: info@humankinetics.co.nz

To Nancy, Brian, Megan, and Heidi,
who've come to understand:
It's not the trophy but the race,
Not the quarry but the chase.

And to Barbara and Kathy,
the perfect activity and fitness companions
who have kept us moving and healthy.

CONTENTS

ASEP SILVER LEVEL SERIES PREFACE

The American Sport Education Program (ASEP) Silver Level curriculum is a series of practical texts that provide coaches and students with an applied approach to sport performance. The curriculum is designed for coaches and for college undergraduates pursuing professions as coaches, physical education teachers, and sport fitness practitioners.

For instructors of undergraduate courses, the ASEP Silver Level curriculum provides an excellent alternative to other formal texts. In most undergraduate programs today, students complete basic courses in exercise physiology, mechanics, motor learning, and sport psychology—courses that are focused on research and theory. Many undergraduate students are looking for ways to directly apply what they learn in the classroom to what they can teach or coach on the court or playing field. ASEP's Silver Level series addresses this need by making the fundamentals of sport science easy to understand and apply to enhance sport performance. The Silver Level series is specifically designed to introduce these sport science topics to students in an applied manner. Students will find the information and examples user friendly and easy to apply in the sports setting.

The ASEP Silver Level sport science curriculum includes the following:

Sport Mechanics for Coaches—an explanation of the mechanical concepts underlying performance techniques; designed to enable coaches and students to observe, analyze, develop, and correct the mechanics of sport technique for better athletic performance.

Sport Physiology for Coaches—an applied approach to exercise physiology; designed to enable coaches and students to assess, initiate, enhance, and refine human performance in sport participation and to improve sport performance.

Sport Psychology for Coaches—a practical discussion of motivation, communication, stress management, mental imagery, and other cutting-edge topics; this text is designed to enhance the coach–athlete relationship and to stimulate improved sport performance.

Teaching Sport Skills for Coaches—a practical approach for learning to teach sport skills, guided by a practical understanding of the stages of learning and performance, individual differences and their impact on skill acquisition, and the critical elements required to create a learning environment that enhances optimal sport skill development and performance.

A variety of educational elements make these texts student- and instructor-friendly:

- Learning objectives introduce each chapter.

- Sidebars illustrate sport-specific applications of key concepts and principles.
- Chapter summaries review the key points covered in the chapter and are linked to the chapter objectives by content and sequence.
- Key terms at the end of most chapters list the terms introduced in that chapter and remind coaches and students, "These are words you should know." The first occurrence of the word in the chapter is boldfaced, and the words also appear in the glossary.
- Chapter review questions at the end of each chapter allow coaches and students to check their comprehension of the chapter's contents. Answers to questions appear in the back of the book.
- Real-world application scenarios called practical activities follow the review questions. These scenarios provide problem situations for readers to solve. The solutions require readers to describe how the concepts discussed in the chapter can be applied in real-world scenarios. Sample solutions appear in the back of the book.
- A glossary defines all of the key terms covered in the book.
- A bibliography section at the end of the book serves as a resource for additional reading and research.
- A general index lists subjects covered in the book.

These texts are also the basis for a series of Silver Level online courses being developed by Human Kinetics. These courses will be offered through ASEP's Online Education Center for coaches and students who wish to increase their knowledge through practical and applied study of the sport sciences.

PREFACE

The authors met in 1980 when Steve was coach and Brian was sport physiologist for the US Ski Team. Brian became coordinator of the team's Nordic Sportsmedicine Council, and Steve was named liaison between the sport scientists on the council and the coaches and athletes of the team. Eventually their lives took separate paths, but they remained friends and kept in touch. In 1998, after completing his doctorate in exercise physiology, Steve applied for a position at the University of Montana, where Brian was retiring after 30 years of service. Since then they have renewed their professional and personal association, conducting research and development activities in the Human Performance Laboratory and in the field. And while Steve has to wait for Brian to catch up, they continue to hike, ski, snowshoe, and bicycle the mountains of western Montana.

This book takes our experiences as athletes, coaches, and sport scientists and presents the information that coaches need to design and carry out effective training programs. The book combines the experience of successful coaches and athletes with the latest research in sport physiology to provide a practical guide for current and prospective coaches. Coaches need to understand the basics of sport physiology in order to plan effective training. *Sport Physiology for Coaches* provides the background along with concrete steps for the design and evaluation of training programs.

Sport Physiology for Coaches consists of four parts: Each part builds on information given previously, and all the parts lead you through the process of designing sound training programs for your athletes. Part I is an introduction that includes important principles of training and discusses the individual response to training. Part II introduces muscular fitness, what it is, how to evaluate it, and the specific effects of training. Part III deals with energy fitness, anaerobic and aerobic, how each component can be evaluated, and the effects of training. Part IV helps you design effective training programs while avoiding overtraining, injury, and illness.

Unique features of the book include field assessment techniques to evaluate muscular and energy fitness of individual athletes; a simple system to control the intensity of training; well-defined training zones designed to focus training for effective results; a proven system to ensure progression of training with adequate time for recovery (periodization of training); and a way for athletes to avoid overtraining. For example, we describe several training zones, including one we call the no-training zone. This zone includes training that is hard enough to cause fatigue, but not hard enough to improve performance. Athletes are encouraged to avoid excess training in this zone, concentrating instead on the performance zone to improve performance and the easy (EZ) zone to ensure endurance and recovery.

You may want to begin by reading the table of contents, then scanning the chapters to get a sense of the organization and the material. As you begin reading the chapters, use a highlighter to identify important points. Keep a note pad handy to jot down ideas and questions as they come to mind. You will be challenged with new ideas and approaches to training muscles and energy systems. We have tried to present the information as clearly as possible, but some new terms will appear. New terms are explained when they first appear, but if you forget a term, consult the glossary. In time the language of sport physiology will become part of your coaching vocabulary.

So get started and good luck! We hope this book contributes to your success and, more importantly, to your understanding and enjoyment of sport and coaching.

PART I

The Athlete, the Coach, and Sport Physiology

Sport consists of preparation and performance—about 99 percent preparation and 1 percent performance. As a coach you need to make the most effective use of the preparation time so that your athletes can achieve their highest possible levels of performance. This part of the book introduces you to sport physiology, important principles of training, and essential information concerning athletes. It will help you understand how knowledge of sport physiology and training can contribute to your athletes' success and to your understanding and enjoyment of coaching.

1 Introduction to Sport Physiology

This chapter will help you

- recognize how knowledge of sport physiology will contribute to the health, safety, and success of your athletes, as well as to your understanding and enjoyment of sport and coaching;
- appreciate how laboratory and field research contribute to performance, safety, and health; and
- determine how to evaluate the sources and quality of information concerning sport physiology and training.

Don loved basketball, and while he wasn't able to make his college team he had always wanted to be a coach. After college he earned a master's degree in sport physiology. Then, to everyone's surprise, he applied for and got the job of head basketball coach for an inner city high school located in a tough neighborhood. He was young and lacked coaching experience. But somehow he got the job, perhaps because better-qualified prospects saw the program as a perennial cellar dweller. Don saw it as an opportunity. His enthusiasm began to wane when preseason practices revealed an eager but listless group of athletes. His goal of teaching tough defense and an up-tempo offense seemed unattainable unless he could find an answer to the team's energy problem.

As Don sought to find out why these otherwise healthy young men lacked the endurance to complete a vigorous practice, he discovered that most did not eat a good breakfast. And lunch often consisted of an "RC and a moon pie" (a cola and devil's food cake with cream filling—nothing but sugar). He arranged to provide breakfast and required the athletes to eat the school lunch. The results were astounding! The energized team fought their way to the runner-up spot in the city championship, and Don was named coach of the year.

Lemon's article in the *Journal of Sports Nutrition* (1995) reviewed the published studies concerning the protein needs of athletes and summarized the findings. As shown in table 1.1, while a nonathlete needs about 0.8 grams of protein per kilogram of body weight daily (for example, 0.8 grams × 70 kilograms [154 pounds] = 56 grams of protein daily), an endurance athlete requires 1.2 to 1.4 grams per kilogram (84-98 grams/day), and a strength athlete may need 1.4 to 1.8 grams of protein per kilogram of body weight per day (98-126 grams/day for a 70-kilogram body weight).

There are 25 grams of protein in a serving of beef about the size of a deck of cards. Physical training requires sufficient protein to build the muscle, enzymes, and other structures influenced by training, and to repair muscles damaged in vigorous effort.

But basketball is a high-energy sport, with practices that can burn 500 to 600 kilocalories per hour, so the diet needs to include ample carbohydrate and sufficient fat to fuel practices and games. If athletes eat too little carbohydrate, the body turns to protein to provide energy. Therefore, nutritious complex carbohydrates including corn, rice, beans, potatoes, and whole-grain breads and pasta should constitute 60 percent of the athlete's diet to provide energy and to conserve protein. Coaching strategies and motivation are important, but successful coaches don't ignore the basics. As Napoleon Bonaparte said, "An army marches on its stomach."

What Is Sport Physiology?

Physiology is the study of the body and how it functions. Physiologists study the structure and function of the tissues, organs, and systems

TABLE 1.1

Protein Needs per Day for a 70 kg Person

	Grams per kilogram of body weight per day	Total grams per day
Nonathlete	0.8	56
Endurance athlete	1.2-1.4	84-89
Strength athlete	1.4-1.8	98-126

© Empics

Athletes should consume adequate protein, fat, and carbohydrate to fuel their bodies.

of the body. **Sport physiology** deals with the immediate and long-term effects of exercise on the body's muscles and systems. Immediate effects include increases in heart rate, respiration, and temperature during exercise, things you have experienced. Long-term effects refer to how systematic training leads to adaptations in muscles, energy pathways, cardiovascular and respiratory systems, and a host of other functions. This book will help you design exercise programs that lead to the beneficial adaptations that form the basis of improved performance.

Sport physiology can be studied at many levels, ranging from the subcellular and molecular levels to changes in structure and function. Many studies look at measures of strength, endurance, and performance. Some sport physiologists study the effects of exercise and training on systems, such as the cardiovascular, muscular, or respiratory system. Cellular research probes the effects on structures within muscle fibers, and molecular biologists investigate the relationship of molecules to training and performance. Some of the most important recent findings in sport physiology have involved exploration of the genetic factors underlying performance. You know that you inherit your physique and other characteristics from your parents' DNA (deoxyribonucleic acid), but did you know that some aspects of endurance are passed on only by your mother? These genetic studies underscore the importance of picking your parents wisely.

- Training. We will show how a particular exercise serves as a **training stimulus** that signals cells to undergo specific changes or **training effects.** The **training program** is the planned accumulation of training effects, leading to improved performances. The **annual program** provides a blueprint for off-season, preseason, and competitive-season training. And the annual program is one chapter in the long-term or **career plan,** a systematic evolution from youth sport to the elite levels of competition.

- Athletes. When the coach knows how to design and carry out training programs, athletes benefit. They enjoy practices more when the coach employs a variety of training methods. Sport physiology is important to team as well as individual-sport athletes. For example, basketball players need strength and power to rebound and endurance to run the court. Of course they also need to polish their shooting and ball-handling skills, and to learn offensive and defensive variations. Muscle and energy fitness should peak when it counts, in big games or tournaments. When fitness, skills, and tactics are at their best, good things happen.

- Coaches. Coaches benefit when they enhance their knowledge of sport physiology, providing the foundation they need to manage their athletes' physical development. Athletes gain confidence in the coach who conducts an effective training program and is able to explain the reasons for specific types of training. Sport psychologists have found that athletes are more willing to put forth effort when they understand why the work is being done and how it will benefit them. Athletes want to know why they

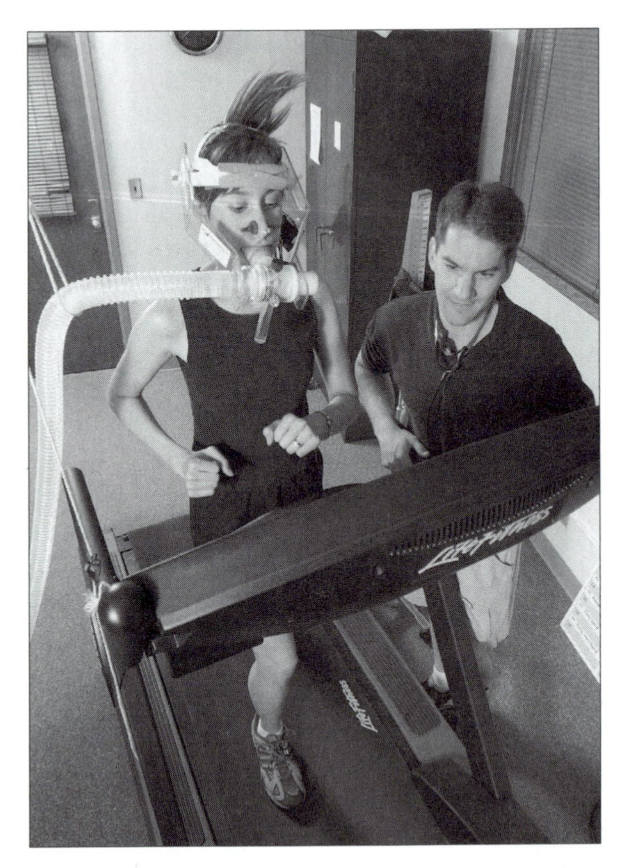

If your facility has the necessary equipment, consider including fitness testing in your training program.

are being asked to do specific workouts such as lifting weights or running intervals. Providing good answers improves compliance with training goals and helps athletes gain a better understanding of their bodies. As athletes gain confidence in the coach, his or her job becomes easier. When athletes are successful, the team is successful, and when that happens the work is particularly satisfying for the coach.

Sources of Sport Physiology Information

Information comes in many forms, ranging from the insights of experienced coaches to the results of controlled research studies. Most coaches are voracious consumers of information concerning their sport. They read books and magazines, attend clinics, and surf the Web for information. We want you to know where to go to get good information, and also to develop a healthy skepticism as you consider the various sources of information.

The introductory text in the American Sport Education Program's coaches education program, *Successful Coaching,* covers the basics of sport physiology, including energy and muscular fitness. In this book we will build on that knowledge and provide the background you need to develop successful training programs. The content of this book is based on laboratory and field studies of athletes conducted throughout the world. The information is blended with the observations and insights of successful coaches, and the material is presented in an easy-to-understand manner.

Lab and field research studies provide a way to determine the true value of a type of training, a diet, or even an advertised supplement. Advertisements and TV infomercials often promote a product with testimonials from an athlete or an attractive model. In all likelihood, neither the athlete nor the model ever used the product to achieve his or her success. When a company makes a claim for a product, we want to see the results of studies that support the claim. Reputable companies are eager to fund research and share their findings. The companies unable to provide independent verification—those that rely on testimonials—should be ignored.

Myths and Traditions of Coaching

Many of us begin our careers coaching the way we were coached, passing on the myths and traditions of the sport. For some that leads to success; for others it means repeating someone else's mistakes. For decades, wrestling coaches allowed the dangerous practice of dehydration weight loss. They liked the flexibility of having a wrestler lose weight to wrestle in a lower weight class. Exercise, voluntary dehydration, rubber suits, and even diuretics and saunas were used to sweat off pounds to make weight. A few state high school associations adopted minimal wrestling weight regulations to avoid this risky practice, but most state and collegiate organizations continued to condone the dangerous tradition. But things changed dramatically in the fall of 1997 when, in the span of two months, three collegiate wrestlers died of heatstroke brought

Minimal Wrestling Weight

The National Collegiate Athletic Association (NCAA) wrestling and sports medicine committees met and designed a minimal wrestling weight program based on early research (Tcheng and Tipton 1973) and the American College of Sports Medicine's position stand on weight loss in wrestling. The minimum weight is calculated using body composition techniques to determine fat-free weight (FFW) and percent body fat. Wrestlers may not compete below a weight equal to FFW plus 5 percent body fat, measured in the hydrated state. The program eliminates the need for dehydration weight loss as well as the unhealthy "starve and stuff" practices once common to the sport. Studies have documented the health and safety benefits of this scientific approach to determination of the competitive weight. The National Federation of State High School Associations is moving to implement a minimal wrestling weight program in all the states.

Myths and traditions can be changed. College wrestlers must now meet minimum weight standards rather than starving and dehydrating to "make weight."

on by dehydration weight loss. Only then did coaches admit the risks and accept the need for changes in the system.

Sport Physiology Research

Carefully planned and controlled studies help us separate fact from fiction. Sport physiologists conduct many types of studies in order to understand the immediate and long-term effects of exercise and training (table 1.2). In **status or comparison studies** we are able to determine differences between groups, such as athletes and nonathletes.

These comparison studies show differences, but they don't prove why athletes are stronger or faster. Chances are the athletes' heredity and environment predisposed them to become successful in a particular sport.

Another type of study, the **correlation study,** looks at the relationship of one factor, such as a training method, and another factor, such as performance. Correlation studies show the relationship of training volume (e.g., kilometers/week) to marathon running performance, but they do not prove the need for high-volume training. Why? A relationship does not prove cause and effect. Marathon runners may be successful in spite of the high-volume training, or because they were able to survive the ordeal while others became injured.

TABLE 1.2

Types of Research Studies Used in Physiology

Type of research	How done?	Prove cause and effect?
Status	Compares groups	No
Correlation	Determines relationships	No
Experimental	Manipulates variables	Yes

The type of study that does allow cause-and-effect conclusions is called an **experimental study.** Athletes are pretested and randomly assigned to experimental or control groups, and their progress is monitored until the end of the experiment, when they are posttested. Using an experimental study, a sport physiologist is able to determine the physiological effects of a particular training method with more confidence. For example, if an experimental training technique improves performance more than the traditional approach, the result is called a "significant difference" when the statistical analysis indicates there is less than a 5 percent probability that the outcome is due to chance. "Highly significant" refers to less than a 1 percent probability, or less than 1 chance in 100, that the results were due to chance. When the preponderance of studies show significant differences and agree on the effect of a type of training, the technique moves from a theory to a proven fact.

We don't expect you to conduct research or even read research journals in order to answer questions in sport physiology. That is our job. Using computerized literature searches, we scoured the world's literature to ensure the best and most up-to-date information for this book. Then we distilled the essence of the findings in order to present the information in simple, easily followed steps. Whenever possible we utilize results gathered on athletes, in conditions that simulate actual training and competition. When that isn't possible, we rely on laboratory studies to fill in the gaps. Sometimes the only studies available were conducted on subjects who were nonathletes, or on laboratory animals. In these cases, we let you know that the evidence isn't based on athletes. Finally, when we pass on the suggestion of a successful coach, one that makes sense physiologically, we label the concept as a plausible theory, not a fact proven in experimental studies on competitive athletes.

Other sources of information include sport and training books from reputable publishers, such as our publisher, Human Kinetics. Many sports have one or more magazines that cover aspects of technique and training. Read the articles, but please don't believe everything you read in the articles and especially in the advertisements. Among the many Internet sites that promise information on sport and training, some market training programs, equipment, diets, and even supplements. Be wary of equipment, diets, or supplements that promise amazing results. If it sounds too good to be true, it probably is. Throughout this book we will give examples of good sources of information. We'll even direct you to some books and journals. And you can search topics to see what is available on the Web. Now let's begin our study of sport physiology by exploring the principles of training.

Experimental Studies

Of the three types of research, only experimental studies allow a cause-and-effect conclusion. Experimental studies involve the manipulation of a variable, such as the number of sets of resistance training, to determine if the variable improves an important measure, such as strength. A recent study suggested that a single set of weight training was the optimal way to gain strength in a muscle. Would you accept that conclusion for athletes? A statistical review of 140 experimental studies indicated that strength increases with the number of sets. Three sets were optimal for previously untrained individuals, while four seemed best for trained subjects. World-class weightlifters do even more sets in their quest for extraordinary strength.

How many sets are optimal? Research has helped find the answer.

SUMMARY

The chapter shows how knowledge of sport physiology can contribute to the health and performance of your athletes as well as your enjoyment and success as a coach. It describes how lab and field studies can be used to help answer important questions concerning training. And it provides advice to help you select good sources of information concerning sport physiology and training.

KEY TERMS

annual program (p. 5)
career plan (p. 5)
correlation study (p. 7)
experimental study (p. 8)
sport physiology (p. 5)
status (or comparison) study (p. 7)
training effects (p. 5)
training program (p. 5)
training stimulus (p. 5)

REVIEW QUESTIONS

Fill in the blank with the appropriate word or words to complete the sentence.

1. _________ is the study of the body and how it functions.
2. _________ studies provide a way to determine the true value of a type of training, diet, or even an advertised supplement.
3. In _________ studies we are able to determine differences between groups such as athletes and nonathletes.
4. An _________ study allows for a cause-and-effect conclusion, as a sport physiologist is able to determine the physiological effects of a particular training method with more confidence.
5. _________ is the percentage of carbohydrates that an athlete's diet should include.

Match the terms in each group on the left with the appropriate answer on the right.

6. Training stimulus ____
7. Training effects _____
8. Training program ____
9. Annual program _____
10. Career plan ____

a. Planned accumulation of training effects, leading to improved performance
b. Systematic evolution from youth sport to the elite levels of competition
c. A particular exercise
d. Blueprint for off-season, preseason, and competitive-season training
e. When cells undergo a specific change

What is the amount of protein each group of athletes needs per day?

11. Nonathlete _____
12. Endurance athlete _____
13. Strength athlete _____

a. 1.4 to 1.8 grams per kilogram body weight
b. 0.8 grams per kilogram body weight
c. 1.2 to 1.4 grams per kilogram body weight

Determine if the following statements are true or false and circle the correct answer.

14. T/F A relationship does not prove cause and effect.
15. T/F Theories do not require experimental verification.

Write a short answer to each of the following questions.

16. How can improved knowledge of sport physiology contribute to your success in coaching?
17. How do sport physiologists verify theories of training?
18. What are some trustworthy sources of information for coaches?

PRACTICAL ACTIVITIES

1. A friend has heard that football linemen need to eat large quantities of meat and eggs to gain size and strength. How would you determine the protein needs for an athlete in this type of sport who weighs 220 pounds (100 kilograms)?
2. A famous athlete endorses a dietary supplement that is reported to improve strength and performance in sport. If you had the opportunity, what questions would you ask concerning the product?
3. An assistant coach has heard that very intense training is correlated to the performances of world-class athletes. He wants to use the training approach on your high school athletes. How would you respond?

2 Principles of Training

This chapter will help you

- appreciate principles of training related to the athlete,
- understand the principles related to the training process,
- identify the principles related to the long-term development of the athlete, and
- be aware of some of the common fallacies of training.

The Martin sisters had been competitive swimmers since they were 6 and 8 years of age. Both had been successful in age-group swimming and now were on the high school team. Although the girls performed all four strokes, Amy, a senior, was a backstroke specialist, while Jen, a sophomore, specialized in butterfly and free. But things were not going as well this year, the first for their new coach. Fresh out of college and eager to make an impression, he set out to double the training volume expected of the team. During the first few weeks the extra hours of training presented a challenge that the girls were determined to meet. But as time went by, the daily grind began to wear on the athletes and to affect other aspects of their lives. Eventually their parents began to notice the changes.

At first the changes were subtle, like sleeping late or being moody—not unusual behavior for adolescents. Then the girls became listless and caught colds that further affected their ability to perform in the pool and the classroom. They started to complain to their folks and actually discussed dropping out of swimming. Fortunately, just as their spirits and energy approached new lows, the coach announced a drastic reduction in training volume known as a taper. During the taper, practices were shortened, intervals were cut back, and athletes were not pushed to exhaustion, allowing time for recovery. Days later at the conference championship, both sisters recorded personal bests.

Competitive swimmers, like any serious athlete, can benefit from pre-event tapers.

Studies of the **taper** process have documented the value of relative rest prior to important competitions (Kenitzer 1998). In a study of female collegiate swimmers, the author concluded that a taper of approximately two weeks represented the limit of recovery and compensation time before detraining became evident. He used lactate measurements during a series of daily sprints to determine when detraining took place. When the taper is optimal, competitive performances improve about 3 percent. In general, the taper should be longer when training volume is high and the event is long. The taper can be brief when the volume of training is low and the event short, such as a sprint. The taper can range from two weeks to just a few days. Coaches often forgo tapering in the early season, preferring to "train through" early meets. But later, when the important events are scheduled, it is wise to engage in an appropriate taper. Incidentally, the taper should be used in team as well as individual sports. It is part of a carefully planned program designed to prepare athletes for peak performances.

The Martins' new coach was just beginning to learn about sport physiology. Fortunately he remembered how his college coach had used the taper before the conference championships. This actual example (names have been changed) illustrates how knowledge of sport physiology may assist a coach in helping athletes be the best they can be. We could use the example to discuss the merits and pitfalls of high-volume training, especially for young athletes. We could use it to document the effects and symptoms of overtraining, or to demonstrate the need to "periodize" training to allow for recovery. We'll discuss these topics in later chapters.

Sport scientist Dr. Ned Frederick has described training as "a gentle pastime by which we coax a slow continuous stream of adaptations out of the body" (Frederick 1973). This description is worth remembering. Frederick goes on to say that "training is a slow, subtle process that cannot be rushed." Done properly, training leads to impressive changes in muscles, systems, and structures—changes that lead to improved performance. Overtraining does not accelerate progress, but inhibits it.

Training is a systematic process. To train athletes properly you must observe certain principles. If you've been through the American Sport Education Program's Coaching Principles course and have read *Successful Coaching,* you have read about most of these principles. We'll review them and provide additional information to help you appreciate their importance. We've organized the principles into three groups that pertain to athletes, training, and longevity.

The Athlete

While all athletes have potential, each responds individually to training, and only when his or her body is ready to respond. We'll start with the principle of readiness, then discuss potential, and then finish this section with a discussion of individual response.

Readiness

The value of training depends on the physical and psychological **readiness** of the individual. Because readiness comes with maturation, physically immature individuals lack the physiological capability to respond completely to muscular or energy fitness training. For example, while prepubertal boys may gain some strength in a resistance training program, they will gain far more after puberty, when the level of the growth-inducing hormone testosterone is considerably elevated. Readiness also implies the need for adequate nutrition and rest if athletes are to benefit from training. Psychological readiness refers to a level of maturation required to delay gratification and to make the commitment and sacrifices required to achieve high levels of performance.

Potential

Each of us has an immense capacity for improvement, but few ever come close to reaching that **potential.** Given the right genes, nutrition, training, and a supportive environment, many of us are capable of achieving success in sport. Some have the capacity for national or even world-class performances, yet many never achieve their potential. Studies show that Olympic medals are correlated to a country's average daily caloric intake. Athletes can't excel without the energy to train. Nor can a potential star emerge without good coaching and competition or the will to sacrifice for a goal. Each young athlete has a potential level of performance. For most, the highest potential performances are still to be achieved. Your job is to help these athletes achieve their potential.

Individual Response

Athletes respond differently to the same training. Heredity, maturity, diet, sleep, and other personal and environmental factors influence the **individual response** to training. Coaches who fail to recognize and account for individual responses risk undertraining the most mature and capable athletes while overtraining the less mature or less gifted participants. We'll discuss the many factors influencing an individual's response to training in chapter 3. In later chapters we'll show ways to assess individual responses, as well as ways to adjust training to accommodate individual differences.

Training

Training is a process governed by principles that emerged from the observations of perceptive coaches and that have been verified in laboratory and field research. Here we'll discuss the training principles of **adaptation, overload, progression, periodization, variation, warm-up** and **cool-down,** and **specificity.**

Adaptation

Day-to-day changes are too small to measure; weeks and even months of patient progress are required to achieve measurable adaptations. Rush the process and you risk illness, injury, or both. Typical adaptations include the following:

- Protein synthesis leading to increased contractile protein, energy system enzymes, or both
- Increased muscular strength, power, or endurance
- Improved blood volume, blood distribution, and related cardiovascular and respiratory adaptations
- Tougher bones, ligaments, tendons, and connective tissue
- Improved neuromuscular coordination and skill

The principle of adaptation tells us that training can't be rushed. The best you can do is follow a sensible program, keep good records, make adjustments when necessary, and be satisfied with the results. Try to do everything in one short season and you may do more harm than good.

Overload

The tale of Milo, a warrior in ancient Greece, illustrates the overload principle. According to the legend, young Milo started lifting a calf soon after it was born. He continued every day, and as the calf grew, so did Milo's strength. Eventually, according to the legend, he was able to lift the full-grown animal.

Training must place a demand on the muscle if the desired adaptations are to take place. At the start, training must exceed the typical demand. As you adapt to increased work, it is necessary to increase the size of the load. The rate of improvement is related to three factors, which you can remember with the acronym **FIT:**

Frequency
Intensity
Time (duration)

We use the overload principle in all types of training. We gradually add more weight (intensity) to the barbell and increase the number of sets (duration) to achieve continued increases in strength. Endurance athletes increase training time and intensity to improve race performances. The overload stimulates changes in the

A lot of bull!

Don't Try This at Home!

We called a rancher friend to put Milo's alleged feat in perspective. A black Angus calf weighs about 75 pounds (34 kilograms) at birth and gains at least 2 pounds (0.9 kilograms) a day until it is weaned at 205 days, weighing 500 to 600 pounds (227 to 272 kilograms). The animal continues to gain weight until it goes to market at 13 to 14 months, weighing in at an impressive 1,200 pounds (544 kilograms). While it is true that modern techniques produce larger animals, it is hard to understand how one could lift an animal half that size.

muscles used in the training, stimulating the production of new protein to help meet future exercise demands. Additional changes occur in systems that supply and support the exercise: The nervous system learns to recruit muscle fibers more effectively, and the circulatory system becomes better able to supply oxygen and fuel to the working muscles.

Progression

To achieve adaptations using the overload principle, training must follow the principle of progression. If the training load is increased too quickly, the body cannot adapt and instead begins to break down. Progression must be observed in terms of gradual increases in FIT.

- Frequency—sessions per day, week, month, or year
- Intensity—training load per day, week, month, or year
- Time—duration of training in hours per day, week, month, or year

But please don't get the impression that progression implies inexorable increases, without time for recovery. The body requires periods of rest in which adaptations take place. So remember to make haste slowly! The principle of progression has other implications also. Training should progress from the general to the specific, from the part to the whole, and from quantity to quality. We will say more about these aspects in later chapters.

Periodization

While training must impose an overload, the progression cannot be inexorable. Periodization is the process of dividing training into smaller and larger sections or cycles, ensuring an appropriate training stimulus as well as time for recovery. The process is used for muscular and energy fitness training, and it is applied from the beginning to the end of the training program. In muscular training, periodization calls for periods of strength, power, and endurance. Each period may last from several weeks to months, and each is divided into smaller cycles providing for increasing loads and periods of relative rest and recovery. Periodization is even used to vary the load during the days of each week. Figure 2.1a shows a hypothetical increase in training hours per year for a developing athlete. The dips in the progression indicate years in which training volume was decreased to decrease stress and increase performance. Typical years for reducing volume might include senior years in high school and college and Olympic years. Figure 2.1b shows weekly hours for one training year. Note that there is an overall progression, with total volume peaking around week 25 and then a gradual reduction of training volume as increased high-intensity training is done to prepare for competitions. (Annual periodization is discussed further in chapter 10.) Note also that there is a three-week cycle of medium-, high-, and low-volume weeks to allow for stress and recovery. Figure 2.1c shows the daily periodization of a month of training. Note the three-day cycle of medium, hard, and easy days within the larger weekly cycles.

Periodization provides a framework for gradual increases in training intensity, duration, and frequency with periods of relative rest and recovery. It ensures the inclusion of essential elements and places them in the proper sequence relative to the season. We will say much more

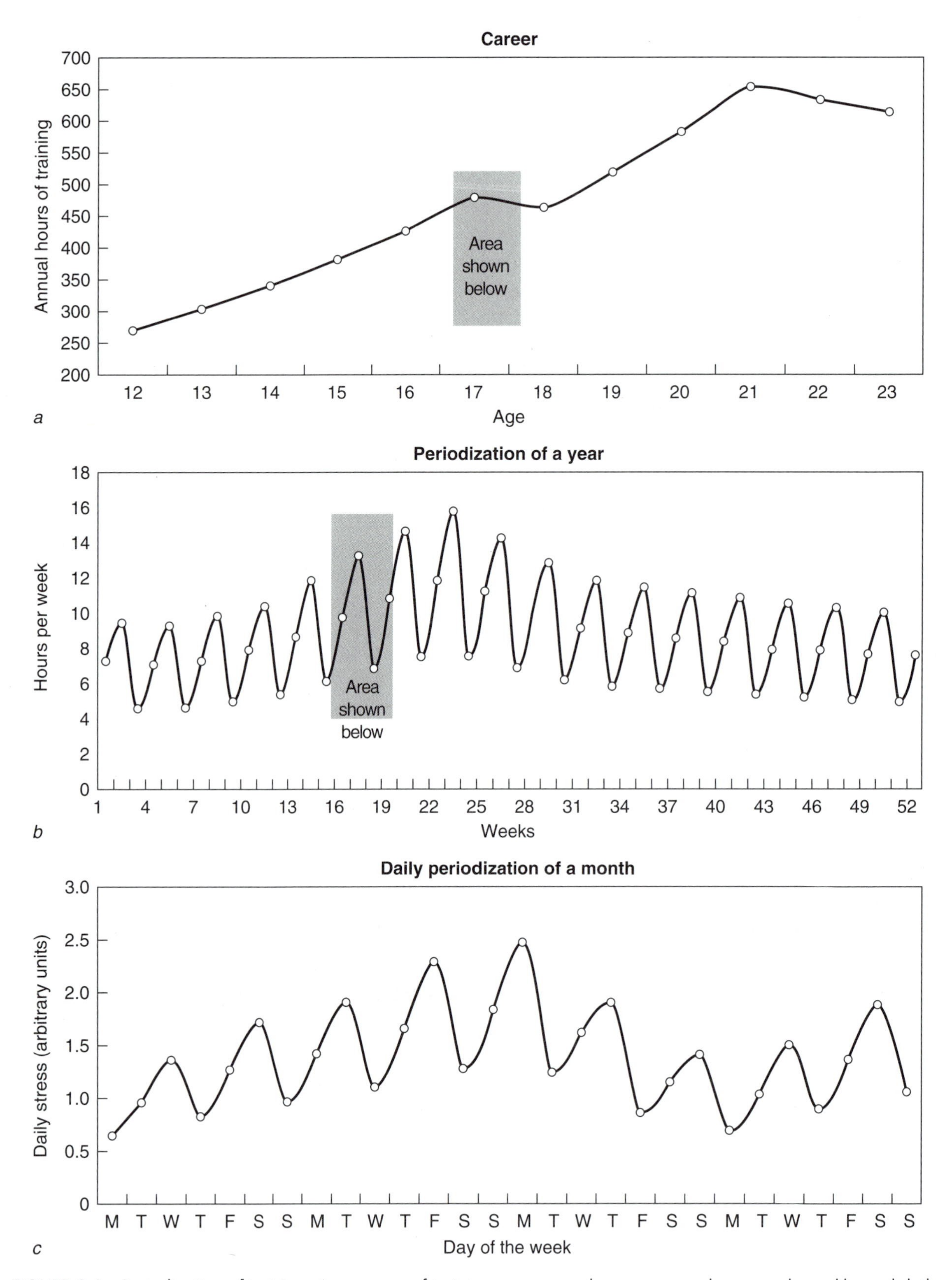

FIGURE 2.1 Periodization of training. As a career of training progresses there are annual, seasonal, weekly, and daily cycles. *(a)* A hypothetical increase in training hours per year for a developing athlete. *(b)* Weekly hours for one training year. *(c)* The daily periodization of a month of training.

about this exciting new principle of training in chapters 6, 9, and 10.

Variation

Training must be varied to avoid boredom and to maintain interest. The principle of variation embraces two important concepts:

Work and rest

Variety

Adaptations come when the overload is followed by rest and when hard training periods are followed by relatively easy periods to allow recovery. Failure to include variation leads to boredom, staleness, and poor performance. Successive sessions of hard work, if not followed by adequate time for rest and recovery, are certain to hinder progress in training. In chapter 10 we will teach you how to integrate relative rest into the program using the principle of periodization.

Coaches achieve variety by changing the training routine and drills. When possible, conduct workouts in different places or under different conditions. Follow a long workout with a short one, or high speed with easy distance. When workouts become dull, do something differ-

Variation: Just Add Mud!

I'll never forget that rainy day at our high school football practice when the coach decided to lighten the mood of a tough week with an unorthodox drill. A drenching downpour had created a series of huge puddles. We were instructed to sprint toward a puddle, do a flat dive, and body surf as far as possible. Best of all was when our normally serious coach demonstrated the drill. Soon we were diving and sliding all over the field, soaked, covered with mud, and grinning from ear to ear.

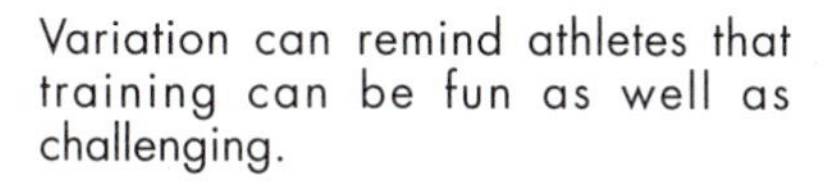

Variation can remind athletes that training can be fun as well as challenging.

ent. Use variety (e.g., cross-training) to diminish monotony and to lighten the physical and psychological burdens of hard training.

Warm-Up and Cool-Down

Warm-up should precede strenuous activity to

- increase body temperature;
- increase respiration and heart rate; and
- guard against muscle, tendon, and ligament strains.

The warm-up can involve calisthenics, skill rehearsal, and a gradual increase in exercise intensity. Stretching may be more effective after the warm-up than before. Warm muscles are more receptive to the benefits of stretching.

The cool-down is just as important as the warm-up because abrupt cessation of vigorous activity leads to pooling of the blood, sluggish circulation, and slow removal of waste products. It may also contribute to cramping, soreness, or more serious problems, such as heart rhythm problems in older athletes and coaches. High levels of the hormone norepinephrine are present immediately after vigorous exercise, making the heart more prone to irregular beats. The cool-down helps remove excess norepinephrine and lowers the body temperature. Light activity and stretching continue the pumping action of muscles on veins, helping the circulation in the removal of metabolic byproducts such as lactic acid.

Specificity

Specific training brings about specific results, so the type of training you do must be related to the desired outcome. The best way to develop the muscular and energy fitness required in your sport is to participate in the sport itself or in closely related activities. The principle of specificity applies to the muscles involved and to the energy systems within those muscles. For example, when you run, you recruit certain muscle fibers, energy pathways, and energy sources. If you run every day, you are training and the adaptations take place in the muscle fibers used during the exercise.

The adaptations to endurance training are different from the adaptations to strength training. Endurance training elicits improvements in oxidative enzymes and increases the particular muscle's ability to burn fat and carbohydrate in the presence of oxygen. Strength training leads to increases in the contractile proteins actin and myosin, but only in the muscle fibers used in training. Endurance training won't make you stronger, and strength training does little to improve endurance.

Cycling is not the best preparation for running, and vice versa. Why? Because the two activities use different muscle fibers. But won't cycling or running bring about improvements in circulation or respiration, changes that could improve or maintain performance in the other activity? Yes, both may improve the heart's ability to deliver oxygen, but the most important changes take place in the muscle fibers, so the transfer of training is limited. Studies have shown that the best results always come with specific training.

Of course, every rule or principle can be taken to the extreme. Specificity does not mean you can't use a mix of activities for variety or in order to minimize the risk of overuse injuries. Nor should you avoid training opposite or adjacent muscles. In fact, you should train other muscles to avoid muscle imbalances that predispose the body to injury. And you can train adjacent muscles to help adapt to changing conditions, as well as to provide a backup for when the primary fibers become fatigued. So some cycling could be good for a runner; it provides muscle balance, trains adjacent fibers, and provides relief from the pounding of running.

Bompa (1990) views specificity as the essence of specialization. Thus sport-specific exercises and training increase as an athlete progresses from the stage of youthful beginner toward the mastery of the mature athlete. Bompa points out that specific adaptations do not refer only to physiological changes, since specialization also applies to technical, tactical, and psychological adaptations (see "Career Development" in chapter 12).

Longevity

In some sports, such as gymnastics, female athletes reach elite status at a relatively early age, before maturation alters the ratio of strength to body weight. In other sports, in which aerobic fitness and muscular fitness

©Empics

© Empics/SportsChrome

While crew teams can benefit from cross-training by running or cycling occasionally, the best training for rowers is rowing.

are paramount, athletes must train for years to achieve their potential. This section considers principles related to long-term success in sport. We'll discuss regression, long-term training, and moderation.

Regression

Most of the adaptations achieved from months or even years of hard training are reversible, and may be lost with detraining when training is terminated. As Calvin Coolidge said, "Nothing in the world can take the place of persistence." It takes approximately three times as long to gain endurance as it does to lose it. With complete bed rest, aerobic fitness declines at the rate of 9 percent per week! Strength is lost at a slower rate, but lack of use (disuse) will weaken even the best-trained muscles. In time the muscle undergoes atrophy, or a decrease in size. Atrophy is the opposite of hypertrophy, an increase in muscle cross-sectional area that occurs with appropriate training. When combined with time and age, this loss of muscle mass can become **sarcopenia,** or vanishing flesh. Training is necessary to maintain aerobic and muscular fitness for sport, and regular lifelong activity is imperative if you are to avoid the health risks associated with sarcopenia.

The principle of **regression** helps explain why **overtraining** is such a risk; overtraining can lead to illness or injury, and both put a halt to training, thereby contributing to regression. This principle also explains why off-season training programs are so important. Prolonged periods of inactivity allow hard-won training gains to regress, to go back to a lower level. The serious athlete builds on each season's gains and progresses to a higher level of performance.

Long-Term Training

The changes that result from progressive overload of body systems lead to impressive improvements in performance. However, it takes years of dedication and effort to approach excellence. Long-term training allows for

- gradual physiological development,
- growth and maturation,
- refinement of skills,
- learning of strategies, and
- greater understanding of the sport.

Prolonged training implies specialization. Those who hope to reach their potential eventually concentrate on one sport. Many elite athletes accumulate over 500 hours of training annually, and some world-class athletes approach 1,000 hours per year. That volume of training can be achieved only by physically mature athletes who have trained for years to achieve a solid foundation, gradually increasing the quantity and quality of training. Excellence comes to those who pursue it in a well-planned, long-term training program.

Moderation

The principle of **moderation** applies to all aspects of life; too much of anything can be bad for your physical and psychological health. Dedication is fine, so long as it is tempered by judgment and moderation. Train too hard, too fast, or too long and the body begins to deteriorate. Practice moderation in all things. Don't sacrifice the future to achieve fleeting immediate success. Remember the line from the song; "Fame, if you win it, comes and goes in a minute." Approach training with moderation, knowing that overtraining is far more disastrous than undertraining. More often than not, success comes to those who pace themselves. Some may burn brightly and then fade; others are in the game until the end, practicing self-discipline, judgment, and moderation.

Fallacies of Training

In addition to the proven principles of training, there are a number of popular misconceptions concerning training. These oft-quoted sayings are not true and have no basis in medical or scientific research.

"No Pain, No Gain"

Serious training is often difficult and sometimes unpleasant, but it shouldn't hurt. In fact, well-prepared athletes often perform difficult events in a state of euphoria, free of pain and oblivi-

Examples of Longevity

Peak performances come at different times in sport. Female gymnasts seem to peak in early adolescence while male gymnasts reach elite status in their early 20s. Track athletes often show promise before college, but most reach their peak after graduation. A few have established world records much later in life. Carlos Lopes of Portugal was a successful international runner for many years, but he achieved his personal best when he established a world's record for the 10,000-meter run—at the age of 37! Professional athletes in team sports continue playing into their 30s, and a few continue beyond the age of 40 years. Longevity for elite athletes or for folks like us depends on our ability to follow the principles of training.

ous to discomfort. Marathon winners sometimes seem to finish full of vitality, while others appear near collapse. Pain is not a natural consequence of exercise or training. It is a sign of a problem that shouldn't be ignored. During exercise the body produces natural opiates, called endorphins, that can mask discomfort of the effort. If an athlete experiences real pain during training, he or she should back off. If the pain persists, have the problem evaluated.

Discomfort, on the other hand, can accompany difficult aspects of training, such as heavy lifting, intense interval training, or long-distance work. Discomfort is a natural consequence of the hydrogen ion (acid) buildup that accompanies the anaerobic effort of lifting or intervals, or of muscle fatigue, microscopic muscle damage (microtrauma), and soreness that can come with distance work. Overload sometimes requires working at the upper limit of strength, intensity, or endurance; and that can be temporarily uncomfortable. But when exercise results in pain during the activity, it is probably excessive. Discomfort that comes 24 hours after exercise is called delayed-onset muscle soreness (DOMS). We will say more about DOMS in chapter 4.

"You Must Break Down Muscle to Improve" and "Go for the Burn"

Microtrauma sometimes occurs in muscle during vigorous training and competition, but it isn't a necessary or even desirable outcome of training. Runners exhibit microtrauma at the end of a marathon or after downhill running that requires eccentric muscular contractions (contractions when a muscle is lengthening). Lengthening contractions have been shown to be a major cause of muscle soreness, which is associated with muscle trauma, reduced force output, and several weeks of recovery. So excessive trauma doesn't enhance training; it halts it. Weightlifters can traumatize muscle by overworking with excess weight or repetitions, but that is not a necessary step in the development of strength. Neither pain nor muscle damage is a normal consequence of training, so you should try to avoid both.

The phrase "go for the burn" is often heard among bodybuilders who do numerous repetitions and sets to build, shape, and define muscles. The burn they describe is likely due to the increased acidity associated with elevated levels of lactic acid in the muscle. While the sensation isn't dangerous, it isn't a necessary part of a strength program designed to improve performance in sport.

"Lactate Acid Causes Muscle Soreness"

This fallacy has been around for years, even though it lacks any basis in fact. Although it is true that lactic acid is often produced in contractions that result in soreness, the lactic acid isn't the direct cause of the soreness. Why not? Lactic acid is cleared from muscle and blood within an hour of the cessation of exercise, while soreness peaks 24 to 48 hours after the effort—long after the lactic acid is removed or metabolized.

Delayed-onset muscle soreness follows unfamiliar exercise, vigorous effort after a long layoff,

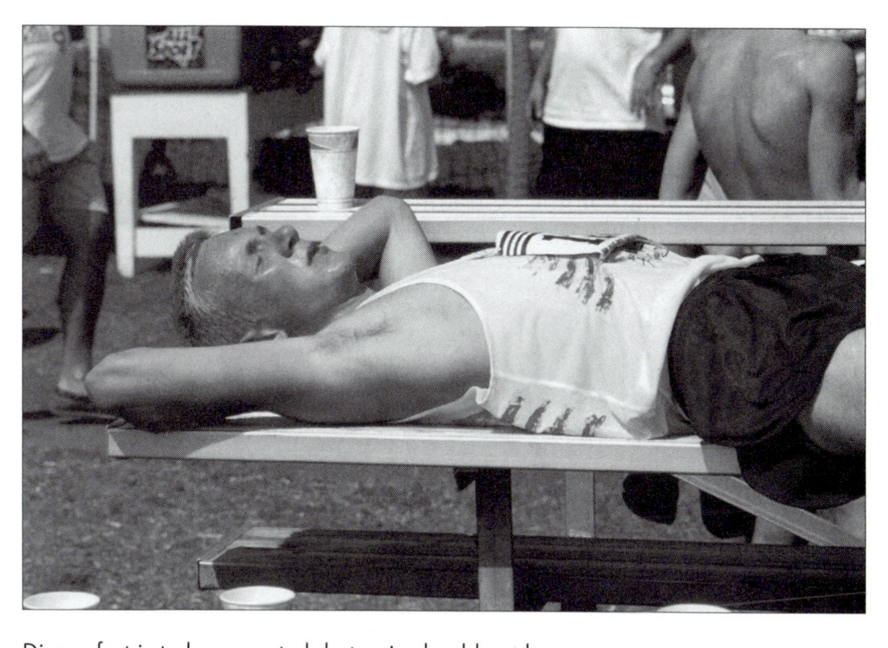

Discomfort is to be expected, but pain should not be.

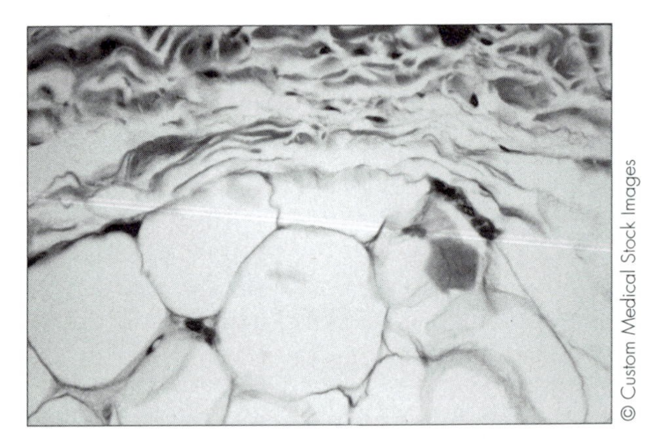

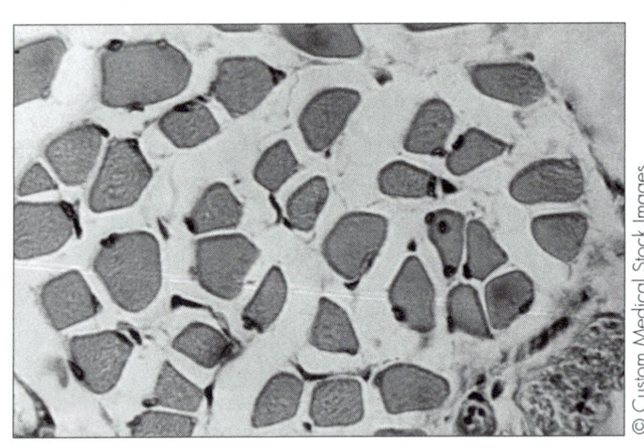

FIGURE 2.2 Fat versus muscle. The two photos show adipose tissue (fat) and a cross-section of skeletal muscle. Fat, an effective form of high-density energy storage, consists of triglycerides, large molecules made up of three fatty acids and a glycerol molecule. Muscle consists mostly of contractile proteins. The two types of tissues do not change from one to the other with training or sedentary activity but independently change based on nutritional and fitness status.

and lengthening (eccentric) contractions. It is associated with microtrauma to muscle and connective tissue, swelling or fluid accumulation (edema), and leakage of substances from the muscle (e.g., the enzyme creatine kinase). After days or weeks of recovery, the athlete is effectively **inoculated** against further soreness, but only from the activity that caused the initial discomfort. Surprisingly, the inoculation lasts for several months.

"Muscle Turns to Fat" (or Vice Versa)

Does muscle turn to fat when an athlete stops training? Muscle and fat are highly specialized tissues, and each has a specific function. Muscle fibers are composed of long, thin strands of contractile protein that are designed to exert force. Training increases the size of muscle fibers (hypertrophy), and detraining reduces their size (atrophy). When training is halted, some fat can be deposited among the fibers if energy intake exceeds energy expenditure; but long, thin muscle fibers do not change into fat cells. Fat cells are spherical blobs designed to store fat. They grow in size when you store fat and shrink when weight is lost. And globular fat cells could never take on the contractile duties of muscle fibers (figure 2.2).

"I Ran Out of Wind"

Run too fast for your level of training and you will experience the sensation of running out of wind, or air. The sensation comes from the lungs and reflects another discomfort of exertion. However, the sensation is more likely to be due to an excess of carbon dioxide (CO_2) than to a lack of oxygen. Carbon dioxide is produced during the oxidative metabolism of carbohydrate, and CO_2 is the primary stimulus that drives respiration. So when CO_2 levels are high, as they are during vigorous effort, they cause distress signals as the lungs attempt to blow off the excess. The respiratory system thinks it is more important to rid the body of excess CO_2 than it is to bring in more O_2.

Intramuscular Fat

A small amount of fat can be stored in endurance-trained muscle fibers, but that fat constitutes less than 0.1 percent of the body's total. Intramuscular fat is the first fat oxidized during exercise. If the effort continues, additional fat will be mobilized from adipose tissue deposits (subcutaneous, abdominal) and transported to the muscle via the bloodstream (circulation).

Excess CO_2 is a sign that you have exceeded your lactate threshold and that you are working above the sustainable level you achieved in training. Become familiar with the sensation and what it means; ignore it during a race and you will soon become exhausted. We will tell you how to use respiratory and other sensations to gauge your effort and avoid exhaustion.

In the next chapter we'll examine the factors that influence an individual's response to training.

SUMMARY

This chapter reviews important principles of training related to the athlete, the training process, and long-term success in sport. Individual response is an important concept for coaches to understand. Athletes respond differently to the same training, due to heredity and the environment (nutrition, training, sleep/rest). Athletes' responses to training are specific, so training should be sport specific. And long-term success requires a long-term view on the part of the coach. Moderation is important for the athlete and the coach.

The chapter also reviews some misconceptions about training. Training doesn't have to hurt, you don't have to break down muscle, and you can have gain without pain. As we have said, training is a gentle pastime in which we coax a slow stream of adaptations out of the body. Have athletes set attainable goals, and help them achieve these goals with well-planned programs. Don't sacrifice the future to achieve immediate success.

KEY TERMS

adaptation (p. 16)
adipose tissue (p. 24)
cool-down (p. 16)
FIT (p. 16)
individual response (p. 15)
inoculated (p. 24)
moderation (p. 22)
overload (p. 16)
overtraining (p. 22)
periodization (p. 16)
potential (p. 15)
progression (p. 16)
readiness (p. 15)
regression (p. 22)
sarcopenia (p. 22)
specificity (p. 16)
taper (p. 14)
variation (p. 16)
warm-up (p. 16)

REVIEW QUESTIONS

Match each of the following training principles to its definition.

1. Readiness
2. Potential
3. Individual response
4. Adaptation
5. Overload
6. Progression
7. Periodization
8. Variation
9. Specificity
10. Regression
11. Moderation

a. Measurable changes taking place over time when one follows a sensible training program
b. Process of dividing training into smaller and larger sections or cycles, ensuring an appropriate training stimulus as well as time for recovery
c. Physical and psychological maturation and the capability to respond completely to muscular or fitness training
d. Loss of training gains due to stoppage in training or to overtraining
e. A change in a training routine to prevent boredom, staleness, or poor performance while still allowing for recovery
f. Response to training that can be affected by heredity, maturity, diet, sleep, and other factors
g. Gradual increases in frequency, intensity, and time to go from general to specific training and quantity to quality
h. Training that is tempered by judgments in attempts to eliminate overtraining
i. Training directly related to the desired outcome
j. Capacity available for improvement
k. Placing a higher demand on the body by increasing the size of the load through frequency, intensity, or time in order to produce measurable gains

Determine if the following statements are true or false and circle the correct answer.

12. T/F If you are not feeling pain during training, then you are not making any physiological gains.
13. T/F You must break down muscle in order to improve performance.
14. T/F Muscle can turn into fat when you stop working out, and fat can turn into muscle when you start.
15. T/F Lactic acid is not the direct cause of muscle soreness.
16. T/F The legendary tale of Milo of Crete is probably true.
17. T/F Progression and periodization describe the same principle.
18. T/F Periodization and variation have similar purposes.
19. T/F The cool-down helps clear lactic acid from muscles, reduce excess epinephrine, and lower the body temperature.

Circle the correct choices.

20. Studies show that the taper period should be longer when training volume is (low or high) and the event is (short or long).
21. While prepubertal boys gain some strength with resistance training, they gain far more after puberty, due in part to the hormone (estrogen or testosterone).

22. Aerobic and muscular fitness training each lead to (general or specific) changes in the muscle fibers employed in training. Strength training leads to increases in (contractile or energy) proteins in the fiber.

Select the correct response (more than one may be correct)

23. Detraining or the cessation of training can lead to muscle
    a. atrophy
    b. hypotrophy
    c. sarcopenia
24. Overtraining can lead to
    a. illness
    b. injury
    c. decreased performance
    d. all of the above
25. No pain, no gain: If you feel pain during training you are
    a. getting results
    b. risking injury
    c. causing microtrauma

PRACTICAL ACTIVITIES

1. Describe how you would include the concept of taper into planning for a team sport, such as soccer or basketball.
2. List some ways to assess a young athlete's potential for success in a given sport.
3. Describe how you could use periodization to ensure an adequate ratio of work and rest in training.

3

Individual Response to Training

© Bongarts/SportsChrome

This chapter will help you

- identify individual differences among athletes,
- understand why individual differences occur,
- appreciate how these differences influence one's ability to respond to training, and
- adjust training loads to accommodate important differences.

Mr. Johnson was an experienced coach and a national-caliber athlete in his own right. He was well liked by students and athletes, and he prided himself on being fair, on treating everyone equally. Unfortunately, this laudable trait is not always the best approach for a coach. In practice he expected all his runners to engage in the same training, regardless of their age or level of ability. As a result, practices were too easy for a few and too hard for many others. Mr. Johnson designed training sessions the way he trained, forgetting that many of his high school athletes were less gifted and far less mature.

In a study from the laboratory of the legendary Claude Bouchard (Lortie et al. 1984), 11 men and 13 women were involved in a 20-week cycle ergometer training program. These sedentary subjects trained four to five times per week for 40 and then 45 minutes, starting at 60 percent of maximal heart rate reserve and increasing to 85 percent intensity. Training was standardized for all subjects, and all sessions were monitored. Aerobic capacity increased 26 percent in the group. But when the individual response to training was calculated, the improvements ranged from 7 to 87 percent. The authors concluded that age, gender, and prior training experience did not contribute much to the variation in trainability. The major causes of variation in the response to training included the current phenotype level (described as the pretraining status of the trait considered) and a genetically determined capacity to adapt to training that is unique for each family of characteristics (i.e., the role of heredity in training, or the genotype–training interaction).

Another study from Dr. Bouchard's lab showed improvements in aerobic power ranging from 0 to 41 percent for 10 pairs of identical twins. In this study, 77 percent of the variation in the response to training was genotype dependent. Clearly, heredity plays a major role in our ability to respond to training.

Many factors influence the way an individual responds to the rigors of training, including heredity, maturation, nutrition, sleep and rest, state of training, and more. Keep in mind that, as Craig Venter said, "The wonderful diversity of the human species is not hard-wired in our genetic code. Our environments are critical" (Ridley 2003). But none of these influences is more potent than that of heredity.

Genes influence potential, but they don't ensure it. The 30,000 genes that form the blueprint of the human body are subject to the influence of the environment and behavior. Your genotype is your genetic constitution, while the phenotype is the observable appearance resulting from the interaction of the genotype and the environment. In sport, genetic potential can be realized only when genes are switched on via the process of training.

Genes carry the code for the formation of proteins. When a specific type of training is performed, it turns on a promoter that activates specific genes. Through a process called transcription, an RNA (ribonucleic acid) strand is formed on the template made up of the gene's DNA (deoxyribonucleic acid). The RNA becomes the messenger (mRNA) that exits the nucleus, enters the cytoplasm, and binds to a ribosome, whose function is to synthesize protein. The mRNA translates the genetic code into a sequence of amino acids to form a specific protein. Without training, genetically gifted individuals cannot achieve success in sport. But **heredity** is much more complicated than genes, DNA, and RNA.

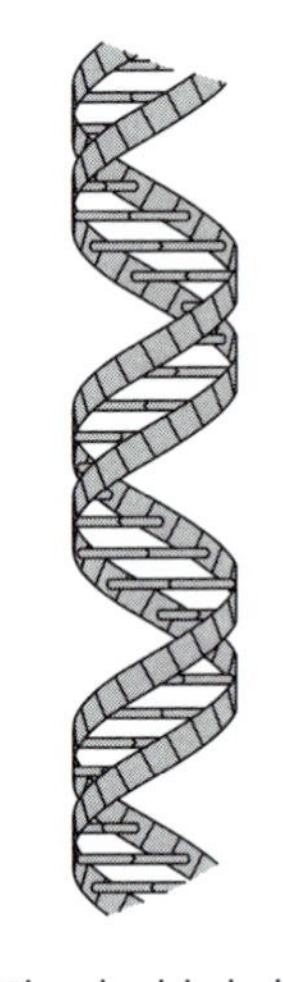
The double helix of DNA.

The genetic response to a specific form of training resides in a cluster of genes, and each can be influenced by the presence or absence of so-called enhancers, regulatory elements that influence the degree of response. That helps to explain why the individual response to identical training is so variable. Factors that may influence the response to training could include maturation (hormones), nutrition (energy, amino acids), adequate rest, and even chemically related emotional factors such as stress (other hormones) or depression (neurotransmitters).

Maturation

Young athletes grow, develop, and mature at different rates; the increases in physical size and functional capacity of body systems are unique to each individual. Because the attributes of size, strength, and endurance are important to success, an athlete's growth and maturation rate directly influence his or her response to training and success in sport. Maturation refers to progress toward the mature state, be it sexual or skeletal. Sexual maturity consists of a fully functional reproductive capability, while skeletal maturity indicates a fully developed skeleton. While **growth** indicates an individual's size (e.g., height), **maturation** focuses on the individual's progress toward full skeletal size. For example, an ungainly, 6-foot (183-centimeter) freshman basketball candidate may grow another 6 to 8 inches (15 to 20 centimeters) before he achieves skeletal maturity. Or a skinny football recruit may add inches and pounds of muscle when he achieves sexual maturity. As a coach, you will work with athletes at various stages of maturation, both sexual and skeletal. In many sports an athlete's full potential emerges only after he or she reaches maturity.

Sexual maturity provides the hormonal environment necessary to support the response to training. Testosterone is an anabolic or growth-stimulating hormone. When it is available in adequate supply it supports a full response to strength training. That is one reason we don't

Growth!

In junior high school my friend Bob didn't have much success in basketball, his favorite sport, because he wasn't very tall. But over the summer before his sophomore year in high school he grew over 5 inches (13 centimeters). When he returned to school in the fall, the coach saw him in a different light. Bob now had the one factor he couldn't practice to improve—height. Bob eventually grew to 6 feet 8 inches (206 centimeters) and enjoyed an exciting basketball career in high school, college, and the European professional circuit.

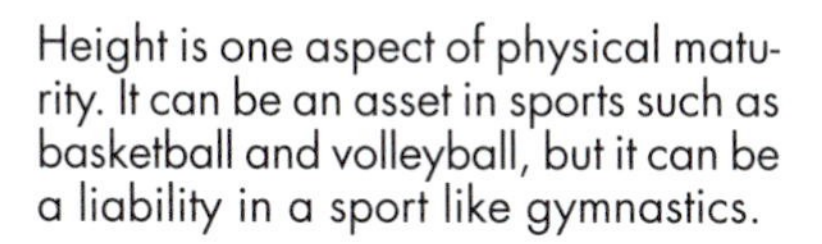
Height is one aspect of physical maturity. It can be an asset in sports such as basketball and volleyball, but it can be a liability in a sport like gymnastics.

recommend heavy weight training for immature athletes. Incidentally, while testosterone is found in males and females, the levels are much higher in mature males.

Assessment

Skeletal maturity may be assessed through comparison of an X ray of the athlete's hand/wrist with standards that indicate progressive levels of skeletal maturity. If the X ray corresponds to that of a 10-year-old, the skeletal age is said to be 10 years. Sexual maturation can be determined through evaluation of the athlete's secondary sex characteristics (pubic hair, breast and genital development) according to the standards developed by Tanner (1962). However, in today's sensitive and litigious climate, we do not recommend that coaches become involved in this assessment. If you are concerned about an athlete's maturation, you should consult with the parents or the team physician.

Development

Development of competence in sport is influenced by the athlete's growth and maturation. Early childhood experiences should emphasize development of basic movement patterns and skills. Then the child should be introduced to a variety of sports and movement experiences.

With age and maturation comes the desire to specialize in one or two sports. Throughout development, regular physical activity is important to the optimal development of an individual's physical capabilities. And training appropriate to the athlete's level of maturation helps the athlete achieve his or her potential (see figure 3.1 and table 3.1).

The body develops from the moment of conception. Development continues after birth as the systems of the body become more refined. A young athlete develops physically in many ways, including aerobic and muscular fitness, intelligence, strategy, and skill. Psychological development can include persistence, concentration, and the ability to defer gratification. Coaches should not be surprised to discover that a workout may be too easy for one athlete and too hard for another. Perceptive coaches quickly realize these differences and adjust practices to account for individual variation. We will provide ways to adjust workouts in later chapters.

Other Factors

Other factors, such as gender, diet and nutrition, rest and sleep, illness and injury, and emotional factors influence athletes' response to training and their ability to succeed in sport. The coach

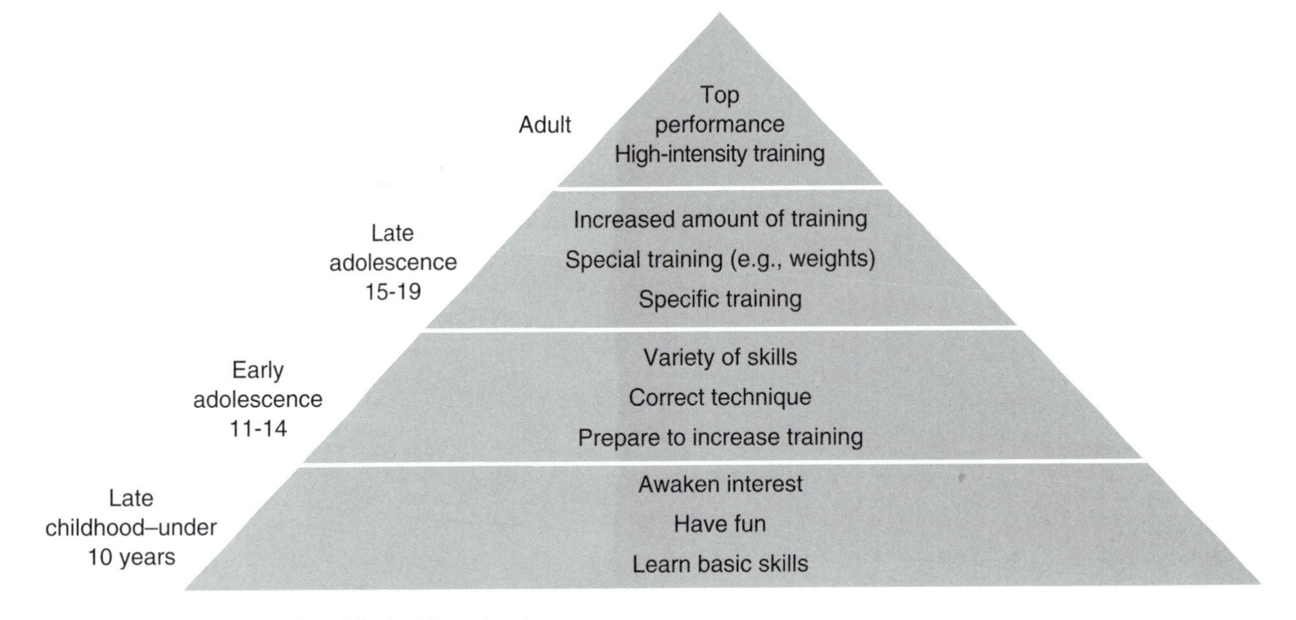

FIGURE B.1 Age-based model of athletic development.
Reprinted, by permission, from B.J. Sharkey, 1986, *Coaches guide to sport physiology* (Champaign, IL: Human Kinetics), 26.

TABLE 3.1

Age-Based Training Guidelines

Growth stages	Muscular fitness		Energy fitness	
	Methods	Time	Methods	Time
Children 6-10	Use body weight as resistance in general conditioning exercises (chin-ups, push-ups, etc.). Maintain flexibility.	15 min 3×/week	Team games with few playing restrictions. Emphasize involvement, play, and free expression. Avoid formal fitness-training methods.	Under 4 hr/week*
Adolescent (early) 11-14	Use moderate resistance and more repetitions (over 10) on weight-training machines to develop endurance. Maintain flexibility.	30 min 3×/week	Continue to develop fitness through the use of team games. Improved aerobic ability is the main training goal. Introduce activities involving long, easy-paced intervals.	4-5 hr/week
Adolescent (late) 15-19	Introduce high-resistance training (under 10 repetitions) and use of free weights to improve strength and develop power. Maintain flexibility.	45 min 3×/week	Increase training intensity. Mix long and short intervals. Train regularly at the anaerobic threshold.	6-8 hr/week
Adult	Advanced muscular fitness training demands depend on specifics of sport specialization.	1 hr 6×/week	Prepare for peak competitive performances by developing a seasonal plan for energy fitness that integrates sport-specific requirements into the annual competitive cycle.	Over 8 hr/week

*Usually 1-2 hr unless children are in organized programs (e.g., swimming) involving slow/easy distance training.

Reprinted by permission from B.J. Sharkey, 1986, *Coaches guide to sport physiology* (Champaign, IL: Human Kinetics), 27.

needs to understand these effects and develop ways to deal with them.

Gender

The authors have both coached boys and girls on the same team. Do males and females respond differently to training? With prepubescent children, it is unnecessary to separate children by gender for sport participation. During childhood, physical differences may be as apparent between individuals of the same gender as they are between the sexes. Coaches working with this age group should focus on individual development and provide the two sexes equal opportunities in practice and competition. Young boys and girls respond similarly to training. However, at the high school level, most athletes will have at least started puberty.

After puberty, body size and related factors tend to favor the young men. On average the males are larger and stronger and have somewhat greater endurance (table 3.2). World records in running favor males by about 10 percent at distances ranging from the 100-meter dash up to the marathon. For these reasons it makes sense to provide separate competitive opportunities for young men and women.

However, in some sports, such as swimming, it is not uncommon for men and women to train together even though they wouldn't compete against each other. In either case, individual differences need to be considered.

Diet and Nutrition; Rest and Sleep

If athletes fail to eat an adequate diet and lack the energy, essential amino acids, vitamins, and

Puberty

Puberty is defined as the period of life when sexual maturation or the ability to reproduce begins. Pubescent means arriving at or having reached puberty, so prepubescent defines the period prior to beginning puberty. The coach should be able to identify the onset of puberty because it signals the time when growth and the response to training change dramatically. Initial development of the breasts and enlargement of the testicles are the first signs of sexual maturation in girls and boys, respectively. However, identification of the adolescent growth spurt is the preferred method for identifying the onset of puberty. The peak height velocity, a measure of height increases per year, occurs at about 13 years of age for females and 14 years for males. During adolescence, the level of testosterone in the blood of boys increases 10 times or more while it remains at prepubescent levels for girls. Training that takes place after the onset of puberty leads to far greater gains in strength, power, and aerobic fitness.

TABLE 3.2

Gender Differences: Male Versus Female

Attribute	Difference
Aerobic fitness	$\dot{V}O_2$max is higher in average and elite male athletes.
Anaerobic threshold	No difference.
Muscular fitness	Strength—men have more upper body strength and somewhat more leg strength (the differences are reduced when expressed relative to body weight or fat-free weight).
Body size	Heart and respiratory systems are scaled to body size.
Other	Men have more hemoglobin (per unit of blood and total), which could influence oxygen delivery to muscles.

minerals they need to support growth and the requirements of training, they will not be able to respond favorably to intense physical demands. As you work with young athletes, remember that nutrition may be one of the reasons some of your players struggle to handle the demands of training. Seek a solution and you may be voted coach of the year!

Growing athletes also need lots of rest, and some need more than others. Many of the adaptations that result from training take place during sleep; so explain to your athletes that those who miss sleep may also miss out on the benefits they've worked so hard to achieve. Recommended sleep guidelines for different age groups appear in table 3.3.

Sleep is important for health as well as performance. Adults who get 7 to 8 hours of sleep are healthier than those who get less than 7 or more than 9 hours of sleep. Naps provide a short-term solution to unavoidable sleep loss. Naps of 20 minutes or 90 minutes seem to work best

TABLE 3.3

Rest and Sleep Guidelines for Young Athletes

Stage	Hours sleep*
Children (6-10)	10
Youth (11-14)	9-10
Young adult (15-19)	8-9
Adult	7-8

*Plus naps as needed.

Reprinted, by permission, from B.J. Sharkey, 1986, *Coaches guide to sport physiology* (Champaign, IL: Human Kinetics), 29.

because they avoid having to wake up during a cycle of deep sleep.

Observe your athletes closely. If they appear overtired, insist that they take time off for rest and recovery. Find out if they are getting enough sleep. Do they have a job? Are they studying long hours, watching too much TV, surfing the Internet, or playing video games? Do brothers or sisters keep them awake? Find the reason for their fatigue and work toward a solution.

Illness or Injury

When an athlete appears flat or listless, this may signal the start of a cold or flu. The biggest enemy of a young athlete in an endurance sport is upper respiratory problems, including viral and bacterial infections, allergies, and asthma. Working athletes too hard when they have an infection may make matters worse. Have them take off a few days of training so they won't ruin weeks of hard work. Allergies have the effect of making athletes tired, thereby making practices more difficult. Be sure you are aware if any of your athletes have allergies or asthma.

While some minor injuries may allow limited practice or training, others require rest and treatment. Be certain that athletes have recovered before they return to practice or competition. And be on the lookout for athletes who try to hide minor injuries for fear of losing their place on the team. See chapter 12 for more information on overtraining and illness.

Emotional Factors

Stress, anxiety, depression, and eating disorders are emotional factors common to young athletes. Some find competition stressful, and the pressure can lead to anxiety. Poor performances related to overtraining can lead to depression. And a small number of athletes may become obsessed with slimness and refuse to eat. Eating disorders are more common among individuals who try to be perfect and please others. As a coach you should be aware of these emotional factors and find ways for athletes to get the help they need.

Now it is time to turn our attention to the important area of muscular fitness and learn the essentials of training for strength, power, power endurance, and speed.

SUMMARY

Athletes differ in their response to training and their ultimate level of performance. These differences, evident in every team, are due to heredity as well as environmental factors such as training, nutrition, and rest. The effects of training are influenced by heredity, along with maturation (hormones), nutrition (energy, amino acids), adequate rest, and emotional factors (stress). Some young athletes do well and then fade, while others emerge after they reach puberty, or even later. Coaches need to be aware of individual differences and to structure practices that yield optimal results for all participants.

Immature athletes cannot profit from the arduous program followed by seasoned athletes. If they try they may face illness or injury, and they are likely candidates for overtraining. Coaches need to be able to assess an athlete's capacity and to utilize age- and maturity-appropriate training. Along with training programs and skills and strategy, coaches provide counsel on nutrition, rest and sleep, illness and injury prevention, and even emotional factors. We never said coaching was easy, but it sure is fun.

KEY TERMS

REVIEW QUESTIONS

Match the age of the athlete on the left with the correct skill development on the right.

1. Late childhood (under 10 years old)
2. Early adolescence (11-14 years old)
3. Late adolescence (15-19 years old)
4. Adult (over 19 years old)

a. Increased amount of training; special training (e.g., weights); specific training
b. Awakening of interest; having fun; learning basic skills
c. Top performance; high-intensity training
d. Variety of skills; correct technique; preparation to increase training

Determine if the following statements are true or false and circle the correct answer.

5. T/F $\dot{V}O_2max$ is higher in average and elite female athletes.
6. T/F There is no difference in anaerobic threshold between males and females.
7. T/F Men have better muscular fitness than females when it comes to strength and muscular endurance.
8. T/F Females have more hemoglobin (per unit of blood and total), which could influence oxygen delivery to muscles.
9. T/F Genetic potential can lead to success at the elite level, even in the absence of training.
10. T/F Sufficiently hard training can lead to success at the elite level, even in the absence of genetic potential.
11. T/F Genes carry the code for the formation of proteins.
12. T/F Your genotype is your genetic constitution, while the phenotype is the observable appearance resulting from the interaction of the genotype and the environment.
13. T/F Genes influence potential but they don't ensure it.

Fill in the blank with the appropriate word or words to complete the sentence.

14. __________ is the most influential factor in the way an individual responds to training.
15. __________ is the increases in physical size and functional capacity of body systems that are unique to each individual.
16. Adults who get __________ hours of sleep are healthier than those who get less or more.

17. Inadequate __________ may be one of the reasons some athletes struggle to handle the demands of training.

Select the correct response (more than one may be correct).

18. Training influences
    a. genes
    b. promoters
    c. DNA
    d. mRNA
    e. all but one of the above
19. The genetic response to a specific type of training resides in
    a. a single gene
    b. a cluster of genes
    c. the chromosome
20. Factors that may influence the response to training include
    a. maturation
    b. nutrition
    c. adequate rest
    d. emotional factors
    e. all of the above
21. Maturation refers to progress toward the mature state, including
    a. sexual maturation
    b. skeletal maturation
    c. both of the above
22. Prepubescent means
    a. during puberty
    b. before puberty
    c. after puberty
    d. peak height velocity

PRACTICAL ACTIVITIES

Assume that you are the supervisor, the athletic director with responsibility over Mr. Johnson, and you become aware of his practice of requiring the same training for all athletes.

1. How would you help him (a) understand the differences among athletes, (b) the reasons for those differences, and (c) how they influence one's response to training?
2. How could he adjust training loads to accommodate important differences? Give examples in energy and muscular fitness training.

PART II

Muscular Fitness Training

Muscles create movement by shortening. Joint angles are changed and movement happens. The specific training of muscles gives athletes the grace and coordination to make complex actions look simple. The combination of muscular strength, power, and endurance that comes through proper training allows great performances. As a coach, you need to understand the principles that guide the development of muscular fitness. This part introduces you to those changes that happen in muscles with proper training. It will help you to differentiate between myth and science and to balance what we have learned from athletes, great coaches, and research data. Here, we focus specifically on muscle. Part III will broaden the scope further to talk about how energy is produced and about the systems that support muscle function.

4

Defining Muscular Fitness

This chapter will help you

- understand how muscles work,
- understand fiber types, and
- become aware of the components of muscular fitness.

After a successful collegiate skiing career, Kim was hired as an assistant coach in charge of Nordic skiing (cross-country skiing and jumping) for the university ski team. He was in a dilemma, as several of the skiers competed in both ski jumping (an agility and power sport) and cross-country skiing (an endurance sport). At that time, it was widely believed that endurance athletes should not do any strength training and that ski jumping performance would be compromised with endurance training. Kim had been a science major and decided to talk with a respected exercise physiology professor, Dr. D., at the university. The professor showed him some new research suggesting that it was possible, and probably beneficial, for athletes in both sports to do some cross-training. Working with Dr. D., Kim designed a resistance training program incorporating many of the principles still in use today, including specificity (for both ski jumping and cross-country skiing), plyometrics, and year-round periodization. That winter, the Nordic skiers helped carry the university to the first of a string of national championships.

We now know much more than we did in the past about resistance training and muscular fitness for both aerobic and power athletes. Surprisingly, few studies relate resistance training to improvements in athletic performance. We often make the leap of faith that improvements in strength and power lead to improvements in performance. Centuries ago, Plato recognized this connection and wrote, "Lack of activity destroys the good condition of every human being, while movement and methodical physical exercise save it and preserve it." One recent study did make the link between improved power and performance in cross-country skiing (Nesser et al. 2004).

Fifty-eight male and female adolescent cross-country skiers were assigned to one of four upper body training methods for a 10-week summer training program. The skiers were evaluated pre- and posttraining for upper body power using an ergometer that simulates the poling action of the sport, along with other tests for strength and endurance. Competitive race results were collected pre- and posttraining along with all training data. Only the group doing ski-specific training on the poling

Appropriate and specific resistance training will help to improve endurance activities such as cross-country skiing, running, rowing, and cycling.

device improved more than the control group in upper body power. Other groups doing more traditional strength training improved in strength but not in ski-specific power or competitive results. Most importantly, improvements in upper body power were significantly related to improvements in competitive race performance over 5- to 15-kilometer (3.1- to 9.3-mile) distances. These results showed that sport-specific, periodized resistance training improved arm power and competitive results. About 25 percent of the improvement in race speed could be attributed to the sport-specific resistance training. The other 75 percent of the improvements were due to other factors (maturity, technique, endurance, equipment, motivation, etc.), demonstrating the difficulty in evaluating the contributions of specific training to performance.

Muscles work by shortening when they are stimulated to contract. Each muscle is surrounded by connective tissues, as are the individual fibers and small groups of fibers within the muscle. All of these layers of connective tissue are continuous with tendons and fibrous sheaths that attach to bones of the skeleton. As muscles contract, the muscle fibers pull the connective tissue that then pulls the tendons and attached bones. Muscles are thus similar to springs that can only pull, not push. When muscles contract, they pull equally on the tendons attached to both ends of the muscle.

Muscle contractions are controlled by the brain via signals transported by nerves. Nerve impulses tell the muscle which fibers to contract, when to contract, and how long to maintain the contraction. The nervous system coordinates contractions of multiple muscles to cause smooth movements (see figure 4.1).

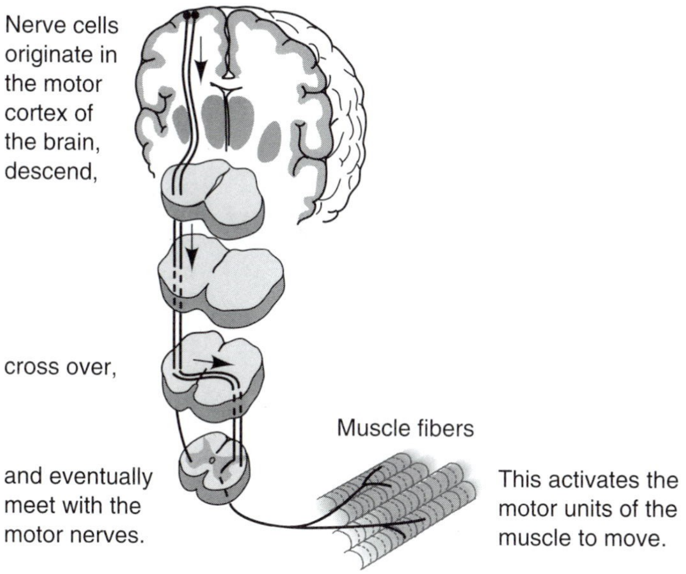

FIGURE 4.1 The control of muscle. Nerve cells located in the brain send impulses down the spinal cord to cause muscles to contract.

Reprinted, by permission, from B.J. Sharkey, 1986, *Coaches guide to sport physiology* (Champaign, IL: Human Kinetics), 35.

When learning new sport skills, athletes must consciously control the movements. This results in poor coordination, as the conscious brain can coordinate only a limited number of movements at a time. With practice and imagery, the resulting contractions become well coordinated. The knowledgeable coach uses a variety of skills (discussed later) such as visual imagery, sequential learning patterns, and specific drills to help athletes learn new skills. Practicing a skill reinforces the nervous pathways between nerves and muscles. Coaches need to be aware that imperfect practice may result in bad habits, which suggests that learning new skills needs to be carefully monitored.

Movement occurs when muscles contract and cause bones to move around joints. Some joints like the knee and elbow allow for movement in only one direction, while others, like the hip and shoulder, allow movement in a large range. See figure 4.2.

Muscle Structure and Contraction

Each skeletal muscle contains **muscle fibers** (cells), connective tissue, nerves, blood vessels, and intramuscular fat. Figure 4.3 shows the structure of the fibers and connective tissue. The continuity of all the connective tissues with the tendons allows the tension developed in individual fibers to be additive such that muscles can generate large forces.

Each muscle contains thousands of spaghetti-like muscle fibers that extend along much if not the entire length of the muscle. Within the fibers are threadlike strands called **myofibrils.** The myofibrils contain bundles of organized proteins that slide over one another to cause muscle shortening (contraction) when stimulated by the nervous system, or relaxation when stimulation stops. The mechanism of shortening is called the **sliding filament theory.** There are numerous proteins involved in muscle contraction. Two of the main proteins are **actin** and **myosin** (figure 4.4). The myosin protein has a headlike structure that can bind to actin and then bend at the neck when stimulated. This bending motion of multiple myosin protein heads bound to actin acts like oars pulling one protein past another. If enough muscle fibers are stimulated, the muscle will shorten and the attached bones will move.

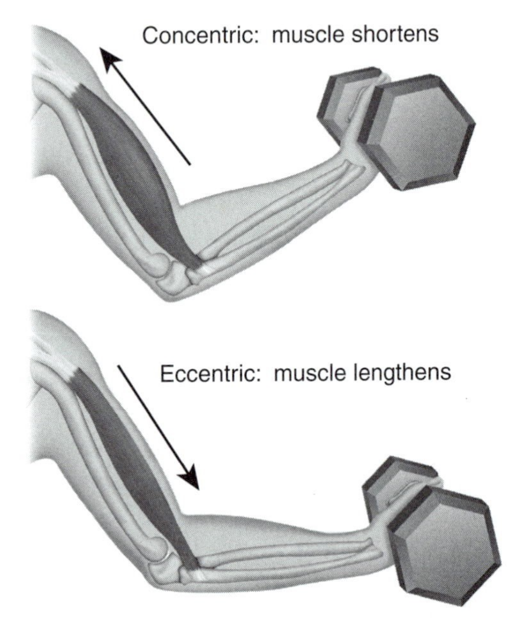

FIGURE 4.2 Movement occurs when muscles contract and cause bones to move around joints. The muscle may either shorten (concentric contraction) against a resistive force, or resist an outside force while lengthening (eccentric contraction), as happens with the biceps muscle when a weight is slowly lowered during a biceps curl.

Adapted, by permission, from J.H. Wilmore and D.L. Costill, 1999, *Physiology of sport and exercise, 2nd ed.* (Champaign, IL: Human Kinetics), 47.

Nerves and Motor Units

Each muscle has numerous slender nerve fibers that enter the body of the muscle. Each nerve splits upon entering the muscle and activates a number of muscle fibers. Some nerves activate only a few fibers in muscles where fine control is needed, such as in the fingers or eyes, while other nerves may activate hundreds of fibers in muscles that generate large forces. The average nerve in skeletal muscle activates about 150 muscle fibers. The nerve and the fibers that

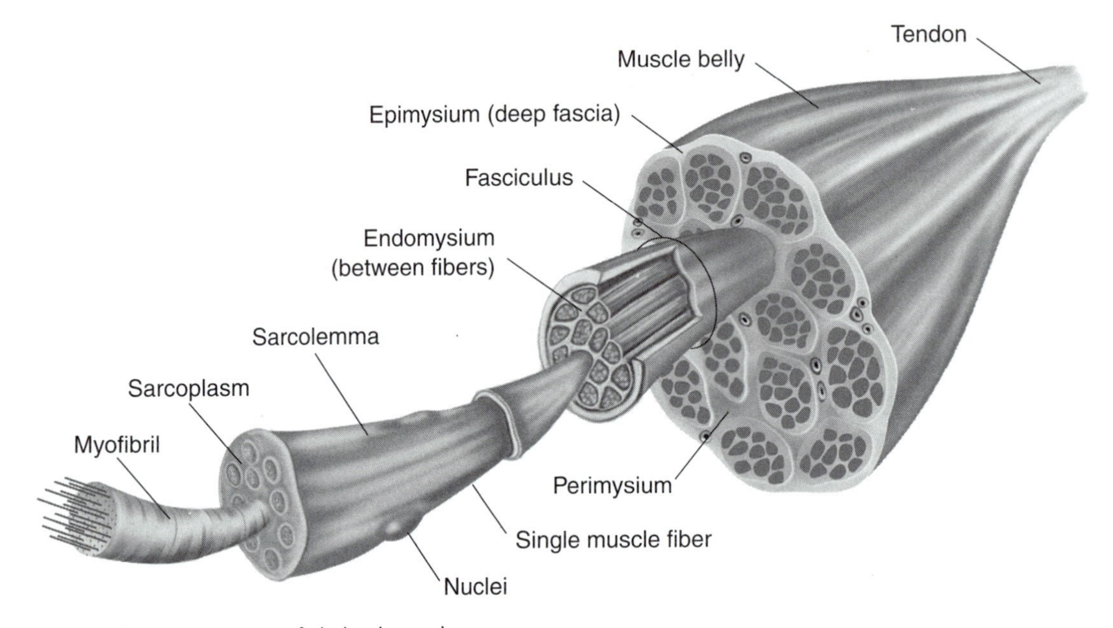

FIGURE 4.3 The organization of skeletal muscle.

Reprinted, by permission, from NSCA, 2000, *Essentials of strength training and conditioning,* 2nd ed. (Champaign, IL: Human Kinetics), 4.

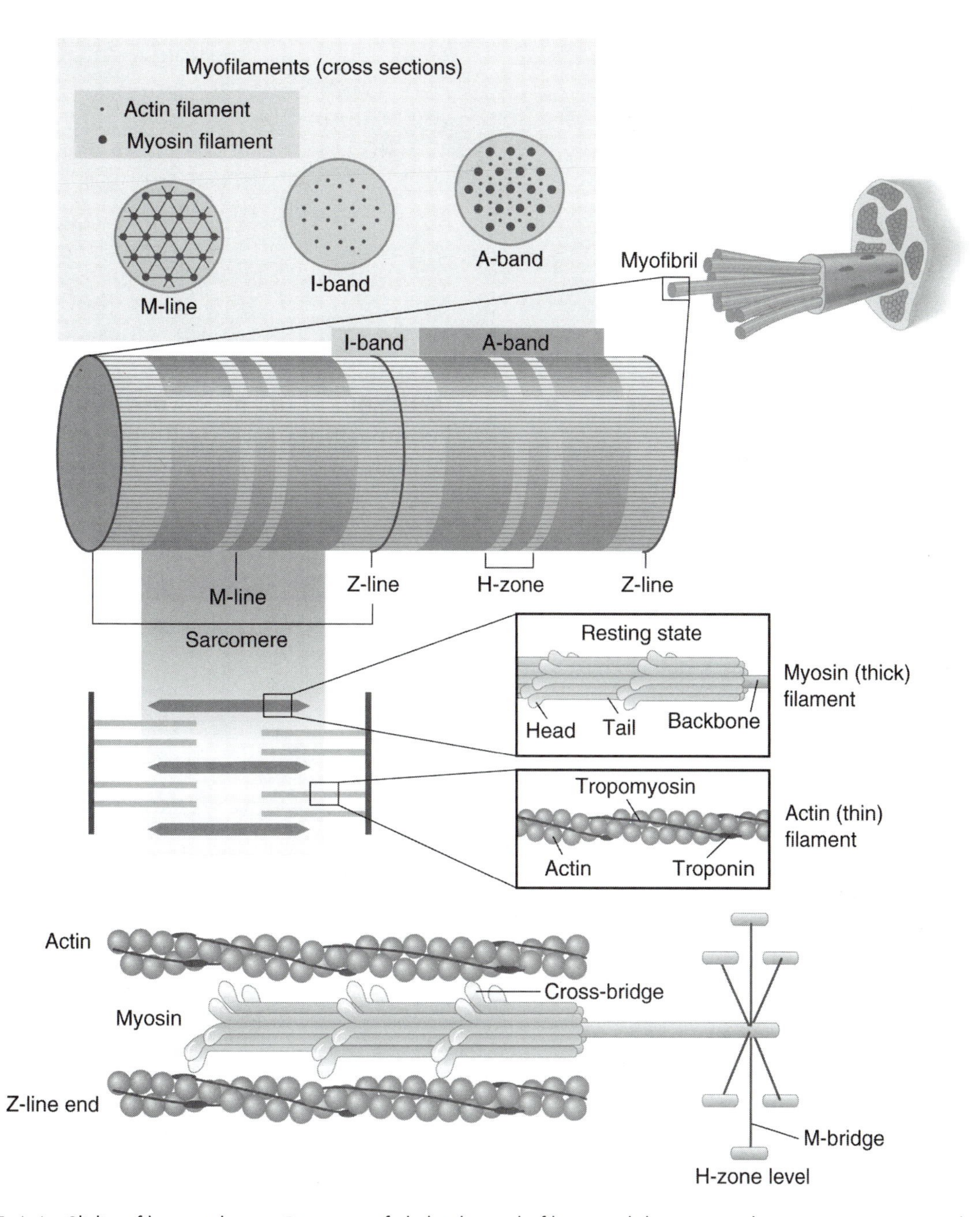

FIGURE 4.4 Sliding filament theory. Diagram of skeletal muscle fibers and the contractile proteins, myosin and actin, that "slide" over one another to cause muscle shortening.

Reprinted, by permission, from NSCA, 2000, *Essentials of strength training and conditioning*, 2nd ed. (Champaign, IL: Human Kinetics), 6.

it commands are called a **motor unit** (figure 4.5). When a motor nerve (nerve going to skeletal muscle) is stimulated, it causes all of the fibers that it commands to contract. Motor units consisting of many muscle fibers produce strong contractions, whereas motor units with only a few muscle fibers result in weak contractions.

Contraction Force and Muscle Fiber Type

The force of a contraction depends on the number of motor units that are activated and the size of the motor units. Additionally, in order for a motor unit to completely contract, continual signals must be sent down the nerve

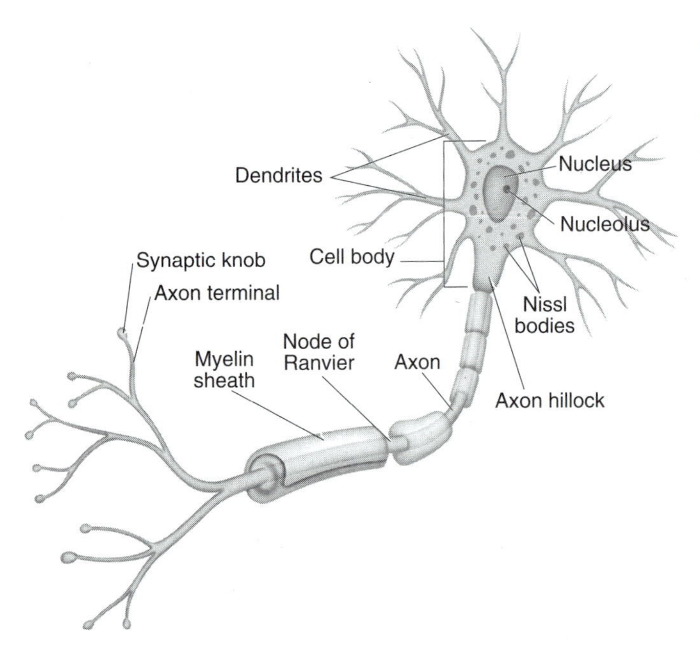

FIGURE 4.5 Each motor nerve stimulates a number of individual muscle fibers to contract. The muscle fibers will all be of the same type (SO, slow-oxidative; FOG, fast-oxidative-glycolytic; or FG, fast-glycolytic) but may not be immediately adjacent to each other. The force of a contraction is determined largely by the size and number of motor units that are recruited.

Reprinted, by permission, from J.H. Wilmore, and D.L. Costill, 2004, *Physiology of sport and exercise*, 3rd ed. (Champaign, IL: Human Kinetics), 62.

to sustain the contraction and keep the fibers fully activated. Thus, the force of a muscular contraction will be the result of the frequency of activation and the number of motor units within a muscle that are stimulated.

Although all fibers within a motor unit function in a similar fashion, there are distinct differences among motor units related to the types of muscle fiber being recruited. There are multiple classifications for skeletal **muscle fiber types.** The easiest and most functional classification is to distinguish fibers as either **slow-oxidative (SO), fast-oxidative-glycolytic (FOG),** or **fast-glycolytic (FG) fibers** (figure 4.6). All muscles have a combination of SO, FOG, and FG motor units. The percentage of fiber types in each muscle varies according to the function of the muscle, training, and, to some degree, genetics. Figure 4.6 shows how SO and FG muscle fibers intermingle within a muscle, and table 4.1 lists characteristics of each fiber type.

Slow-oxidative fibers contract more slowly than FOG or FG, but they are ideally suited for endurance work due to their high aerobic

TABLE 4.1

Characteristics of Muscle Fibers

Characteristics	Slow-twitch or slow-oxidative (SO)	Fast-twitch	
		Fast-oxidative-glycolytic (FOG)	Fast-glycolytic
Average fiber percentage	50	35	15
Speed of contraction	Slow	Fast	Fast
Time to peak tension (sec)	0.12	0.08	0.08
Force of contraction	Lower	High	High
Size	Smaller	Medium	Large
Fatigability	Fatigue resistant	Less resistant	Easily fatigued
Aerobic capacity	High	Medium	Low
Capillary density	High	High	Low
Anaerobic capacity	Low	Medium	High

Oxidative means that oxygen is used to produce energy. Glycolytic means that glycogen (stored glucose) is broken down to provide energy for contractions.

Reprinted, by permission, from B.J. Sharkey, 1986, *Coaches guide to sport physiology* (Champaign, IL: Human Kinetics), 28.

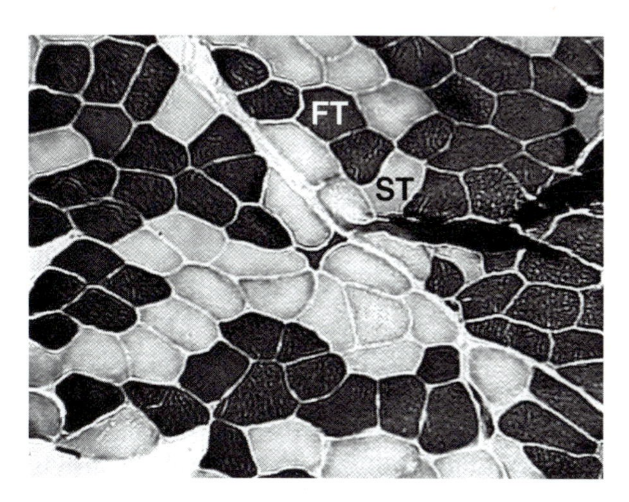

FIGURE 4.6 Fast- and slow-twitch muscle fibers. FT=Fast Twitch; ST= Slow Twitch.

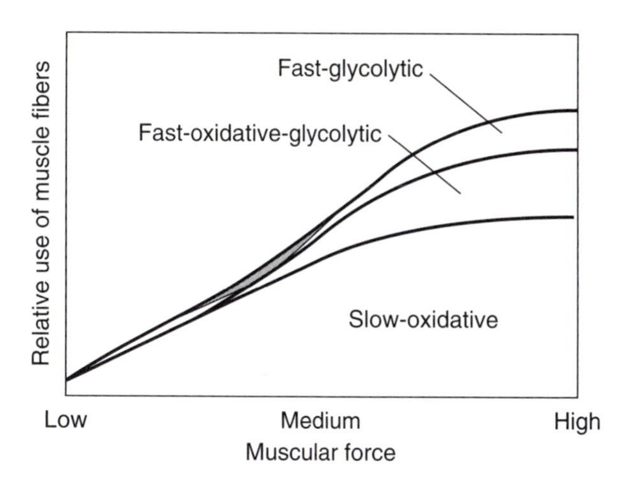

FIGURE 4.7 Additive effect of muscle recruitment. Slow-oxidative (SO) muscle fibers are recruited first for low-intensity muscular force. As intensity increases, fast-oxidative-glycolytic fibers are added; and as intensity reaches high levels, fast-glycolytic fibers are added. Note that SO fiber recruitment continues to increase as muscular force increases.

capacity (their ability to use oxygen). Fast-oxidative-glycolytic fibers share some of the characteristics of both SO and FG fibers and represent FG fibers that have been aerobically trained. Fast-glycolytic fibers are more suited for short-term, intense anaerobic (nonoxygen consuming) work.

The nervous system generally recruits SO muscle fibers first and, as more force of contraction and speed are required, adds additional FOG and then FG motor units. For slow movements like walking, the nervous system recruits SO fibers. As force of contraction is increased, for example when one goes from walking to running or from easy cycling to an uphill sprint, more fibers are continually added. Additional SO fibers are added first, then FOG fibers. Finally FG are recruited; they add greatly to the speed and force of contraction but are quick to fatigue as depicted in figure 4.7.

Muscles and Movement

Muscles generate force by shortening. This shortening causes bony lever systems to articulate, resulting in movement of limbs or segments. Since muscles can only pull, muscles come in groups, with groups of muscles working to allow movement in multiple directions. In addition, the shape and form of muscles, along with the fiber characteristics, determine the speed, force, and endurance of specific movements.

Biomechanics of Movement

In general, the attachment of muscles to bones is such that a slight shortening within a muscle may cause a large movement at the end of a limb. Visualize the bicep muscle of the upper arm, which can cause flexion (bending) at the elbow. The bicep muscle is attached to the scapula (shoulder blade) and, at the other end, to the radius (a long bone of the forearm) about 1 inch (2.5 centimeters) from the elbow. During some movements of the elbow, such as during a biceps curl (figure 4.8), the shoulder is stabilized. As the biceps contracts it pulls on the radius, causing the elbow to flex. Since the distance from the elbow to the hand is about 12 inches (30 centimeters) and the distance from the elbow to the biceps attachment is only about 1 inch, slight shortening of the biceps causes large hand movements. This type of lever arrangement is common in the muscular-skeletal system and allows us to generate high speeds of limb movement.

Generating Speed of Movement

The trade-off for these fast movements is that the bone and muscle lever systems require large forces. The elbow is an example of a typical lever system. During a biceps curl (figure 4.8) when the elbow is at 90 degrees, it takes about 600

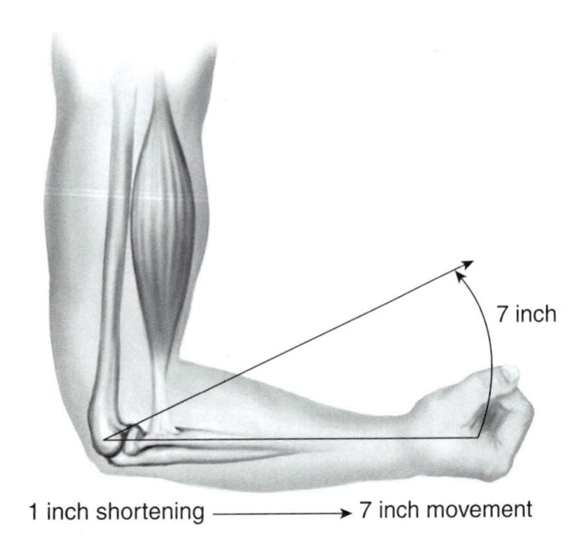

FIGURE 4.8 Basic biomechanics. High movement speeds are possible due to the normal lever arrangement of muscles and bones. In the case of the biceps, a very small shortening of 1 inch (2.5 centimeters) will result in a much larger (7-10 inch) movement by the hand.

Reprinted, by permission, from NSCA, 2000, *Essentials of strength training and conditioning*, 2nd ed. (Champaign, IL: Human Kinetics), 33.

pounds (272 kilograms) of muscular force to lift 50 pounds (23 kilograms). As the joint angle increases or decreases from 90 degrees, the required muscular force must further increase. In the skier study discussed at the beginning of the chapter, the latissimus and triceps muscles were used to generate poling forces resulting in high speeds. Increases in strength and power of these muscles greatly improve cross-country skiing speeds.

Shortening and Lengthening Contractions

As mentioned earlier, muscles do work by pulling. Joints move in both directions because we have opposing muscle groups. These opposing muscles are often called flexors or extensors based on their function. Consider an individual doing a biceps curl as shown in figure 4.8. The bicep is the muscle that flexes the elbow and raises the bar. During elbow flexion the muscle is shortening while generating force. This type of muscle action, shortening during contraction, is called a **concentric contraction.** Now consider the motion to slowly lower the bar back down. Gravity provides the force to lower the bar, and the biceps muscle must now resist gravity to maintain the slow downward motion. In this case the muscle is contracting while lengthening. This type of muscle action, lengthening while contracting, is called **eccentric contraction.**

Muscle lengthening and shortening contractions are important in sport. During running, muscle shortening contractions in the legs and hips generate most of the push-off force. When the front foot strikes the ground, muscles absorb the impact and contract to stop the leg from collapsing. This brief muscle lengthening contraction also serves to store energy that enhances the subsequent push. Imagine doing a vertical jump without first doing a short downward lengthening contraction or "elastic recoil."

Elastic Recoil

In nearly every athletic motion, the concentric power production is preceded by a short prestretching motion that elongates the muscles and acts much like a rubber band to increase the final force output. The increase in force production is often called the **elastic recoil** or stretch–shortening cycle. Elastic recoil is important as it occurs in nearly every sport. Running, throwing, striking, and jumping (figure 4.9) all make use of stored energy potential resulting from the quick stretch followed by a shortening contraction. The potential energy is caused by elastic energy in the muscle fibers resulting from stretching of the cross-bridges in the muscle proteins, activation of stretch reflexes, and possibly another muscle protein called elastin, which seems to have recoil properties. In chapter 6 we provide you with information about how to do plyometric exercises that train elastic recoil.

During the early stages of training, or when one is beginning a new exercise, excessive lengthening contractions, either from elastic recoil or from activities like downhill running, may cause severe muscle soreness beginning 12 to 24 hours after exercise. This soreness is called delayed-onset muscle soreness.

Muscle Soreness

There are two primary types of soreness associated with exercise training: short- and long-term. Short-term soreness may occur during or soon

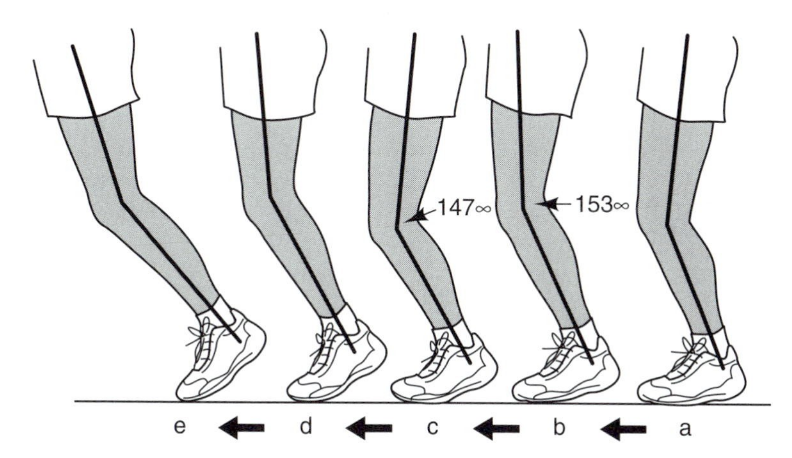

FIGURE 4.9 Elastic recoil: the stretch–shortening cycle during running. Just before the power phase and push-off during running, there is a brief moment in which gravity causes knee flexion (b-c) and eccentric (lengthening) contraction of the quadriceps. This brief stretch results in elastic recoil, which adds to the force of muscular contraction.

after extended or intense exercise. This temporary soreness is probably due to acid–base imbalances, slight swelling, and the buildup of metabolic waste, all of which are relieved within a short period. Long-term soreness that occurs the next day and lingers for several days is often due to the muscle lengthening (eccentric) contractions that have been done during the exercise. This residual pain that begins 12 to 24 hours after exercise and may last for up to two weeks is termed **delayed-onset muscle soreness (DOMS).** The pain from DOMS can often be severe and can limit physical activity. Muscle lengthening contractions that are greater than normal, such as in running downhill for the first time during a training season, cause microscopic tears in muscle membranes, releasing chemicals from muscle fibers. These chemicals stimulate free nerve endings, and the damage stimulates inflammatory responses. Sometimes the initial swelling may induce further damage and delayed healing.

Heredity Versus Training

While each of us inherits a certain percentage of SO and FG muscle fibers, the way the fibers are habitually used also influences the percentage of each fiber type. Research indicates that elite distance athletes have higher percentages of SO fibers, while sprinters and high jumpers have a higher proportion of FG fibers. Figure 4.10 shows these differences in the vastus lateralis muscle of athletes from a variety of sports.

Delayed-Onset Muscle Soreness

Remember the early spring day when you went out and tossed the ball around for an hour? Everything felt fine until the next day, when your throwing muscles were sore. Delayed-onset muscle soreness occurs about 12-24 hours after a vigorous effort that you haven't done for many months. Typically the effort involves eccentric contractions, which occur when muscles are involved in slowing vigorous movements (e.g., throwing a ball, lowering a weight, running downhill). Muscle soreness is correlated with submicroscopic muscle damage, accumulation of fluid (edema), leakage of enzymes (creatine kinase), inflammation, and diminished strength that may last up to two weeks.

Soreness can be viewed as something to cope with during the early season. Delayed-onset muscle soreness can be avoided or minimized if you increase training slowly, especially early in the season. The best rule is to avoid an excess of activities, such as downhill running, that cause lengthening contractions. It is advisable to begin an exercise program with concentric or muscle shortening contractions and only slowly add muscle lengthening work. For athletes who experience DOMS there is a small benefit. Following a period of DOMS, an individual will be protected for up to six months from DOMS. Modest vitamin E and C supplementation, taken prior to muscle lengthening exercise, may reduce the tissue damage following muscle lengthening contractions and may lessen the DOMS symptoms.

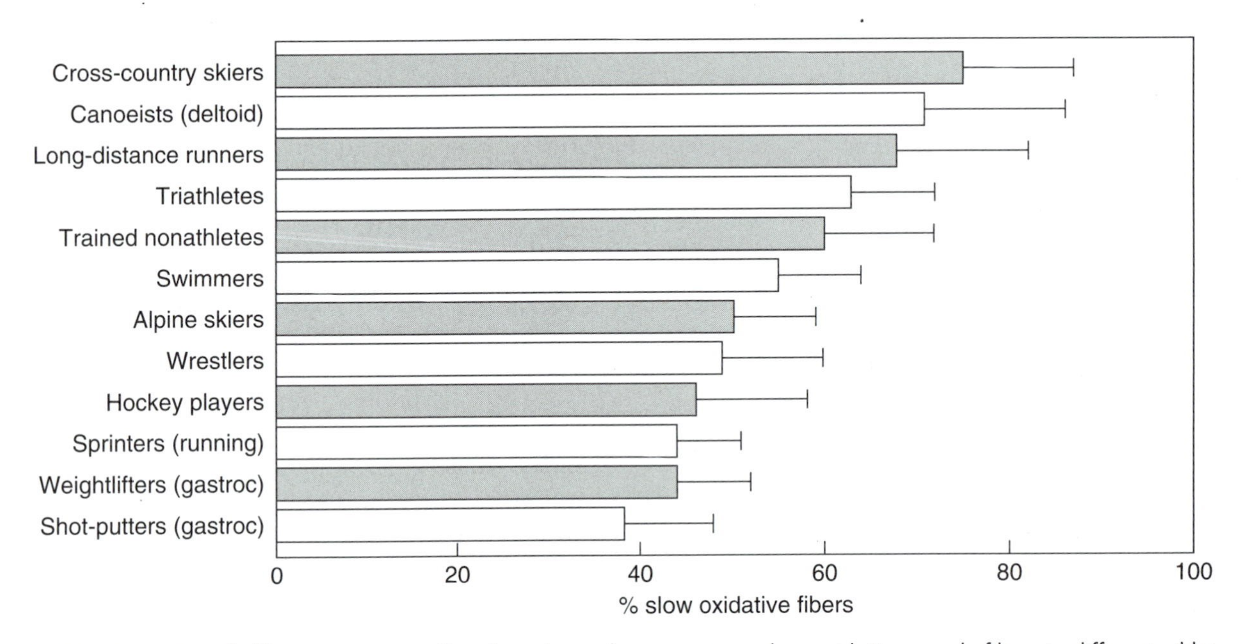

FIGURE 4.10 Muscle fiber composition. The chart shows the percentage slow-oxidative muscle fibers in different athlete groups. The balance of the muscle will be fast-twitch (fast-oxidative-glycolytic or fast-glycolytic) fibers. Unless noted, the muscle is the vastus lateralis portion of the quadriceps muscle of the thigh. The narrow bar represents the variation in individual subjects. As expected, endurance athletes have a higher percentage of slow-twitch fibers than other athletes. Data were assembled from a variety of sources.

Training can have an influence on muscle fibers. The fibers will respond specifically to the demands placed on them. Resistance training will result in increases in the size of muscle fibers (muscle hypertrophy) by stimulating an increase in the number of myofibrils (bundles of contractile proteins) within each cell. Fast-glycolytic fibers increase in size more than do SO fibers during heavy resistance training, but both will respond to the training stimulus. It is generally accepted that we increase the size of fibers rather than growing new muscle fibers (muscle hyperplasia) during strength training. Slow-oxidative, FOG, and FG fibers respond to endurance training by increasing their ability to use oxygen and resist fatigue. Slow-oxidative fibers respond to a greater degree, but FG fibers also undergo remarkable metabolic changes with endurance training. Fast-glycolytic fibers appear to maintain most of their fast characteristics as they increase oxidative enzymes and become FOG fibers. Many multisport athletes who require both power and endurance have a high percentage of FG fibers that have been trained to become FOG fibers. Muscle fibers are capable of considerable adaptation when exposed to the stress of training.

Components of Muscular Fitness

Now that you have learned something about the structure and function of muscles, let's take a look at the components of muscular fitness and see how training influences them.

Strength

Strength is synonymous with force, where force is the ability to move an object. **Strength** is the maximal force that can be exerted in a single effort. All sports require a certain amount of strength, but strength becomes a priority in sports in which heavy weights such as your body must be lifted, carried, or thrown. The strength of a muscle is directly related to the cross-sectional area of the muscle. Resistance training, if done properly, can increase the number of fibers recruited for work. Training also increases the cross-sectional area of muscles by increasing contractile protein, thereby increasing contractile strength.

Strength is unrelated to muscle fiber type. Slow-oxidative, FOG, and FG fibers are equally strong; when equal cross-sectional areas are compared, all three generate similar forces. The

How Much Is Enough?

As a coach, the challenge is for you to determine how much strength is required for specific skills and how to best train your athletes to develop and maintain the strength needed. As the duration of an event increases, the force per contraction that can be sustained is reduced. Trained endurance athletes may be able to generate nearly 30 percent of maximal force with every contraction for several hours, while 100-meter sprinters may be able to recruit nearly 95 percent of their maximal force at every stride for 10 seconds. In endurance events such as distance running, cycling, cross-country skiing, and distance swimming, strength gains are less important to improved performance than in sports such as weightlifting or football, in which athletes never seem to have enough strength. For every sport and individual athlete there will be optimal levels of strength. More is not always better, and each coach and athlete must decide when a strength program is warranted. Athletes with little strength will generally profit the most from strength training. Those who are already quite strong may see little improvement in performance, already having sufficient strength for the demands of their sport.

difference between fiber types is that the FG and FOG fibers can generate their force more rapidly than SO fibers.

Power

Have you ever wondered why a smaller defensive lineman in football may be able to outplay a larger offensive tackle? Obviously technique plays a large role; but if the two players are equally skilled, the one who has the greatest momentum (mass times velocity) will be able to stop his opponent. The smaller player must move at a higher speed than the larger player to have equal or greater momentum. How does the successful athlete develop greater velocity? When the ball is snapped in football, both players begin to accelerate. The player who can accelerate most quickly will attain the greatest velocity before the two collide. What allows one player to accelerate more quickly? The answer is **power.**

Power is defined as the rate of doing work (see sidebar "Work"). The relationships of strength, velocity, and work show that there are many ways in which one can think about power. Power during sport is best thought of as strength multiplied by velocity. The key concept is that athletes are universally concerned with speed and acceleration. Timed sports require covering a known distance in the shortest time. Ball sports require accelerating an object (ball) to high velocities. Boxing requires accelerating the hand and glove to maximal velocity when a punch is landed. A soccer play requires acceleration and the ability to change directions quickly, as well as the ability to impart momentum (using high foot

Soccer is an example of a sport that requires endurance to perform continuously for 90 minutes but also relies on powerful movements to take a powerful shot, to quickly change direction, and to outsprint an opponent.

Work

Work is defined as force multiplied by distance. In a practical sense this means moving some object, such as your body, a given distance. Another way to consider work is to think of overcoming the resistive forces (gravity, friction, or wind or water resistance) to move an object or body a fixed distance. In resistance training, lifting a 50-pound (23-kilogram) weight from the floor to over your head requires the same work whether done slowly or quickly. Another simple way to think of human work is to evaluate the number of calories that are expended. Two activities that require similar caloric (energy) cost are similar in the total work performed. The concept of work is important to our understanding of movement during training and competition, as work provides the link between strength and power.

Relationships of Force (Strength), Work, Power, and Velocity

work = force × distance

power = work / time

velocity = distance / time

Thus:

power = [force × distance] / time

or

power = strength × velocity

Selected Units

kilogram-meters, calories, foot-pounds

kilogram-meters/second, watts, joules

kilometers/hour, miles/hour, meters/minute

velocity) to the ball. All competitive sports rely on power for success. In general, the shorter the duration of the activity, the greater the power that is required.

It is useful to understand that the components of power include strength (force) and velocity. In order to become more powerful, athletes must be strong enough to generate the necessary power for success. Once they have adequate strength, athletes must train to use that strength at the velocities necessary in competition.

Muscular Endurance and Power Endurance

The ability to sustain submaximal contractions is usually called **muscular endurance.** It allows a barber to use scissors for many hours, whereas we would become fatigued in minutes. Muscular endurance is different from whole-body aerobic endurance since it usually involves small versus large muscle groups. Moreover, repeated small-muscle contractions do little to improve whole-body endurance. Nonetheless, muscular endurance is important in many occupations and recreational sports. When muscular endurance is required in sport, it must be achieved in sport-specific training, and that is where power endurance comes into play. In sport, speed is the factor that separates winners from losers, so training must be accomplished at or near competitive speeds.

Power endurance is the ability to repeatedly perform a skill with adequate speed. Power endurance is usually measured by the number of repetitions an athlete can perform within a time limit. Sports that involve multiple repetitions with a medium load over a few minutes require medium-term endurance. Wrestlers, for example, have to overcome an opponent's body weight. Because the competition lasts only a few minutes, wrestlers need short- to medium-term power endurance. Other endurance sports such as cycling and running involve numerous repetitions with a light load, requiring long-term power endurance. Sports such as the shot put, football, and sprinting require very short intense muscular work and involve only short-term power endurance. Many sports require that athletes can demonstrate explosive power, but also that they have long-term endurance to complete the match and to meet the changing demands of the activity. Examples might be team sports such as soccer, volleyball, basketball, and lacrosse. Individual sports such as tennis also require a combination of explosive power and quick recovery; tennis utilizes long-term power

Strength–Endurance Continuum

Strength is the ability to produce force. Endurance is the ability to sustain that force production or to repeat it a few or many times. Each sport requires a balance of strength and endurance. The Olympic powerlifter requires tremendous strength and power but little endurance. A wrestler needs to be strong and powerful to take down or move an opponent, but also needs good muscular endurance to control an opponent for three periods. During ski skating, the cross-country skier needs long-term power endurance to repeatedly produce small bursts of power. Where does your sport fit on the strength–endurance continuum? Table 4.2 shows the differing muscular fitness training across the strength–endurance continuum.

TABLE 4.2

The Strength–Endurance Continuum

			Power endurance		
	Strength	Power	Short-term (anaerobic) endurance	Intermediate endurance	Long-term endurance
Force	Maximal force	Explosive force	Brief (2-3 min) persistence with heavy load	Persistence with intermediate load	Persistence with lighter load
Prescription	2-8RM 1-3 sets	5-15RM 1-4 sets	10-25RM 3 sets	30-50RM 2 sets	Over 100RM 1 set
Improves	Contractile protein (actin and myosin) ATP and PCr Connective tissue	Strength and speed of contraction ATP and PCr Connective tissue	Some strength and glycolytic metabolism (glcolysis)	Some endurance and anaerobic metabolism Slight improvements in strength for untrained	Aerobic enzymes Mitochondria Oxygen and fat utilization
Doesn't improve	Oxygen uptake, endurance	Oxygen uptake, endurance	Oxygen uptake		Strength and power

RM = Repetition Maximum, or how many repetitions an athlete can do with a given resistance before failure.
ATP = Adenosine Triphosphate
PCr = Phosphocreatine

Adapted, by permission, from B.J. Sharkey, 2002, *Fitness & Health*, 5th ed. (Champaign, IL: Human Kinetics), 153.

endurance in the dominant arm while aerobic endurance is required overall. A cyclist generally relies on long-term endurance for most of a ride, but may need additional power to climb a short but intense hill or sprint to the finish. During these short periods of high-intensity activity, athletes need to be able to produce and use energy rapidly.

Power endurance training increases the muscles' ability to generate power by enhancing the ability of the metabolic pathways to produce energy for contractions. Long-term power endurance training improves the muscles' ability to utilize oxygen, thereby increasing endurance and recovery. High-intensity power endurance training enhances the anaerobic (nonoxidative) portion of the energy production system. How you train determines what you get, demonstrating the principle of specificity—you get what you train for.

Reaction Time, Quickness, and Speed

The terms reaction time, quickness, and speed are often used interchangeably, but they are really quite distinct in their meanings and underlying physiology. Reaction time is the time elapsed from the stimulus until movement starts. Quickness is a combination of reaction time plus the power of the initial movement. And speed is the final velocity attained. A few examples may help you to understand these concepts and explain them to your athletes.

Reaction Time

Consider the batter in baseball. Once a pitch is thrown, the batter receives stimuli that he or she must process in order to decide where the ball is headed, whether to swing, when to swing, and where to swing. The batter processes the information and sends signals from the brain down nerves to the muscles that will be used to start the swing. The time spent processing and signaling is called reaction time. We probably can't change the signaling time, but with practice and experience we learn to process the information more efficiently. With experience, the batter learns how to identify different pitches and to quickly decide if, when, and where to swing the bat. In this way, athletes improve that portion of reaction time requiring the processing of information.

Quickness

Quickness can most easily be thought of as the way an athlete reacts within a small area. Volleyball is an example of a sport requiring quickness. When we say that a volleyball player is "quick," we are referring to both the player's reaction time to stimuli (genetic and learned) and the speed of his or her initial movements. The player's initial speed depends on muscular power. Athletes with high levels of muscular power can accelerate quickly to make the appropriate movements within a small area of one or two steps. This initial acceleration relies on muscular force (trained) and the number of FG and FOG fibers (genetic) that the athlete can recruit. A third component of quickness, firing the appropriate motor fibers in the most efficient sequence, is another aspect that is greatly improved through practice and experience (learned). Thus, quickness has learned, trained, and genetic components, suggesting that while all can improve, some individuals will be able to increase quickness to a higher level than others.

Speed

Speed can be thought of as the final product of reaction time and quickness. Athletes who are said to be "fast" or to have good "speed" are those who are able to attain high velocities once they get moving. Sprinters need to have excellent reaction time, quickness, and speed. Middle-distance runners (400-1,600 meters or 437 to 1,750 yards) do not depend so much on their reaction time or quickness, but they must have adequate speed to be competitive for their distance. Most team sports (hockey, tennis, volleyball, linemen in football) require good reaction time and quickness, while other team sport athletes (soccer, lacrosse, receivers and quarterbacks in football) require excellent speed as well.

In general, each athlete can improve in reaction time, quickness, and speed. However, due to genetic influences, some athletes may have an edge in different areas. Athletes tend to find those sports that complement their abilities and to balance deficiencies with strengths. The very fast forward in soccer may be able to overcome a slower reaction time with good anticipation and by outrunning the defense. It becomes the coach's job to help each athlete use his or her talents and improve important areas as much as possible.

Balance

Dynamic balance is the ability to maintain equilibrium during vigorous movements. **Static balance** is the ability to maintain equilibrium in a stationary position. Both types of balance depend on the ability to use visual cues and input from receptors located in the inner ear, muscles, and joints. Participation in a variety of sport and movement experiences improves balance. Because balance is largely task specific, it is best improved by task-specific practice. In many sports, specific techniques to improve balance

Flexibility is sport specific, and a general rule is that athletes should be flexible through a slightly greater range of motion than that necessary for the unhindered performance of their sport.

are routinely recommended by knowledgeable coaches. In gymnastics, for example, widening the base, lowering the center of gravity, and focusing on a spot are useful tips to improve balance.

Flexibility

Flexibility is the range of motion through which joints are able to move. The natural range of motion of each joint depends on the design of the joint and the associated tendons, muscles, and ligaments. Furthermore, the flexibility of each joint is influenced by gender, habitual use, and stretching of the joint and associated structures.

Because many injuries occur when a limb or muscle is forced beyond its normal limits, flexibility training that gradually increases a joint's range of motion can help to reduce the risk of injury. In addition to limiting range of motion and increasing the risk for injury, tight muscles may impede optimal performance. Well-stretched muscles that flow easily through the range of motion require less energy and may facilitate better skill performance. Flexibility is sport specific, and a general rule is that athletes should be flexible through a slightly greater range of motion than that necessary for the unhindered performance of their sport.

As in many areas of training, more is not always better. Flexibility is inversely related to joint stability. This means that as athletes exceed the flexibility needed for their sport, joints become less stable and may become more prone to injuries. Unless you are coaching athletes in sports such as gymnastics, dance, or diving, a moderate degree of flexibility is probably adequate.

Agility

Agility, the ability to change speed and direction rapidly with precision and without loss of balance, depends on all of the factors already discussed: strength, power, power endurance, speed, quickness, reaction time, balance, and flexibility. Practicing sport-specific movements improves agility in that sport by enhancing learned movements, reducing cognitive processing time, training specific muscle fibers, and developing power endurance specific to the movement patterns required for the sport. Most coaches incorporate agility drills into early-season practices. If agility is a problem for any of your athletes, have them participate in games and sports requiring rapid changes in direction, such as racquetball, handball, and tennis, during the off-season.

Excess weight may hinder agility and balance. If any of your athletes are overweight, you may consider talking with the athlete's parents and possibly recommend that they consult with a nutritionist to develop a sensible weight loss program.

With a good understanding of the components of muscular fitness, you are now ready to learn more about assessing your athletes and preparing them for competition. The next two chapters will provide you with the information you need to evaluate and train for muscular fitness.

SUMMARY

In this chapter we have reviewed how muscles work, the differences in muscle fiber types, and the components of muscular fitness. Muscles create movement by shortening and moving bones around joints. The brain controls all stimuli for coordinated contractions that create movement. It is thus necessary for athletes to be able to visualize and understand movements in order to learn proper technique. Muscle fibers are the individual cells within a skeletal muscle that contract when the proteins actin and myosin slide over one another resulting in a shortening of the muscle. Nerves called motor neurons carry electrical impulses from the brain to the muscles. Each nerve stimulates a group of muscle fibers within a specific muscle to shorten. These motor units, ranging from a few slow-twitch fibers to a larger number of fast-twitch fibers, are the increments by which we recruit more muscular force. Three major muscle fiber types are generally recognized in humans: slow-twitch endurance fibers, fast-twitch glycolytic anaerobic fibers, and intermediate fast-oxidative-glycolytic fibers. These fiber types each respond to specific training and are recruited as needed for specific movements. Each individual has a genetic distribution of fiber types.

Muscle fibers can provide force by shortening (concentric) or by resisting lengthening (eccentric). Muscles provide the greatest force when they are stretched quickly before shortening; this is similar to the quick downward movement prior to a vertical jump. Stretching a muscle under tension (eccentric contraction) is often responsible for the delayed-onset muscle soreness (DOMS) experienced during the early periods of a training program.

Strength is the ability to generate force and is gained quickly at first through better recruitment of fibers, then more gradually through increases in muscle size. Work is the ability to apply force to move objects such as the body or external objects. Power is the speed at which we can do work. Thus power training requires both improvements in strength and improvements in movement speed. Muscular endurance is the ability of a specific muscle to resist fatigue and continue providing force for long periods of time. Muscular endurance is different from whole-body aerobic endurance since it usually involves small versus large muscle groups. Power endurance is the ability to repeatedly perform a skill with adequate speed for the sport or activity. Each sport requires a balance of strength and endurance, allowing the various sports to be categorized on the strength–endurance continuum.

Reaction time is the elapsed time from stimulus until movement starts. Quickness is a combination of reaction time plus the power of the initial movement, and speed is the final velocity attained. Dynamic balance is the ability to maintain equilibrium during vigorous movements. Static balance is the ability to maintain equilibrium in a stationary position. Flexibility is the range of motion through which joints are able to move. Well-stretched muscles that flow easily through the range of motion require less energy and may facilitate better skill performance. Flexibility is sport specific, and a general rule is that athletes should be flexible through a slightly greater range of motion than that necessary for the unhindered performance of their sport. Agility, the ability to change speed and direction rapidly with precision and without loss of balance, depends on strength, power, power endurance, speed, quickness, reaction time, balance, and flexibility.

KEY TERMS

actin (p. 44)
agility (p. 55)
concentric contraction (p. 48)
delayed-onset muscle soreness (DOMS) (p. 49)
dynamic balance (p. 54)
eccentric contraction (p. 48)
elastic recoil (p. 48)
fast-glycolytic (FG) fibers (p. 46)
fast-oxidative-glycolytic (FOG) fibers (p. 46)
flexibility (p. 55)
motor unit (p. 45)
muscle fibers (p. 43)
muscle fiber type (p. 46)
muscular endurance (p. 52)
myofibrils (p. 44)
myosin (p. 44)
power (p. 51)
power endurance (p. 52)
sliding filament theory (p. 44)
slow-oxidative (SO) fibers (p. 46)
static balance (p. 54)
strength (p. 50)

REVIEW QUESTIONS

Match the terms in the left column to the correct definition in the right column.

1. Slow-oxidative muscle fibers (SO)
2. Fast-oxidative-glycolytic muscle fibers (FOG)
3. Fast-glycolytic muscle fibers (FG)
4. Strength
5. Muscular endurance
6. Quickness
7. Speed
8. Agility
9. Reaction time

a. Maximal force that can be exerted in a single effort
b. Ability to change speed and direction rapidly with precision and without loss of balance
c. Slow contraction; lower force of contraction; small in size; fatigue resistant; high aerobic capacity; low anaerobic capacity
d. Elapsed time from stimulus until movement starts
e. Final velocity attained
f. Fast contraction; high force of contraction; large in size; easily fatigued; low aerobic capacity; high anaerobic capacity
g. Combination of reaction time plus the power of initial movement

h. Fast contraction; high force of contraction; medium in size; less resistant to fatigue; medium aerobic capacity; medium anaerobic capacity

i. Ability to sustain submaximal contractions

Fill in the blank with the correct word or words to complete the following statements.

10. Movement against a resistive force that shortens the muscle during contraction is called a(n) __________ contraction.
11. Resisting an outside force while lengthening the muscle during contraction is called a(n) __________ contraction.
12. The two main proteins involved in muscle contraction that "slide" over one another are __________ and __________.
13. Increase in force production due to a prestretching motion that elongates the muscle is called the __________.
14. Residual pain that begins 12 to 24 hours after exercise and sometimes lasts up to two weeks is __________.

Determine if the following statement is true or false and circle the correct answer.

15. T/F The force of muscle contraction depends on the number and size of the motor units that are activated.

PRACTICAL ACTIVITIES

1. George, a head college track coach, has a meeting with his assistant coaches to discuss the training needs of different groups of athletes. One of the first topics involves categorizing each sport in terms of its requirements for endurance, strength, power, and power endurance. The sports of interest include 10,000 meters, 5,000 meters, 1,500 meters, 800 meters, 400 meters, 200 meters, 100 meters, pole vault, high jump, shot put, and javelin. How would you categorize the requirements of these sports?
2. Mary, a high school history teacher, has been asked to step into the head softball coach position based on her past athletic background. The current season has just wrapped up, and the current coach has stepped down after 20 years at the helm of a successful program. Mary would like to start a training program including muscular fitness for the athletes. Mary starts by analyzing the needs of her different players, trying to pay attention to the specific skills needed for each position. How would you go about doing this?

5 Assessing Muscular Fitness

This chapter will help you

- identify appropriate tests for your athletes,
- understand the criteria for good tests,
- use tests that coaches and sport scientists have already developed and tested,
- learn how to use the test data to evaluate your team and the improvements made by individual athletes, and
- appreciate the value of long-term record keeping in improving your coaching and your athletes' performances.

Kathy had just passed the test for her soccer coaching license. The same day, she was offered a teaching and coaching position at the local high school. Kathy had been a successful college player, and she looked forward to sharing her passion for the sport with another generation. That fall, when soccer practice started, nearly 100 girls showed up to try out for the 36 spots on the varsity and junior varsity teams. Kathy had the JV coach and two assistant coaches to help evaluate the candidates. The athletes were scored subjectively for individual skill and knowledge of the game. Additionally, each player completed a series of physical tests. When the coaches' scores were tallied, Kathy was surprised by how well the ratings agreed, with 15 players appearing in the top 18 for all four coaches. The next 18 were also quite close. She spent most of that night comparing the physical test results with the subjective order. She was surprised to find that the rankings for the maximum vertical jump, repeated 40-yard (37-meter) sprint, and 1.5-mile (2.4-kilometer) run were quite similar to the coaches' subjective list. Kathy decided to use those test scores to pick players who were not unanimous choices of all the coaches.

The next thing that Kathy did was to create an intramural league for players not making varsity or JV teams. Volunteer coaches were found, and no players were cut. Kathy gave players a copy of their ratings and physical test scores. She challenged every player to improve to meet the top 10 standard for the physical tests. Players who had not made the varsity or JV team were included in the varsity or JV practice sessions when they met the standards. Kathy led her team to 10 straight winning seasons and never cut a player from the team.

Kathy understood assessment as well as motivation and development. Her methods followed guidelines similar to those commonly accepted and recommended for coaches. One recent study (Siegler, Ruby, and Gaskill 2003) had the purpose of evaluating changes in the soccer-specific power endurance of female high school soccer players throughout a season, either with or without an intermittent high-intensity exercise protocol. Thirty-four female high school soccer players were tested prior to the fall season and again 10 weeks later. The athletes first completed a game simulation and then did an increasing-intensity shuttle test for evaluation of match fitness. Other tests were hydrostatic weighing, vertical jump, 20-meter (22-yard) running-start sprint, and 30-second Wingate leg power test. The experimental group (n = 17, age 16.5 ± 0.9 years) completed a 10-week in-season plyometric resistance training and high-intensity anaerobic program. The control group (n = 17, age 16.3 ± 1.4 years) completed only traditional aerobic soccer conditioning. The experimental group showed significant improvements in the shuttle test and 20-meter sprint. The nonspecific Wingate test (often used for soccer players) did not change over the training season. This study indicates that a strength and plyometric program improved power endurance and speed over aerobic training. The value of the soccer-specific power endurance test was evident from the results. The soccer study shows that sport-specific tests are a valuable tool and may give very different results than standardized, nonspecific tests.

Kathy intuitively understood that some tests were related to soccer performance, and she used those tests to motivate athletes to improve in areas of weakness. Kathy further recognized that future potential is difficult to assess in developing athletes, so she found a method to keep everyone playing. She challenged the lesser-skilled players to improve. Many who might have normally been cut eventually made the varsity team.

What is the value of athlete assessment? The answer is straightforward: We test (1) in order to evaluate strengths and weaknesses in areas that might be important to performance and (2) to see if the training plan is working. Put another way, "If you don't know where you are going, you may end up someplace else" (Anonymous).

Appropriate assessment tests need to give the coach and the athlete useful information that can be used to improve performance. To be useful, an assessment method must meet several basic criteria. It must

be sport specific,

be valid (measure what you want to measure),

be reliable (give a similar result when repeated),

be objective (not be affected by coach bias),

be gender and maturity appropriate,

be easy to administer,

be easy for athletes to learn, and

give results that can be used to monitor progress.

The ultimate assessment is performance. Performance is the sum of many factors. Great athletes must be strong in nearly all areas required for their sport. The purpose of athlete testing is to identify factors of performance and to give the athlete feedback on each factor so that specific training, if needed, can be designed to improve weaknesses while maintaining strengths. The beauty of assessment is that it allows measurement of the effectiveness of training plans in terms of both overall performance and individual skills required for that performance.

The purpose of athlete testing is to identify factors of performance and to give the athlete feedback on each factor so that specific training, if needed, can be designed to improve weaknesses while maintaining strengths.

Sport Specificity

Each sport has a wide range of factors dictating success. Top athletes require good mental, physiological, technical, tactical, time management, social, and other skills to be successful. This chapter focuses specifically on assessment of muscular fitness. However, the basic rules for assessment can be applied to all other areas.

In the area of muscular fitness, coaches need to identify specific movement patterns (skills) and the associated muscular fitness that is required for their sport. These specific areas of muscular fitness need to be prioritized for each team and athlete so that testing can give meaningful data that will help each athlete. Chapter 4 covered the various aspects of muscular fitness. Later in this chapter, we provide examples of possible tests in each category—tests for strength, power, power endurance, flexibility, and agility. Given the wide variety of sports, it is not possible, or desirable, to define appropriate testing for all athletes. Coaches should use the examples and guidelines in this chapter to guide them in developing appropriate tests for their athletes. To get you in the right frame of mind, consider the high school girls' basketball team discussed in the sidebar "Testing for Basketball."

Test Design

Once the sport-specific muscular fitness attributes have been identified, appropriate tests need to be chosen or designed. When you design or choose tests, there are four main characteristics of each test that you should consider: validity, reliability, objectivity, and appropriateness to the group. More detail on designing appropriate tests is included in appendix D.

Validity

The most important aspect of testing is **validity:** Does a test measure what you want it to measure? Most coaches primarily use tests that are generally accepted as tests of the desired attribute, such

Testing for Basketball

Mary was a high school coach for the girls' basketball team. When designing her muscular fitness testing program, she identified jumping ability, running speed, agility, balance, upper body strength, and leg endurance as important to success. She prioritized the list and realized that in some areas such as sprinting speed, she needed to further evaluate attributes of speed, such as acceleration (leg power) and strength of the quadriceps and hamstring muscles. Here is what she designed for her team: repeated vertical jump test for leg power and power endurance (p. 67); 10RM hamstring curl and knee extension tests to measure the strength and the ratio of the hamstring and quadriceps muscles (p. 68); stair run for leg power and acceleration (p. 142); repeated 40-yard (37-meter) sprints for power endurance, speed, and power (p. 71); agility run (p. 76); and the sit and reach test for hamstring flexibility (p. 72). She administered these tests at the beginning of the season. She then set up testing so that athletes completed all six tests every two weeks. The players were always given feedback on their improvements. Mary used the tests to help athletes set goals and to evaluate the training program. She had great success with her teams.

All sports and athletes can benefit from a program of regular testing to evaluate changes in sport-specific fitness.

as a laboratory test that measures maximal aerobic fitness ($\dot{V}O_2max$). Other tests estimate a performance factor; for example, a 1.5-mile (2.4-kilometer) run can be used to estimate maximal aerobic fitness. These estimates are usually related to an accepted standard (such as $\dot{V}O_2max$). Some tests might have predictive value by relating closely to performance. The key is that coaches can identify what they desire to test and find the appropriate test. Many existing tests are available that have been scientifically tested for validity.

Reliability

Reliability is the repeatability of the test. If you were to give the same test on two consecutive days, a perfectly reliable test should give identical scores. Of course, no test will give exactly the same score each time, but coaches should use only tests that always give similar results if an

Validity and Reliability

Tests must be both valid and reliable in order to be useful to coaches. It is somewhat obvious that in order to be valid a test must also be reliable, as without reliability we don't really know what the results mean. However, it is a common mistake for coaches to believe that reliable tests are therefore valid. Coaches in many sports have been using the reliable power endurance tests of pull-ups and push-ups to evaluate athletes. Though these tests may be appropriate for some activities (rock climbing would be obvious), many sports have no related muscular fitness needs. Unless a coach can show that one type of validity is fulfilled, a test should not be administered.

athlete is retested within a short period of time. The differences in test results, which make tests unreliable, may be the result of either inconsistent performance or measurement error. Measurement error may occur for many reasons (see appendix D). To improve reliability, tests should be simple and easy to understand. Coaches are encouraged to do the following:

Use simple standardized instructions.

Have the same trained person make the same measurement on each test.

Be sure athletes have adequate practice time if a "learning curve" is involved in performing the test.

When possible, repeat the test until consistent measurements are recorded.

Always try the test first before administering it to athletes.

Objectivity and Bias

Objectivity is the ability of separate scorers to independently agree. Problems in objectivity can be demonstrated using the example of evaluating a vertical jump test. The athlete jumps up and touches a tape on the wall. Three observers record the point he or she touches. One observer is watching from a chair on the floor. A second observer is standing on a platform just behind the athlete with his eyes close to the touch level. The third observer, the coach, is also on the platform. The first observer, because of his or her angle of view from the floor, may have difficulty seeing how high each athlete reaches. The first observer's score may thus disagree with that of the second observer, who has a good view of the highest point reached by each athlete. The second observer and the coach may also disagree based on the coach's (or the second observer's) bias. One may want the athlete to succeed or fail and may record a score appropriate to that bias. A better method is to use a vertical jump measurement system such as the Vertec system (see figure 5.1), in which athletes move the highest bar that they touch, so there is no subjective judgment.

FIGURE 5.1 The Vertec vertical jump measurement system removes the need for subjective judgment in the vertical jump test.

In order to improve objectivity, it is essential to have a clearly defined and consistent scoring method for each test. When using a stopwatch, be sure the timer is well trained and unbiased. Use the same timer each time you run the tests. If possible, use an electronic timing system. Use scorers who have no personal interest in the athletes (are unbiased). It is best to use scorers who do not personally know the athletes.

Some tests are subjective by nature. In the story at the beginning of this chapter, Kathy based her team selection primarily on subjective coach ratings. She did this only after comparing the ratings of all the coaches to verify that the selections were similar. Where she felt that there was inadequate agreement (objectivity), she based her final decisions on objective assessments.

Criteria for Good Tests

As we have said, the most important criterion in evaluating athletes is how well they perform. All other measurements help evaluate factors that may contribute to an athlete's success. There are 11 basic criteria for athlete assessment. Good tests should have most or all of the following characteristics:

- Relevant: Does the test measure factors related to athletic success?
- Objective: Can the test be objectively scored or, if subjective, are the scores reliable across evaluators?
- Specific: Is the test sport specific and is proper technique controlled?
- Appropriate: Is the difficulty, skill, strength, speed, or distance appropriate for the age and ability level? Have the athletes been training for this attribute?
- Learning curve: Does the test have a learning curve, and, if so, are adequate trials and practice allowed before the test is used for scoring?
- Gamelike: Does the testing involve game simulations, or can it be done at game intensity?
- Meaningful: Are the results meaningful and useful to the coach and the athlete in designing and evaluating training programs?
- Individual: Can the test be done individually?
- Simple: Is the test simple and straightforward?
- Challenging: Will the test challenge the athletes and differentiate between them?
- Sensitive: Can the test measure small training differences that can be evaluated statistically? (see appendix D)

Administering Tests

In order to properly administer assessments, coaches should adhere to a number of principles. These are meant to increase the validity, reliability, and objectivity of the testing. Some of the topics in this section revisit points made previously.

- Test order. Order your tests starting with those that will require the shortest recovery and proceeding to those that require the longest recovery. Always allow adequate rest time between tests. Start with low-fatigue tests of skill, agility, and coordination and then move on to tests more likely to result in high levels of fatigue. Next administer moderate-fatigue strength and power tests, allowing at least 4 to 5 minutes' rest after each test. Finally, high-fatigue tests such as those for aerobic or anaerobic capacity should be run last. If you have both anaerobic capacity (1- to 3-minute maximal capacity) and aerobic capacity (>4-minute maximal) tests, these should be administered on separate days.

- Warm-up. Athletes need to warm up for each test requiring physical performance. Allow the same warm-up that you would for a practice or competition. Start with a general aerobic warm-up and finish with activities specific to the test.

- Instruction. Athletes should be given clear instructions and should be shown the test procedures. Allow the athletes to ask appropriate questions, and provide appropriate practice time. Multiple trials of low- or moderate-fatigue tests, with the best or average score being recorded, often improve reliability.

• Oversight. As the test administrator you are responsible for safety and accuracy of the testing. Be sure the facilities for testing will provide a fair trial for all athletes. Be sure that the area and the exercises are safe and appropriate for the athletes to be tested. Observe your test administrators, ensuring that they are paying close attention and following procedures. Stress that good form should be used during all testing. When appropriate, supply spotters during strength and power testing.

• Encouragement. Offer equal encouragement to all athletes. If possible, simulate game intensity and motivation.

• Rest. Be sure to allow for adequate rest between tests or trials. This requires that the administrator have a solid understanding of the tests. In general, skill and agility trials require only a 30- to 60-second rest; short-term maximal-effort tests require 4 to 5 minutes between trials while longer maximal-effort tests may require 1 to 3 hours between tests.

• Feedback. Athletes should be given feedback as soon as possible during or after the testing. Be sure that athletes understand that the results are for their personal benefit and that the tests are not a competition. Giving immediate feedback often improves motivation for subsequent trials. Timely feedback after the testing, including how to improve in areas of weakness, is also a great motivational tool in that it provides specific goals for each athlete to achieve.

Sample Tests

The following tests are meant to be examples of what coaches might use to measure different aspects of muscular fitness. Some of the tests might be applicable for your sport, but more often you will need to develop your own tests or find sport- and position-specific tests that have already been scientifically established (valid, reliable, and objective). Table 5.1 gives a basic overview of practical tests for each area of muscular fitness.

Strength Tests

Strength is considered the maximum weight that an individual can move slowly through a specific range of motion. To test for strength, coaches first need to identify the sport-specific movements that require strength and then find or design the tests that will measure the move-

TABLE 5.1

General Testing Guidelines for Muscular Fitness for Coaches

Attribute	Typical tests	Necessary equipment	Rest between trials
Strength	1-3RM	Free or machine weights	2-3 min
Power	Standing long or vertical jump, Wingate cycle test, <5 sec stair run	Tape measure, vertical jump scale, stationary cycle, stopwatch	1-5 min
Power endurance	Chair dips, push-ups, sit-ups, bench hop, 20-30RM weights, repeated vertical jump	Chairs or benches, stopwatch, free or machine weights, vertical jump scale	10-15 min
Speed	5 to 40 yd sprint	Track, electronic timer, stopwatch	2-5 min
Flexibility	Sit and reach test, shoulder rotation test, full-body rotation test	Sit and reach box, measuring stick, tape measures, 6 ft pole	20-60 sec
Balance	Balance board, one-legged balance	Balance board, stopwatch	20-60 sec
Agility	Ladder run, agility run	Agility ladder, cones, stopwatch, gym floor	2-3 min

Necessary equipment refers to equipment to which most coaches might have access. Laboratory measures are not covered in this table.

ment. Generally, a 1RM (1-repetition maximum) lift is used for these tests.

The actual measurement of maximal 1RM force is never done in beginners and is generally used only after the athlete has trained the lift for several weeks and has the necessary skill, balance, coordination, and supporting musculature to allow for safe lifting. Be sure to use spotters if warranted. Here is the procedure for estimating strength from 1RM effort:

- Have the athlete warm up aerobically.
- Have the athlete perform the lift with progressively higher resistance of about 50, 70, and 85 percent of estimated 1RM resistance. Have the athlete rest after each set.
- Estimate the 1RM weight for the athlete and have the athlete attempt the lift.
- If the athlete is unable to lift the weight or is able to do more than one repetition, stop and give a 3- to 5-minute rest. Depending on the effort of the previous lift, adjust the resistance and repeat. Continue until the athlete fails at 1RM. Be sure to allow adequate time for the athlete to recover between trials.

With younger or less skilled athletes, a resistance that causes failure within 10 repetitions (10RM) can be used to estimate maximal strength. Here is the procedure for estimating strength from a 2- to 10RM effort:

- Have the athlete warm up aerobically.
- Have the athlete perform the lift with light resistance to warm up specific muscles.
- Estimate a 10RM weight for the athlete.
- If the athlete is able to complete only 2 to 10 repetitions, use table 5.2 to estimate his or her maximum strength for that activity.
- For an athlete able to complete more than 10 repetitions, stop the athlete at 11 and give a 3- to 5-minute rest. Depending on the ease with which the athlete completed the previous set of 10, adjust the resistance and repeat. Continue until the athlete fails between 2 to 10 repetitions.

Use table 5.2 to estimate maximal strength and measure changes in the strength of your athletes. Divide the amount of weight lifted by the percentage in the left-hand column (as a decimal) to get an estimated 1RM value. Example: The athlete is bench pressing 75 pounds (34 kilograms) and completes seven repetitions. From the table you find that 7RM represents 83.5 percent of maximum strength. Divide 75 by 0.835 to get an estimated maximum strength of 89.8 pounds (75 pounds ÷ 0.835 = 89.8 pounds).

Athletes can be tested for strength using many different lifts, including the bench press.

Muscular Power Tests

Power is force times velocity. The resistance in sport may be an athlete's body weight during human-powered propulsion (e.g., road cycling, mountain biking, running, swimming, cross-country skiing, long jump, pole vault) or another object as in throwing or pushing sports (e.g., shot put, baseball, football, wrestling) or Olympic powerlifting. The range of required power for different sports covers a continuum from repetitive power endurance, such as the power required for each

TABLE 5.2

Estimating Maximal Strength (1RM) From 2- to 10RM Sets

RM to failure	Percentage of maximum
2	0.935
3	0.91
4	0.885
5	0.86
7	0.835
8	0.785
9	0.76
10	0.735

Use of this table is an estimate of maximal strength, and the accuracy will vary for different exercises. It does provide an adequate method for measuring increases in strength.

From Fleck and Kraemer 1987.

push-off during running or cross-country skiing, to the single maximal effort required for a vertical jump or a shot-put throw. The coach must understand the requirements of the activity in order to design appropriate tests. Here we give two examples to help demonstrate tests that can be used at different points on this continuum.

Single and Repeated Vertical Jump Test

Sports that require the ability to jump vertically range from explosive (high jump) to endurance or intermittent (basketball and soccer). Measuring a single maximal vertical jump may be less useful for basketball players who do repeated jumps in a game than for the high jumper. The following test can help you to evaluate both explosive power and power endurance. Power endurance will help you evaluate how an athlete is likely to perform during a game when he or she is tired. See table 5.3 for information on the relationship between vertical jump results and fast-twitch fiber composition.

Tools

- Stopwatch
- Clipboard with recording sheet and pencil
- Vertical jump measuring tape on a high wall (landings in stairwells often work well)
- Step ladder or 36-inch (91-centimeter) bench
- Three test administrators: (1) judge standing on step ladder or riser just behind the test area, (2) timer, and (3) recorder (see figure 5.2—if not enough people are present, the judge on the step ladder can also be the timer)

Procedure

After an aerobic warm-up and several easy practice jumps, the athlete will be ready to start.

- Standing with the dominant hand and arm by the wall and feet flat on the floor, the

TABLE 5.3

Relationship Between Vertical Jump Results and FT Fiber Composition

Age and sex	Low % FT fibers	Average (50%) % FT fibers	High % FT fibers
UNDER 14			
Males	Under 13 in.	13-18 in.	Over 18 in.
Females	Under 6 in.	6-10 in.	Over 10 in.
OVER 14			
Males	Under 15 in.	15-21 in.	Over 21 in.
Females	Under 8 in.	8-13 in.	Over 13 in.

FT = Fast-twitch muscle fibers

FIGURE 5.2 Repeated vertical jump test.

athlete reaches up as far as possible with the dominant hand. This height is recorded.

- The judge stands on the ladder or bench so that his or her eyes are at about the height that the athlete is expected to reach.
- Standing in place, keeping the dominant arm in the air and the nondominant arm by the side, the athlete jumps as high as possible and taps the wall. The timer starts the watch. The judge notes the highest position reached by the fingertips regardless of where the wall is tapped. The judge orally says the height for the recorder to note.
- At 10 and 20 seconds, the athlete repeats the jump on the timer's signal, "Ready-jump," and the judge relays the height achieved to the recorder.
- Starting at 30 seconds, the athlete is instructed to complete one jump for maximum height every 2 seconds on the timer's signal, "Ready-jump." The athlete continues to jump 15 times (once every 2 seconds for 30 seconds). The judge notes the height of each jump and informs the recorder.
- Each jump must be completed starting with the reaching arm vertical and the nonreaching arm remaining by the athlete's side. This isolates leg power by eliminating the use of the arm swing. The athlete's feet must be stationary at the start of each jump.

Scoring

Subtract the reach height from each of the 18 jumps. The highest of the first three jumps is scored as the maximal vertical jump (see figure 5.3). The averages for the next three sets of jumps (4-8, 9-13, and 14-18) can help coaches evaluate power endurance. Athletes who remain within 95 percent of the maximal height for jumps 14 through 18 have good power endurance, while those who drop below 85 percent have poor power endurance.

Stair Run

The stair run test is shown in chapter 8 (p. 142). Variations can be made to match the demands of specific sports. These variations include one-legged stair hops instead of running and increases in the number of stairs (hops or running). For one-legged hops, test right and left legs individually with adequate rest between trials. Be sure to allow athletes time to practice the test. Hopping is usually done two steps at a time. Tests that are very short may require timing pads and will measure maximal muscular power, while tests as long as 30 seconds (using stadium steps) can be used to test power endurance.

Power Endurance Tests

Power endurance might more appropriately be called "local power endurance," the ability of specific muscles or muscle groups to repeatedly produce adequate power to maintain a high level of performance. Power endurance is similar to aerobic endurance (chapter 7) in that both require the muscle to be able to sustain energy production via aerobic (oxygen) pathways. The difference between the two is that aerobic endurance places a large demand on the cardiovascular system by recruiting many muscle groups simultaneously, resulting in a large total oxygen demand. Cross-country skiing is an example of a sport requiring a high degree of aerobic energy

Name____

Standing reach height ________ **inches**

Time	Trial #	Jump height (to 1/2 inch) (1)	Vertical jump (2)	Scoring
0	1	________	________ inches	
15	2	________	________ inches	Maximal vertical jump
30	3	________	________ inches	Highest: trials 1, 2, 3_________
32	4	________	________ inches	
34	5	________	________ inches	
36	6	________	________ inches	
38	7	________	________ inches	
40	8	________	________ inches	Average: trials 4-8____________
42	9	________	________ inches	
44	10	________	________ inches	
46	11	________	________ inches	
48	12	________	________ inches	
50	13	________	________ inches	Average: trials 9-13___________
52	14	________	________ inches	
54	15	________	________ inches	
56	16	________	________ inches	
58	17	________	________ inches	
60	18	________	________ inches	Average: trials 14-18__________

FIGURE 5.3 Recording sheet for repeated vertical jump test.

From *Sport Physiology for Coaches* by Brian J. Sharkey and Steven E. Gaskill, 2006, Champaign, IL: Human Kinetics.

fitness. During long-duration events of 10 minutes to 3 hours, cross-country skiers use both their arms and legs. This large recruitment of multiple muscles requires skiers to have a high aerobic capacity (ability to take in, transport, and use oxygen) and stresses the cardiovascular system. Cross-country skiers also need to have adequate power endurance to allow each muscle to produce the necessary bursts of energy for each ski stride repeatedly for many minutes. Some activities, such as repeating 100 sit-ups, also require local power endurance but do not

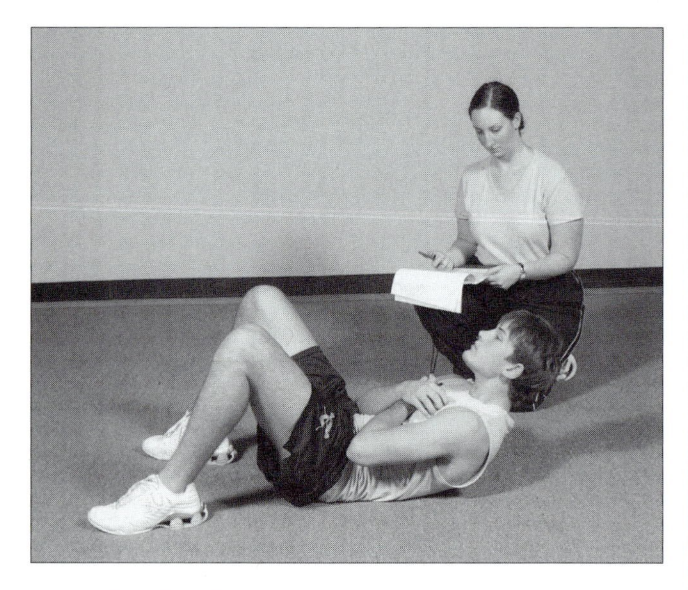

Three of the most popular tests for power endurance are timed sit-ups, push-ups, and dips. To be effective, tests should be specific to the muscular actions used in a sport.

activate enough total muscle mass to demand high levels of aerobic energy fitness.

Tests for power endurance thus are similar to tests for aerobic energy fitness as discussed in chapter 8, but focus on specific muscle groups. Timed sit-ups, push-ups, and dips are common examples of power endurance. The repeated vertical jump, previously discussed, is also a good example of short-term power endurance. Power endurance can be tested using free or machine weights. These tests are typically performed using standard weights that an athlete can lift about 15 to 30 times to exhaustion. Improvements are evaluated either when the athlete lifts more weight the same number of times or when the athlete lifts the same weight more times within a given time period. With computerized cycle ergometers, cycling-specific power endurance can be tested through measurement of average watts over 1 to 15 minutes. As you recall from the previous chapter, power can be defined as the rate of doing work. Whenever possible, tests for power endurance should include the number of repetitions (work) within a set time. Here is the procedure for the bench hop (figure 5.4), used to estimate power endurance.

Tools

- Bench or gymnasium bleacher, 16-1/4 inches (41 centimeters) high
- Stopwatch

FIGURE 5.4 Bench hops.

Procedure

- Athletes should warm up aerobically.
- Allow athletes a short practice and recovery period.
- On the signal "Ready–go," a stopwatch is started and the athlete starts a two-footed jump onto the riser and back to the floor. Athletes who cannot continue for a full minute may step up and down. A repetition is counted only when both feet are on the floor.

Scoring

Use table 5.4 to evaluate scores. The values are norms for the age groups shown. Individuals at the lower end of an age group might be expected to perform slightly less well. Maturation is not considered in the norms; and for athletes 10 to 16 years old, maturation may be a significant factor in individual values. The best way to use test values is to use the criterion scores only as guidelines and use individual athlete scores as baselines against which to measure improvements.

Speed Tests

Speed requires power; and the greater the power that can be generated by a muscle, the greater the potential for high speed. Speed tests must be specific to the requirements of the sport. Chapter 8 on assessment for energy fitness covers tests for short-term anaerobic and longer-term aerobic speed. Here we discuss the issue of assessing both speed and recovery. Many events require high levels of speed (muscular power) that must be achieved frequently and repeatedly following varying periods of rest or light activity. Many intermittent sports such as volleyball, soccer, lacrosse, football, and basketball fall into this category. The coach must design tests that evaluate speed not only in the rested state, but also in game simulations. In order to design a test to evaluate speed during game situations, a coach must first evaluate specific positions and have an understanding of the average sprint length and the average rest period. Design standardized tests that require athletes to sprint an appropriate

TABLE 5.4

Criterion Scores for the 16-1/4 Inch Bench Hop

Group	Excellent	Good	Average	Below average
MALES				
10-13	>52	44-51	38-44	<38
14-18	>57	51-57	46-56	<46
19-30	>63	57-63	51-56	<51
FEMALES				
10-14	>50	40-50	34-39	<34
14-20	>52	42-52	36-41	<36
20-30	>54	44-54	39-43	<39

distance and then rest an appropriate length of time before repeating.

An example of testing during a gamelike simulation is a 45-minute soccer-specific shuttle test that was used during a research project conducted in the Human Performance Lab at the University of Montana. (A brief summary was given at the start of this chapter.) The test, called the shuttle test, starts with a game simulation using standardized sprint and rest periods similar to those analyzed during soccer games. Once the athletes are tired, they are challenged to perform increasingly fast sprints with standardized rest periods until they can no longer meet target goals. We have provided a more generic test, using an aerobic run followed by progressively faster 40-yard (37-meter) sprints. Here is the procedure for a repeated 40-yard sprint test.

Tools

- 400-meter or 440-yard (1/4-mile) running track
- Measured 37-meter or 40-yard section on the track, or a field with 40 yards marked by lines or cones
- Stopwatch
- Personnel: timer and recorder
- Good running shoes are needed by athletes

Procedure

- Standardized warm-up: Each athlete is asked to jog one lap of the track in the following times:

  Lap 1 (start at time 0): 2:00-minute lap followed by 30 seconds rest

  Lap 2 (start at time 2:30): 1:55-minute lap followed by 35 seconds rest

  Lap 3 (start at time 5:00): 1:50-minute lap followed by 40 seconds rest

  Lap 4 (start at time 7:30): 1:45-minute lap

  Athletes now get 45 seconds' recovery, during which they jog to the start of the 40-yard sprint, if different from the start of each lap.

- Athletes now run 40-yard sprints and jog back to the start according to the time schedule shown in figure 5.5. The starter stands on the 40-yard finish line and yells "Ready" 5 seconds before the start and signals ready with the arm held vertical. The "Go" signal is a rapid downward arm movement. At the allotted time for the 40-yard sprint noted in figure 5.5, the timer yells "End." Any athletes who clearly do not cross the finish line by the "End" signal are pulled out by the assistant. Their best time is noted on the recording sheet, and they begin a cool-down lap. Athletes have 30 seconds to jog back the 40 yards to the start line. The sprints continue to get faster until each athlete no longer crosses the finish line in time. (Note: You may use shorter age-appropriate distances for this test with younger or less skilled athletes.)

Scoring

No criterion values are given for this test, as age, gender, skill level, and maturation vary between groups and individuals. Setting baseline standards for each individual relative to your team average is the best use of the results. Improvements over time will help you to evaluate your program. Athletes with low initial tests should be encouraged to work on improving power endurance.

Flexibility Tests

Flexibility refers to the range of motion through which a joint can freely move. One of the most popular flexibility tests is the sit and reach test for hamstring flexibility. This is a valuable test since poor hip flexibility is often related to lower back pain in athletes. Other flexibility testing can be done that is specific to the needs of each sport. Shoulder flexibility is valuable for swimmers and athletes involved in throwing sports. A useful test for general shoulder flexibility is the shoulder rotation test. Here is the procedure for the sit and reach test (figure 5.6).

Tools

- Yardstick or ruler
- Area for sitting with back to wall

Procedure

- Have the athlete sit with legs flat on the floor and buttocks, back, and head touch-

40 yd time	Check row when completed	5 sec warning time			5 sec warning time			5 sec warning time			
sec	√	min	:	sec	min	:	sec	min	:	sec	Athletes not finishing in time
15		0	:	00	0	:	05	0	:	20	
15		0	:	50	0	:	55	1	:	10	
14		1	:	40	1	:	45	1	:	59	
14		2	:	29	2	:	34	2	:	48	
13		3	:	18	3	:	23	3	:	36	
13		4	:	06	4	:	11	4	:	24	
12		4	:	54	4	:	59	5	:	11	
12		5	:	41	5	:	46	5	:	58	
11		6	:	28	6	:	33	6	:	44	
11		7	:	14	7	:	19	7	:	30	
10		8	:	00	8	:	05	8	:	15	
10		8	:	45	8	:	50	9	:	00	
9		9	:	30	9	:	35	9	:	44	
9		10	:	14	10	:	19	10	:	28	
8		10	:	58	11	:	03	11	:	11	
8		11	:	41	11	:	46	11	:	54	
7		12	:	24	12	:	29	12	:	36	
7		13	:	06	13	:	11	13	:	18	
6		13	:	48	13	:	53	13	:	59	
6		14	:	29	14	:	34	14	:	40	
5		15	:	10	15	:	15	15	:	20	
5		15	:	50	15	:	55	16	:	00	

FIGURE 5.5 Repeated 40-yard sprint recording and time schedule sheet.

From *Sport Physiology for Coaches* by Brian J. Sharkey and Steven E. Gaskill, 2006, Champaign, IL: Human Kinetics.

ing the wall. (If no wall is available, the test can still be done. Just make sure the athlete sits up straight.)

- In this position, have the athlete extend the arms forward over the legs, parallel to the floor.
- Check to make certain that the athlete's legs are flat and that the buttocks, back, and head are touching the wall (if using).
- Adjust the ruler so the fingertips are at 0 inches. The ruler must now be held in place without moving. (You can place it on top of a box or other item of appropriate height, as in figure 5.6.)
- Have the athlete reach forward, one hand flat on top of the other, and reach as far down the ruler as he or she can and hold for 1 to 2 seconds. Repeat for a total of three

FIGURE 5.6 The sit and reach flexibility test.

trials. The score is the distance reached on the third trial. Record the distance to the nearest 1/2 inch (1.3 centimeters).

Scoring

Scoring is shown in table 5.5.

Here is the procedure for the shoulder rotation test (figure 5.7).

Tools

- 5-foot (1.5-meter) broom handle or stiff 3/4-inch (1.9-centimeter) dowel with a cloth measuring tape attached starting with 0 inches on one end.

Procedure

- Athletes should warm up the muscles of the back, neck, and shoulders before doing this test.

TABLE 5.5

Flexibility Ratings From Sit and Reach Test Results

Group	Low	Average	High
Males	<7 in.	7-12 in.	>12 in.
Females	<9 in.	9-14 in.	>14 in.

Adapted from Sharkey 1986.

- Measure the distance between the acromial processes of the shoulder. These are the bumps on the front side of the shoulders where the clavicle attaches to the shoulder.
- Record this distance to the nearest 1/2 inch (1.3 centimeters).
- With the bar behind the athlete's back, have the athlete hold the bar with an overhand grip so that the thumbs are pointing away from the body. The index finger of the right hand should be at the end of the bar, even with the start of the tape measure. The other hand can be placed as far to the left as necessary.

FIGURE 5.7 The shoulder rotation flexibility test.

- Standing, the athlete now brings the bar forward over the head, keeping elbows straight. If the athlete is unable to bring the bar over the head, start over with the athlete's hands farther apart.
- Repeat, with the athlete moving the left hand in about 1/2 to 1 inch (1.3 to 2.5 centimeters) each trial until he or she is unable to bring the bar directly overhead without undue pain or bending of the elbows. Measure by reading the number under the left index finger to the nearest 1/2 inch. Record this number.
- The final score is the total width (between index fingers) minus the distance between the acromial processes.

Scoring

The scores for this test are adjusted to the biacromial width. Use table 5.6 to evaluate scores. Please note that these are average criterion scores and that each sport will have different shoulder flexibility requirements. Swimmers, gymnasts, and athletes in throwing sports may require better shoulder rotation flexibility than runners or cyclists do.

Balance and Agility Tests

Balance is the ability to maintain equilibrium. Dynamic balance is important in team sports such as basketball, soccer, and football. Static balance is important in sports requiring static equilibrium, for example when a diver stands poised on the edge of the board or a gymnast holds a handstand position. Agility is the ability to change direction quickly while maintaining control of the body and balance. Agility depends on strength, power, speed, balance, anticipation, and coordination. Both balance and agility are sport specific. They improve with participation in the sport, but they also can improve with participation in a variety of activities. Included in this section is one test of static balance, the stork test, as well as the agility run and a shuttle run. Here is the procedure for the stork stand test (figure 5.8).

Tools

- Stopwatch
- Level floor

Procedure

- Athlete stands in bare feet on the dominant leg, then places toes of the other foot against the knee of the dominant leg and puts hands on hips as shown in the photo.
- At the commands "Ready" and "Go," the athlete raises the heel of the dominant foot and tries to maintain balance without the heel touching the floor, the foot moving, or the hands coming away from the hips.
- Give each athlete three attempts.

Scoring

Compare the best of each athlete's three scores with the norms in tables 5.7 (males) and 5.8 (females).

Here is the procedure for the agility run.

TABLE 5.6

Criterion Scores for the Shoulder Rotation Flexibility Test

Age	Excellent	Average	Low	Poor
MALES				
13-18	<19	19-27	28-29	>29
19-30	<14	14-22	23-28	>28
30-50	<24	24-28	29-32	>32
>50	<31	31-33	34-34	>34
FEMALES				
13-18	<12	12-20	21-25	>25
19-30	<10	10-21	22-24	>24
30-50	<19	19-25	20-27	>27
>50	<21	20-27	28-30	>30

Scores are given in inches (multiply by 2.54 to calculate in cm).

Adapted from Hoerger and Hoerger, 2004, *Principles and labs for physical fitness,* 3rd ed. (Wadsworth Publishing).

FIGURE 5.8 Stork stand position.

TABLE 5.7

Static Balance Rating for Males As Estimated From Time in Seconds Performing the Stork Stand Test

Age	Low	Average	High
Under 10	15	30	45
10-15	25	40	55
Over 15	35	50	65

TABLE 5.8

Static Balance Rating for Females As Estimated From Time in Seconds Performing the Stork Stand Test

Age	Low	Average	High
Under 10	10	20	35
10-15	15	30	45
Over 15	25	40	55

Adapted, by permission, from B.J. Sharkey, 1986, *Coaches guide to sport physiology* (Champaign, IL: Human Kinetics), 139.

Tools

- Stopwatch
- Four cones
- Gym floor with basketball free-throw lanes marked

Procedure

- Cones are placed at the four corners of the free-throw lane (12 × 19 feet or 3.7 × 5.8 meters) as shown in figure 5.9.
- During the entire shuttle run, the athlete remains facing the far wall of the court.
- Athlete stands in the starting position just to the right of the right cone on the baseline as noted in the figure.
- At the commands "Ready" and "Go," athletes begin the run following the pattern shown on the diagram. They must remain facing the far wall at all times. The stopwatch is stopped when they return to the starting position. Time to the nearest 1/10th second.
- Record the time.

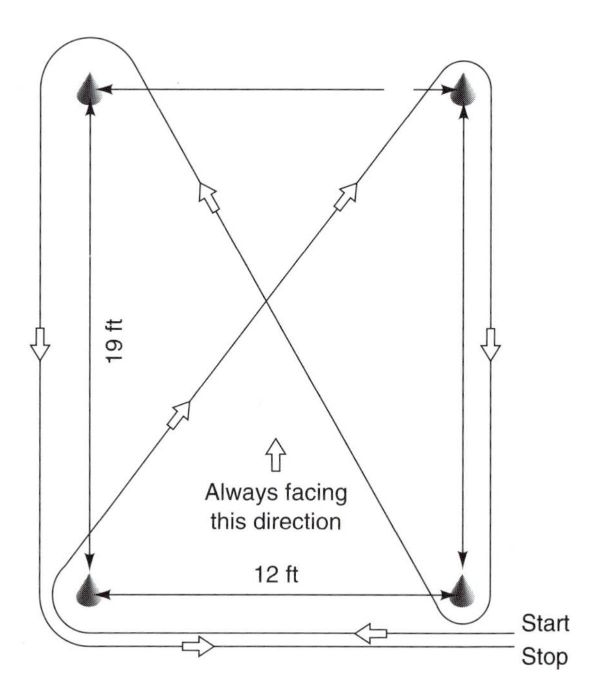

FIGURE 5.9 For this agility test, place cones as shown in the diagram or use the free-throw lane of a basketball court. The athlete must face forward at all times. Follow the pattern indicated in the diagram, having the athlete side-step and run backward as indicated by the arrows.

Adapted from R. Kirby, 1971, "A simple test of agility," *Coach and Athlete*. June: 30-31.

- During the test, the athlete must always go around each cone.
- Give each athlete three attempts.

Scoring

Compare the best of each athlete's three times with the norms in table 5.9.

Here is the procedure for the shuttle run.

Tools

- Four tennis balls
- Two shoe boxes
- Two stopwatches
- Gym floor with two parallel lines, 30 feet (9 meters) apart (width of a regulation volleyball court).

Procedure

See figure 5.10 for setup diagram.

- Have the athletes warm up thoroughly, then compete in pairs.
- The athletes stand behind one line, facing the other line 30 feet away. Four tennis balls are placed behind the 30-foot line (two in front of each athlete), and a box is placed on the start line next to each athlete.
- From the start line, on the command "Ready," then "Go," athletes run to the 30-foot line, pick up one ball, run to the start line, and place the ball in the box.
- Athletes return to get the second ball and finish by placing the ball in the box as they cross the start line.

TABLE 5.9

Agility Run Evaluation

Age	Low	Average	High	Excellent
Under 10	>14.1	14.1-12.2	12.3-13.7	<12.3
10-15	>13.3	13.3-12.5	11.1-12.6	<11.1
Over 15	>12.1	12.1-11.4	10.3-11.5	<10.3

Values shown are times in seconds.

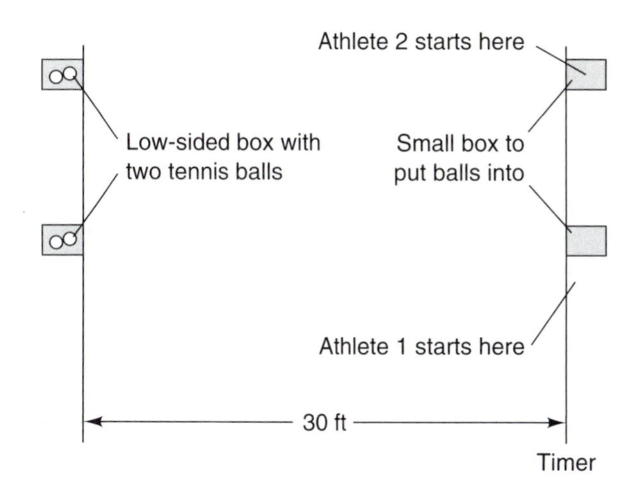

FIGURE 5.10 Setup diagram for shuttle run.

- The athlete's score is the time from start to finish (to the nearest 1/10th of a second).
- Give each athlete two to three trials.

Scoring

Use each athlete's best score and compare with the norms in tables 5.10 (males) and 5.11 (females).

TABLE 5.10

Agility Rating for Males As Estimated From Time in Seconds on Shuttle Run Test

Age	Low	Average	High
9-10	12.0	11.2	10.3
11	11.5	10.9	10.0
12	11.4	10.7	9.9
13	11.0	10.4	9.6
14	10.7	10.1	9.2
15	10.4	9.9	9.1
16	10.4	9.9	8.9
17+	10.4	9.8	8.9

Reprinted, by permission, from B.J. Sharkey, 1986, *Coaches guide to sport physiology* (Champaign, IL: Human Kinetics), 137. Data from AAHPERD Youth Fitness Test.

TABLE 5.11

Agility Rating for Females As Estimated From Time in Seconds on Shuttle Run Test

Age	Low	Average	High
9-10	12.5	11.8	10.6
11	12.1	11.5	10.4
12	12.0	11.4	10.3
13	12.0	11.2	10.2
14	12.0	11.0	10.0
15	11.8	11.0	10.1
16	12.0	11.2	10.3
17+	12.0	11.1	10.0

Reprinted, by permission, from B.J. Sharkey, 1986, *Coaches guide to sport physiology* (Champaign, IL: Human Kinetics), 137. Data from AAHPERD Youth Fitness Test.

SUMMARY

To be useful, an assessment method must meet several basic criteria. It should be sport specific, valid (i.e., should measure what you want to measure), reliable (i.e., should give a similar result if repeated), objective (should not be affected by coach bias), gender and maturity appropriate, easy to administer, easy for athletes to learn, and provide results that can be used to monitor progress. The ultimate assessment is performance. Once the sport-specific muscular fitness attributes have been identified, appropriate tests need to be chosen or designed. When designing or choosing tests, coaches should consider four main characteristics of each test: validity, reliability, objectivity, and appropriateness to the group. Basic criteria for athlete assessment include relevance, objectivity, specificity, and appropriateness. Tests should have an easy learning curve; should be gamelike, simple, and challenging; and should measure small training differences. Before testing, athletes should be well rested, as they would be on a competition day. After tests, athletes should be given feedback as soon as possible regarding strengths, areas of weakness, and how to improve. The chapter presents a number of sample tests for strength, power, power endurance, speed, flexibility, balance, and agility.

In this section, we have defined muscular fitness and given you the tools and information to design muscular fitness tests to determine a starting point for your athletes. Coaches are encouraged to read appendix D on testing criteria and evaluation. In the next chapter, we will provide up-to-date information on how to develop muscular fitness.

KEY TERMS

objectivity (p. 63)
reliability (p. 62)
validity (p. 61)

REVIEW QUESTIONS

Answer the following questions by filling in the blank with the correct word or words.

1. Identify the 11 basic criteria for a good athlete assessment.
2. _________ , _________ , _________ , and _________ are the four main characteristics that a coach should consider when designing or choosing a fitness or skill test.
3. Skill and agility tests generally require a _________-second rest; short-term maximal-effort tests require _________ minutes between trials; and longer maximal-effort tests require _________ hours between tests.

Match the physical attribute on the left with the test commonly used to assess the attribute.

4. Strength
5. Power
6. Power endurance
7. Speed
8. Flexibility
9. Balance
10. Agility

a. Chair dips; push-ups; sit-ups; bench hops; 20- to 30RM weights; repeated vertical jump test
b. 1- to 3RM lifts using weights
c. Sit and reach; shoulder rotation test; full-body rotation test
d. Ladder run; agility run
e. Standing long or vertical jump; Wingate cycle test; less than 5-second stair run
f. Balance board; one-legged balance
g. 5- to 40-yard sprint

Determine if the following statements are true or false and circle the correct answer.

11. T/F Test order should start with those that require the longest recovery and end with those that require the shortest recovery.
12. T/F Athletes should be given feedback on their testing a few days after they have had time to reflect on their performance.

PRACTICAL ACTIVITIES

1. In chapter 4, George, the head coach for a college track team, evaluated the types of training that would be necessary for athletes in different events. The next task that he and his assistants tackled was to design tests to evaluate the gains in muscular fitness for each group of athletes. The events of interest included 10,000 meters, 5,000 meters, 1,500 meters, 800 meters, 400 meters, 200 meters, 100 meters, pole vault, high jump, shot put, and javelin. What tests of muscular fitness would you recommend for each sport? List them in the order in which you would administer them.

2. John coaches U12 girls' soccer. The club that he coaches for recommends a standard series of tests for all of the girls to be used during tryouts, including 100-yard (91-meter) dash; shooting stationary balls from the penalty mark and from 15, 20, and 25 yards (14, 18, and 23 meters) from the goal; throw-ins for distance; distance covered during five continuous two-legged hops; kicking a stationary ball (on the ground) for distance and accuracy; and dribbling a soccer ball through a series of cones for time. John would like to develop a series of tests that will show improvement in soccer-specific fitness and that he can use throughout the season. Which of these tests might be good to use repeatedly, and what different tests might you propose?

6

Developing Muscular Fitness

This chapter will help you

- understand the types of training that are necessary to develop strength,
- appreciate the types of training that are necessary to develop power,
- understand the types of training that are necessary to develop power endurance,
- recognize what types of training are necessary to develop speed,
- recognize what types of training are necessary to develop flexibility, and
- understand the types of training that are necessary to develop agility and balance.

John played JV junior high football, filling different offensive and defensive line positions as needed. He was small for his 14 years and seldom started a game. Both John and his dad hoped that he would be able to make the high school team. John's dad talked to the high school coach, asking for advice. He was told that John needed to start lifting weights and get bigger, stronger, and faster. John's family already had a membership to the local YMCA. Nearly every day John met with a group of boys who lifted on their own with no coaching. Most sessions were a contest about who could lift the most weight and do the most reps and sets. Without knowing what they were doing, most of the boys began to look more buff, but were often sore and tired. After the first month, they did not gain much strength. John began to lose weight. He was irritable, didn't sleep well, and was driving his parents nuts. After two months, his father decided that John needed professional guidance and set him up with a certified strength coach. During the first few meetings, the strength coach listened to John's goals, heard about what the boys had been doing for training, recognized the signs of overtraining in John, and decided on a long-term course of action. When John went to high school a year later, he had increased both his bench press and squat strength by nearly 60 percent, put on nearly 20 pounds (9 kilograms), and increased his speed and power. Having learned the lesson "Train, don't strain" (author unknown), John made the high school JV team the first year and varsity the next.

Designing proper resistance training programs requires an understanding of the basic training principles. His father's decision to hire a certified strength coach helped John improve by providing

In order to stimulate adaptation toward specific training goals, appropriate sport-specific progression in the resistance training protocol is necessary.

guidance in designing and implementing a training program with proper overload, rest, progression, and periodization. The development of muscular fitness has been well studied, as shown in the following research.

A large body of scientific research has evaluated resistance training for strength, power, and power endurance. The American College of Sports Medicine (ACSM) frequently asks panels of experts to summarize current research in areas concerning exercise and health. Dr. Kraemer, a leading expert in strength and power development, led a distinguished group to write the ACSM's position stand on resistance training (2002). The article is summarized here. For the full text, go to the ACSM website (www.acsm.org).

In order to stimulate adaptation toward specific training goals, progression in the resistance training protocol is necessary. It is also recommended that strength programs sequence exercises to optimize the quality of the exercise intensity (large before small muscle group exercises, multiple-joint exercises before single-joint exercises, and higher-intensity before lower-intensity exercises). For initial resistances, it is recommended that loads corresponding to 8- to 12-repetition maximum (RM) be used in novice training. For intermediate to advanced training, it is recommended that individuals use a wider loading range, from 1- to 12RM, with eventual emphasis on heavy loading (1-6RM), and that they use at least 3-minute rest periods between sets performed at a moderate contraction velocity. For training at a specific RM load, the recommendation is to apply 2 to 10 percent increases in load when the individual can perform the current workload for one to two repetitions over the desired number. The recommendation for training frequency is two to three days per week for novice and intermediate training and four to five days per week for advanced training.

Progression in power training entails two general loading strategies: (1) strength training and (2) use of light loads (30-60 percent of 1RM) performed at a fast contraction velocity, with 2 to 3 minutes of rest between sets with multiple sets per exercise. It is also recommended that emphasis be placed on multiple-joint exercises, especially those involving the total body.

For developing local muscular endurance, the recommendation is that people perform light to moderate loads (40-60 percent of 1RM) for high repetitions (>15) using short rest periods (<90 seconds). In the interpretation of this position stand, as with prior ones, the recommendations should be viewed in context of the individual's target goals, physical capacity, and training status.

Training programs follow fundamental rules and use specific language. This section reviews the basic training principles that were presented in chapter 2.

- Overload and recover principle. One can state the overload principle as follows: In order to stimulate physiological changes in muscles, you must overload them more than they are accustomed to, then allow them time to recover and respond. In resistance training, we can overload specific muscles or multiple muscle groups in a number of ways: by adding more resistance, by increasing the number of lifts done in a workout, by changing the speed of movement, by increasing the frequency of workouts, or through some combination of these additive stresses. The commonly used FIT principle of overload can be modified to suit resistance training: Frequency equals the number of sessions per week; intensity equals the amount of resistance and speed of movement; and time can be thought of as the **volume of training** (number of lifts × resistance) that an athlete completes.

- Specificity principle. Your athletes' muscles and physiological systems will respond specifically to overload stress. For resistance training, this means that only the specific muscle or muscles that are trained will respond. These muscles will respond according to how the overload is applied. The range and speed of motion, resistance, recovery time, and other factors will all determine the specific response. Conditioning programs should involve the necessary muscle groups in ways specifically related to their use in the sport. The task of the coach is to design training programs that stimulate and overload the muscles in a sport-specific manner, but that also sequentially improve strength, power, and power endurance.

- Individuality principle. Since each athlete has different needs and different fitness levels, each athlete will respond somewhat differently, and coaches are continually challenged to adjust training to meet individual needs.

- Maintenance and reversibility principles. Once a new fitness level is achieved, it will take less training to maintain the fitness than it did to achieve the fitness level. The higher the level of fitness, the greater the amounts of training that will be required to maintain that fitness level. If training stops, or if maintenance of training drops below necessary levels, the athlete will begin to detrain and fitness will decline. Many training programs incorporate a plan for a period of reduced training, called a taper, leading into critical competitions. Tapers are designed to maintain fitness while increasing rest and recovery so that athletes are able to give a maximal effort for the competition. Maintain a taper too long and competitive performance will decrease.

- Progression, variation, and periodization. Progression is the idea that as an athlete improves, the overload needs to be increased in order for additional gains to be realized. During resistance training, a general guideline is as follows: When an athlete can lift the current resistance for one to two repetitions over the desired number, it is time to increase the load by about 2 to 10 percent. This basic concept is built into nearly all packaged training programs. Though the concept of progression is correct, the application can be somewhat difficult. It has been well established that variation in training programs is necessary for optimal improvements. Biological systems seem to respond to cycles of rest and variation in overload. Additionally, as training progresses, changing its focus is often necessary. The concept of periodization accommodates the need for both progression and variation. **Periodization** refers to planned training variation over years, seasons, months, weeks, and even days. In this way, the routines of athletes are varied. Different parts of a training season, weeks within sections of a program, and even days within a week may have different foci. In resistance training, we often design training programs with a preparatory period in which athletes start with very light loads and progressively increase training intensity and volume as they learn better technique and their body adapts. Following the preparatory period, athletes may apply progression to a period of strength training, then focus on power and eventually move on to power endurance training.

Resistance Training Terms

Discussion of training for muscular fitness requires a common language. In this section, we define and discuss a number of key terms and concepts. These terms and concepts will be used in the remaining chapters of the book.

Muscle Balance

Resistance training requires that we maintain a balance in the strength between opposing muscles. A classic example is the necessary balance between the hamstring and quadriceps muscles of the leg. Each sport has primary movers (agonists), the muscles that are the main power generators for the required movements. It is easy to design training programs for these muscles, but it is vital that coaches have their athletes also train the opposing (antagonist) muscles. We say more about muscle balance later in the chapter.

Repetition Maximum

Based on research by DeLorme and Watkins in 1951, the concept of repetition maximum (RM) evolved. A repetition maximum is the number of times an athlete can lift a load before exhaustion. If an athlete can lift a weight 10 times before failing, this is referred to as a 10-repetition maximum (10RM) load. The use of repetition maximum is a practical method for designing resistance programs, as it allows for program development in the absence of knowledge about an athlete's maximal strength (1RM load). Many books and articles continue to define resistance training using **percent maximum strength (% max).** Table 5.2, in the previous chapter, relates % max to RM. In general, a 5RM lift will be about 85% max, and a 10RM lift will be about 70% max.

Resistance Training Methods

Resistance training methods are usually divided into three distinct categories:

- Isometrics, or static contractions
- Isotonics, or same-weight resistance training
- Isokinetics, or same-speed, variable-resistance training

Each of these methods may have some use in athletic strength and power development. Let's consider some of the key features of each method.

Isometrics or Static Contractions

Isometrics, or **static contractions,** are contractions against an immovable object such as a wall. In the 1960s, research showed large strength gains using isometric contractions. Coaches and athletes were quick to adopt the new method. Unfortunately, the researchers who found large gains from isometric strength programs also used isometric measurements to evaluate the strength gains. Athletes in dynamic sports soon found that there were few benefits to be gained with isometric training. Researchers had ignored a basic principle of specificity: Most sports require dynamic, not static, strength. The use of static, contractions is not suited to developing the type of strength necessary for most sports. Static contractions are still valued only in sports that use static contractions (e.g., archery, shooting, and possibly rock climbing [finger strength]) or in rehabilitation programs when limbs cannot be moved.

Isotonics or Weight Training

Isotonics refers to resistance training using either free weights or weight machines. The word isotonic means "same weight," thus referring to the method in which the same weight is lifted through a range of motion. Current literature often refers to this training as **dynamic constant external resistance (DCER)** training, since the external resistance remains constant but—due to biomechanics of muscle motion—the actual muscular force changes over the range of motion, creating "sticking points." Isotonic training is the resistance training method preferred by serious athletes throughout the world. Typically, isotonic weight training exercises involve both concentric and eccentric muscle contractions. The concentric contraction occurs as the muscle shortens during the "lift," and the eccentric contraction occurs when the muscle lengthens as the load is lowered. Current research supports a combination of concentric and eccentric lifting. Since a greater load can be lowered than can be lifted, there has been interest in increasing the eccentric load to further overload the muscle. Eccentric contractions are known to produce excessive muscle soreness when the overload is increased too rapidly, and care must be taken to

Isometric training remains valuable in sports such as rock climbing that require periods of static contraction. Rock climbers may use sport-specific exercises to develop isometric finger and wrist strength while also developing dynamic leg and arm power.

increase the resistance gradually during isotonic training.

Free weights provide the most inexpensive form of weight training. With the purchase of a bar and a selection of weights, every coach can institute a strength training program. Machines are generally easier and safer for beginners, allowing the completion of exercises that are difficult with free weights, such as lat pulldowns. Machines also help to stabilize the body, reduce balance problems, and isolate movement to the main muscle groups. Free weights often allow for more sport-specific training and activate synergistic muscles involved in balance and body stabilization. A general recommendation is that during the early stages of weight training and for youth and elderly persons, machine weights be primary, with added free weights for sport specificity. As athletes become more advanced, free weights will generally predominate, with machines added as necessary for specific muscle actions. The advantages of weight machines and free weights are compared in table 6.1.

© Bongarts/SportsChrome

© Rick Polke

Free weights often allow for more sport-specific training and activate synergistic muscles involved in balance and body stabilization. It is generally recommended that during the early stages of weight training and for young and elderly people, machine weights be primary, with added free weights for sport specificity. As athletes become more advanced, free weights will generally predominate over machines.

Isokinetics or Variable Resistance

Isokinetics or variable-resistance training controls the speed of movement and allows for variable resistance by the athlete. During concentric contractions, the slower the speed, the greater the possible force that can be applied. Sports come in all speeds. It seems to make sense that isokinetic resistance training, done at the speeds required for one's sport, would provide an optimal sport-specific overload. Research with isokinetic training has shown that power generation is specific to the training speed, and athletes who train at high speeds can generate better power at high movement speeds than athletes who have been training at low speeds. Other research has shown that training at moderate speeds with high resistance improves power over a large range of speed, though probably not quite to the extent that speed-specific training does.

Isokinetic training devices are effective in a number of ways:

TABLE 6.1

Free Weights and Weight Machines Compared

Advantages	Disadvantages
FREE WEIGHTS	
Greater variety of exercises available	Less safe than machines and often require spotters
Require balance	Greater possibility of theft
Can train accessory (synergistic) muscles	More time-consuming
More sport specific	Require more technique
Low initial cost	Difficult to move with high speed
WEIGHT MACHINES	
Safe (self-spotting)	Limited number of exercises
Low likelihood of theft	Balance not learned
Time efficient to use	Do not train accessory muscles
Easy for circuit training and large groups	High initial cost
Can isolate muscle groups	May favor stronger limb

Adapted, by permission, from B.J. Sharkey, 1986, *Coaches guide to sport physiology* (Champaign, IL: Human Kinetics), 53.

- They provide high resistance and a training effect throughout the range of movement trained.
- The machines are ideally suited to performance of exercise at velocities similar to those required in specific sports.
- Isokinetic contractions are generally concentric and do not produce muscle soreness.

Although isokinetic resistance training has advantages for improving sport performance, it is not without disadvantages. The high cost of variable-resistance devices limits their widespread availability, and their design often limits the exercises and range of motion that can be trained. One of the main limitations of free weights, that they do not permit moving quickly, is being overcome with new modified machines that allow for explosive, multijoint lifts. Most strength and conditioning coaches, including those working with well-endowed college and professional teams, continue to focus their training using a combination of free weights and isotonic (DCER) machines specially developed to be sport specific. Practical experience, the ability to isolate and exercise sport-specific muscles, and the ability to train accessory muscles continue to make isotonic resistance training the preferred method for athletes. Finally, the use of free weights trains concentration—a quality of tremendous value in sport, which is difficult to teach athletes by any other means. When athletes have to balance heavy weights, they must focus their undivided attention on the exercise. Few other activities require this degree of concentration.

Sets, Reps, Circuits, and Other Terms

The terminology of resistance training is generally quite straightforward. In the following sections, in which we start to organize and then prescribe resistance training, the terms set, rep, set-rep, and circuit are frequently used.

- A **rep (repetition)** is one complete cycle of an exercise, such as one bench press or one sit-up.
- A **set** is the number of reps completed continuously without a rest period, such as a

set of 10 bench press reps or 30 sit-ups. A set is always followed by a rest period.

- **Circuit training** is a method of putting sets of exercises together such that an athlete does one set of each exercise before repeating sets of any exercise. As an example, the athlete could complete one set of dips, sit-ups, and then push-ups, which may be repeated two or more times. Circuit training is popular for power endurance development using light weights or body weight resistance.
- **Set-rep training** organizes training into multiple repeated sets of one exercise before the next exercise, such as three sets of 10 bench presses followed by three sets of 10 squats. The resistance and number of repetitions may be adjusted for each set.
- **Intensity (of resistance training)** refers to the amount of weight, generally in terms of RM (repetitions to failure) or percent of maximum (% 1RM). Since the speed of movement also influences the amount of weight that can be lifted, the best definition of intensity is the amount of work per time (power) per repetition.
- **Rest period (for strength training)** refers to the amount of time allotted for recovery between sets. In the following sections, the rest periods are generally given as minimum time for recovery. For the development of strength and power, longer rest periods are often beneficial, as they allow for greater recovery and better effort during each set.

Organizing a Strength Training Program

Using a systematic approach to developing the resistance training program will help each coach design an appropriate program specific to his or her athletes. By following the steps outlined here, you can design a program that will benefit your athletes by matching the program to their needs and goals. Start with defining your goals and the needs of the sport; progress to general core exercises, and specialize with sport-specific exercises.

Goals and Needs Assessment

To get somewhere, it is necessary to know where you are and where you want to go. In chapter 5 we covered assessment—which tells you where your athletes currently are. If your fitness tests were well designed, you have already defined the goals of your sport and sport-specific activities. By defining team and athlete goals and needs, you have determined the destinations that you would like to arrive at. After testing your athletes you will also have the starting points and can define the route for each athlete. With good information and goals, you can design a program to improve areas of weakness while maintaining strengths.

The amount of information that organized coaches accumulate may seem overwhelming, especially if they are thinking about having different programs for each athlete. It really isn't that complex. You can design a basic training program for your team, then teach your athletes how to adjust the program to fit their needs.

Core Stability

A long-understood but little applied fundamental of resistance training is the role of core body muscles (trunk, hips, and shoulders) in the stabilization of the primary movers in the limbs. **Core stability** is a part of the chain in power development in throwing and other sports that require transfer of power, starting with the legs and moving upward to accelerate the arms. Athletes in throwing sports start their power development in the legs, then proceed to the trunk, and finally finish with their arm musculature. Throwing is generally ineffective if the trunk and leg muscles cannot first develop power that is transmitted to the shoulder and arm. Core stability is related to strength, dynamic balance, coordination, and balance among core muscle groups. Recently, core training has gained popularity as strength coaches have begun to recognize the importance of having a strong base from which to develop sport power. Many other full-body sports such as wrestling, tennis, boxing, rowing, and cross-country skiing also require transferring power through the central core. Core training is a great addition to the preparatory training period. Adding core training to the overall training plan will be discussed in chapters 10 and 11. By first stabilizing the abdominal, back, and shoulder muscles, your athletes will increase their training gains, improve performance, and reduce low back and other injuries. Core exercises are generally done with a resistance that allows for 15 to 30 repetitions. The recent emphasis on core training appears to be beneficial and helps to improve the response to sport-specific resistance training for strength, power, and speed. Coaches should evaluate each athlete and choose, or design, appropriate exercises. Appendix B shows a number of exercises for training core stability.

Exercise Selection

Shown in table 6.2 is a list of weight training and resistance exercises for different sports. Some of these exercises are fully described and illustrated in appendix B. Although suggested exercises are indicated for each sport, your final selection should be determined by seasonal priorities and the unique abilities and needs of your athletes. When organizing a resistance program, many coaches distinguish the following four types of exercises: major, assistant, supplementary, and specialty.

Major exercises are those that have the greatest influence on strength development. The value of squats for leg strength and the bench press for upper body strength is evident from table 6.2. In many programs these are considered the major exercises. **Assistant exercises** are those that have also been identified as having a significant training effect for a particular sport. Most resistance training programs include one to three major and one to three assistant exercises. **Supplementary exercises** consist of approximately four to six carefully selected, but less vital, exercises. The program is then completed by the addition of one to three **specialty exercises,** selected according to each athlete's individual needs. How to select and integrate exercises into a comprehensive training program is discussed in chapters 10 and 11.

Exercise Specificity

You are probably beginning to notice how the theme of specificity continues to be emphasized throughout this book. Note that table 6.2 suggests specific exercises for different sports. In addition to knowing the recommendations, you will need to evaluate the needs of each athlete. A tennis coach could argue, with some justification, that an exercise like the bench press will not develop the specific strength that a tennis player needs. This is partly true; however, these exercises are still valuable for creating a foundation upon which to build greater training specificity. Also, depending on the abilities of your athletes and the contribution of muscular fitness to performance in your sport, this may be as specific as you need to get in relation to the training process.

For even greater benefits, you might like to consider how you could modify existing exercises or develop new training methods to more closely simulate sport-specific movements for your athletes. With a little imagination, you will probably find that the training principles for developing most or all of the components of muscular fitness can be applied to sport-specific skills. For example, a shot-putter could use shots of varying weights: a heavier shot to develop strength and a softball to train for speed. To improve for power endurance in cross-country skiing, a lat machine can be used, or pulleys attached to weights (such as buckets of rocks) could be set at different heights to exercise different patterns of the arm movement. By varying the resistance and speed, the skier can determine the training effect. Many companies are now beginning to develop sport-specific resistance training equipment that can be used to mimic the motions and speeds required by specific sports.

The introduction of sport-specific training methods becomes increasingly beneficial as athletes enter the competitive season. This periodization of training is discussed later in this chapter and integrated into training programs in chapters 10 and 11.

Exercise Order

The order of exercises may have a large impact on the efficacy of strength, power, or power endurance development. Multiple-joint exercises, such as the squat, have been shown to be effective for increasing general muscular strength, and one

TABLE 6.2a

Recommended Weight Training and Resistance Exercises for Different Sports*

Exercises	**Body area**	Aikido	Archery	Backstroke	Badminton	Baseball/Softball	Basketball	Bobsled	Boccie	Bowling	Boxing	Breaststroke	Butterfly	Canoeing	Cricket	Cycling	Discus/Shot put	Distance running	Diving	Equestrian
Neck flexion and extension	Upper and lower arm	√	√		v	√	√	√	√	√	√	√		v	v	√	√	√		√
Back extension	Lower back	√	√	√	√			√			√	√	√	√	√	√	√		√	√
Bench press	Chest	√	√	√	√	√	√	√			√	√	√	√	√	√	√	√	√	√
Bent-arm pullover	Chest	√		√	√	√	√				√	√	√		√		√		√	
Bent-knee sit-ups	Abdomen	√	√		√	√	√	√	√	√	√			√	√	√	√	√	√	√
Bent-over rowing	Shoulder girdle	√	√					√			√	√	√	√		√			√	√
Fingertip push-ups	Grip	√	√					√			√	√	√	√		√			√	√
Heel (toe) raise	Lower leg	√		√	√	√	√	√			√	√	√		√	√	√	√	√	
Incline press	Upper arm	√		√	√	√		√			√	√	√	√	√		√		√	
Knee extension	Upper leg	√		√	√	√	√	√			√	√	√		√	√	√	√	√	√
Lateral arm raise	Shoulder	√		√	√		√				√		√		√		√		√	
Leg curl	Upper legs	√			√	√	√	√			√	√	√		√	√	√	√	√	√

Leg raise	Trunk	√	√	√	√	√		√			√	√	√	√		√		√	√	√
Military press	Shoulder, upper arms				√	√	√							√	√		√		√	
Neck flexion and extension	Neck	√									√								√	
Parallel bar dip	Shoulder, upper and lower arm	√	√	√	√	√	√	√			√	√	√	√						
Power clean	Trunk, shoulder girdle	√		√		√	√				√	√	√		√		√	√	√	
Press behind neck	Shoulder			√									√							
Pulldown—lat machine	Shoulder girdle	√	√	√	√	√	√	√			√	√	√	√	√				√	
Reverse curl	Lower arm		√		√	√	√	√	√	√				√	√	√				√
Reverse wrist curl	Forearm		√		√	√	√	√	√	√				√	√	√	√	√		√
Shoulder shrug	Shoulder	√		√	√						√	√	√	√			√		√	
Squat (half)	Lower and upper back, upper legs	√	√	√	√	√	√	√	√	√	√	√	√	√	√	√	√	√	√	√
Straight-arm pullover	Chest				√	√	√							√	√					
Triceps extension	Shoulder, upper arm	√		√	√	√	√				√	√	√	√	√	√	√			
Upright rowing	Shoulder	√	√	√	√	√	√	√			√	√	√	√	√	√		√	√	√
Wrist curl or wrist roller	Forearm	√	√	√	√	√	√	√	√	√	√			√	√	√	√			√

*Adapted, by permission, from B.J. Sharkey, 1986. *Coaches guide to sport physiology* (Champaign, IL: Human Kinetics), 55-58.

TABLE 6.2b

Recommended Weight Training and Resistance Exercises for Different Sports*

Exercises	Body area	Fencing	Field hockey	Figure skating	Football	Off/Def linemen	Off backs	Receivers	Def backs	Kickers/Punters	Freestyle swimming	Frisbee	Golf	Gymnastics	High jump	Hurdling	Ice hockey	Javelin	Judo	Karate
Arm curl	Upper and lower arm		√			√	√	√	√	√		√	√	√	√		√	√	√	√
Back extension	Lower back	√	√	√		√	√	√	√	√	√	√		√		√	√	√	√	√
Bench press	Chest	√	√	√		√	√	√	√	√	√	√	√	√	√	√	√	√	√	√
Bent-arm pullover	Chest		√			√	√	√	√	√	√			√	√		√	√	√	√
Bent-knee sit-ups	Abdomen	√	√	√		√	√	√	√	√		√	√	√	√	√	√	√	√	√
Bent-over rowing	Shoulder girdle		√			√	√	√	√	√	√			√		√	√		√	√
Fingertip push-ups	Grip		√				√	√	√	√		√	√	√			√	√	√	√
Heel (toe) raise	Lower leg	√	√	√		√	√	√	√	√	√		√	√	√	√	√	√	√	√
Incline press	Upper arm					√	√	√	√	√	√		√	√	√	√		√	√	√
Knee extension	Upper leg	√	√	√		√	√	√	√	√	√	√		√	√	√	√	√	√	√

Lateral arm raise	Shoulder	√											√	√				√	√	√
Leg curl	Upper legs	√	√	√		√	√	√	√	√		√		√	√	√	√	√	√	√
Leg raise	Trunk	√	√	√			√	√	√	√	√	√		√	√	√	√		√	√
Military press	Shoulder, upper arms					√	√	√	√		√		√	√				√		
Neck flexion and extension	Neck		√			√	√	√	√	√							√		√	√
Parallel bar dip	Shoulder, upper and lower arm		√			√	√	√	√	√	√	√	√	√			√		√	√
Power clean	Trunk, shoulder girdle	√	√	√		√	√	√	√	√	√		√	√	√	√	√	√	√	√
Press behind neck	Shoulder					√	√	√	√					√						
Pulldown—lat machine	Shoulder girdle		√			√	√	√	√		√	√	√	√		√	√		√	√
Reverse curl	Lower arm	√	√														√			
Reverse wrist curl	Forearm	√	√			√	√	√	√			√	√	√			√	√		
Shoulder shrug	Shoulder		√			√	√	√	√	√	√						√	√	√	√
Squat (half)	Lower and upper back, upper legs	√	√	√		√	√	√	√	√	√	√	√	√	√	√	√	√	√	√
Straight-arm pullover	Chest	√				√	√	√	√		√		√	√	√					
Triceps extension	Shoulder, upper arm		√			√	√	√	√		√	√		√	√		√	√	√	√
Upright rowing	Shoulder	√				√	√	√	√	√	√	√	√	√					√	√
Wrist curl or wrist roller	Forearm	√	√			√						√	√	√			√	√	√	√

*Adapted, by permission, from B.J. Sharkey, 1986. *Coaches guide to sport physiology* (Champaign, IL: Human Kinetics), 55-58.

TABLE 6.2c

Recommended Weight Training and Resistance Exercises for Different Sports*

Exercises	Body area	Kayaking	Lacrosse	Long jump	Luge	Motocross	Mountaineering	Orienteering	Pole vault	Racquetball	Roller skating	Rowing	Rugby	Sailing	Shooting	Skiing/Alpine	Skiing—Cross-country	Ski—jumping	Soccer	Speed skating
Neck flexion and extension	Upper and lower arm	√	√	√	√	√	√	√	√	√		√	√	√	√				√	√
Back extension	Lower back	√	√	√	√	√	√	√	√	√	√	√	√		√		√	√	√	√
Bench press	Chest	√	√	√	√	√	√	√	√	√	√	√	√	√	√	√	√	√	√	√
Bent-arm pullover	Chest	√		√		√	√		√	√		√	√						√	
Bent-knee sit-ups	Abdomen	√	√	√	√	√	√		√	√	√	√	√	√	√	√	√	√	√	√
Bent-over rowing	Shoulder girdle	√	√	√		√	√						√	√			√		√	√
Fingertip push-ups	Grip	√	√			√	√		√	√					√					
Heel (toe) raise	Lower leg		√	√	√	√	√	√	√	√	√	√	√			√	√	√	√	√
Incline press	Upper arm	√	√	√	√		√		√	√			√						√	√
Knee extension	Upper leg		√	√	√	√	√	√	√	√	√	√	√			√	√	√	√	√

Lateral arm raise	Shoulder	√				√	√		√	√			√							
Leg curl	Upper legs		√	√	√	√	√	√	√	√	√	√	√			√	√	√	√	√
Leg raise	Trunk	√	√	√	√	√	√	√		√	√	√	√	√	√	√	√	√	√	√
Military press	Shoulder, upper arms	√	√				√			√			√							
Neck flexion and extension	Neck												√						√	
Parallel bar dip	Shoulder, upper and lower arm	√	√		√	√	√	√	√	√		√	√	√	√		√		√	
Power clean	Trunk, shoulder girdle	√	√	√			√		√		√	√	√			√		√	√	√
Press behind neck	Shoulder	√											√						√	
Pulldown—lat machine	Shoulder girdle	√	√	√	√	√	√	√	√	√		√	√	√		√	√		√	√
Reverse curl	Lower arm		√		√									√		√				
Reverse wrist curl	Forearm		√		√				√	√		√		√		√				
Shoulder shrug	Shoulder	√	√			√	√		√	√			√	√						
Squat (half)	Lower and upper back, upper legs	√	√	√	√	√	√	√	√	√	√	√	√	√	√	√	√	√	√	√
Straight-arm pullover	Chest	√	√	√			√		√	√			√						√	
Triceps extension	Shoulder, upper arm	√	√	√		√	√	√	√	√		√	√	√		√	√		√	
Upright rowing	Shoulder	√	√		√	√	√	√	√	√		√	√	√			√			
Wrist curl or wrist roller	Forearm	√	√		√	√	√		√	√		√	√	√		√				

*Adapted, by permission, from B.J. Sharkey, 1986. *Coaches guide to sport physiology* (Champaign, IL: Human Kinetics), 55-58.

TABLE 6.2d

Recommended Weight Training and Resistance Exercises for Different Sports*

Exercises	**Body area**	*Sprinting*	*Squash*	*Surfing*	*Synchronized swimming*	*Table tennis*	*Tae kwon do*	*Team handball*	*Tennis*	*Triple jump*	*Volleyball*	*Water polo*	*Weightlifting*	*Wrestling*						
Neck flexion and extension	Upper and lower arm	√	√	√		√	√	√	√	√	√	√	√	√						
Back extension	Lower back	√	√	√	√	√	√	√	√	√	√	√	√	√						
Bench press	Chest	√	√	√		√	√	√	√	√	√	√	√	√						
Bent-arm pullover	Chest		√	√	√	√	√	√	√	√	√	√	√	√						
Bent-knee sit-ups	Abdomen	√	√	√	√	√	√	√	√	√	√	√	√	√						
Bent-over rowing	Shoulder girdle	√			√		√	√		√		√	√	√						
Fingertip push-ups	Grip		√			√	√	√	√		√	√	√	√						
Heel (toe) raise	Lower leg	√	√	√		√	√	√	√	√	√		√	√						
Incline press	Upper arm	√	√			√	√	√	√	√	√	√	√	√						

Knee extension	Upper leg	√	√	√		√	√	√	√	√	√	√	√	√						
Lateral arm raise	Shoulder		√		√	√	√	√	√		√		√							
Leg curl	Upper legs	√	√	√		√	√	√	√	√	√	√	√	√						
Leg raise	Trunk	√	√	√	√	√	√	√	√	√	√	√	√	√						
Military press	Shoulder, upper arms		√			√	√	√	√		√	√	√	√						
Neck flexion and extension	Neck	√					√							√						
Parallel bar dip	Shoulder, upper and lower arm		√	√		√	√	√	√		√	√	√	√						
Power clean	Trunk, shoulder girdle	√					√			√		√	√	√						
Press behind neck	Shoulder								√				√							
Pulldown—lat machine	Shoulder girdle	√	√	√	√	√	√	√	√	√	√	√	√	√						
Reverse curl	Lower arm		√			√		√	√			√	√	√						
Reverse wrist curl	Forearm		√			√		√	√			√	√	√						
Shoulder shrug	Shoulder		√			√	√	√	√		√	√	√	√						
Squat (half)	Lower and upper back, upper legs	√	√	√	√	√	√	√	√	√	√	√	√	√						
Straight-arm pullover	Chest		√	√	√	√		√	√	√	√	√	√							
Triceps extension	Shoulder, upper arm		√	√		√	√	√	√	√	√	√	√	√						
Upright rowing	Shoulder		√	√		√	√	√	√		√	√	√	√						
Wrist curl or wrist roller	Forearm		√			√	√	√	√		√	√	√	√						

*Adapted, by permission, from B.J. Sharkey, 1986. *Coaches guide to sport physiology* (Champaign, IL: Human Kinetics), 55-58.

Many companies are beginning to develop sport-specific resistance training equipment that can be used to mimic the motions and speeds required by specific sports. *(a)* Athlete using swim bench and *(b)* athlete swimming freestyle.

a

© Paige E. Kaleita

b

© Empics

should focus on those exercises for optimal muscular fitness improvements. It is most advantageous to start with multiple-joint exercises early in the training season and then progress toward more sport-specific exercises as the competitive season approaches.

There are recommendations for ordering exercise within each workout. The following list was modified from the ACSM's progression models for resistance training for healthy adults (Kraemer et al. 2002).

- Begin all resistance training with light warm-up sets for all muscle groups to be trained.
- When training all major muscle groups in a workout, train large muscle groups before small muscle groups, perform multiple-joint exercises before single-joint exercises, and perform high-intensity exercises before lower-intensity exercises.

Frequency of Resistance Training

How often your athletes train specific muscle groups will depend on a number of variables, including goals, fitness level, other training needs, intensity of training, and the ability to recover. In general, maintenance of strength seems possible with one to two days per week. Development of strength requires more training. For beginners, two to three days per week may be adequate for improvements in strength and power; but as muscular fitness increases, further strength and power gains may occur only with additional days of training—with up to four sessions per week for specific muscle groups. If your sport requires a number of muscle groups for both upper and lower body fitness, your athletes may find it beneficial to alternate upper and lower body days (split training) so that they are doing resistance training six days a week with three sessions for each muscle group. There is some evidence that for very elite weightlifters, up to five days of training each week may be beneficial but that more training does not improve results. Three days a week appears to be the best use of time for most athletes. Pay attention to your athletes, using common sense and realizing that they may have limited time availability. If they seem overly tired, reduce the volume of training. If they seem to have reached a plateau in strength, take a little time off and then adjust what you are doing. Sometimes more rest between sessions and increased intensity during the ensuing sessions will help athletes reach new levels.

Velocity of Lifting

The speed with which resistance training is performed has an effect on neuromuscular coordination, fiber type recruitment, metabolic response to the training, and muscle hypertrophy development. Using isokinetic (same speed, variable resistance) training, researchers have shown that the greatest improvements are made at the speed at which the training is completed. It is important to note that performing resistance training at moderate speeds produces reasonable strength gains across a wide range of tested velocities.

Since most athletic teams do not have access to isokinetic training devices and use isotonic training (DCER training), it is important to understand the relationship between speed and resistance in this type of training. During lifts involving high effort, speed of movement decreases as the resistance is increased. While slow movements due to high resistance are great for strength development, moderate and fast velocities of training have been shown to better develop power. It is recommended that athletes use a high effort with all resistance loads, and that for power development they limit resistance such that moderate and fast movements are possible. During multiple lifts, the speed of movement tends to decrease as the athlete nears failure on the final few lifts of a set. If power is the training focus, it might be better to continue reps only as long as the speed of movement can be maintained and to end the set before failure.

Unfortunately, with free weights and many machines, high velocities are impossible because of the ballistic nature of the weights. In these cases, athletes should use weights that limit the lift speeds to within safe ranges. Fortunately, more equipment is being designed that allows high movement velocities and the sport-specific development of power. It is generally innovative coaches who lead the way in modifying existing equipment or designing new equipment to allow sport-specific power development.

Untrained individuals should start with slow velocities and modest resistance to learn proper technique and develop the neuromuscular coordination necessary for resistance training. As technique and strength improve, athletes may move from slow- to moderate- and higher-velocity training.

Developing Strength

The resistance required to develop strength depends on the goals and initial fitness level of the athlete. Untrained individuals with low strength can see improvements with as little as 50% 1RM or a weight that they can lift 15 to 20 times. However, as muscular fitness improves, it becomes necessary to increase the resistance relative to an athlete's maximal strength. The basic prescription for developing strength in untrained to moderately trained athletes is to use a resistance of about 60% to 70% 1RM for

How your athletes lift will determine the training stimulus and the outcomes. Pay attention to the resistance, the reps, the rest periods, and proper progression.

8 to 12 repetitions, progressing with increases of 2 to 10 percent as athletes are able to do 14 or more reps. For advanced athletes, it is optimal to use 80% to 100% 1RM, or about 2- to 5RM resistances. Recent studies have shown that a variety of loading is more effective in developing strength than when all sets are completed with the same load.

The volume of training (total repetitions × resistance) needed to stimulate improvements also varies with fitness status and program goals. Research has shown that single and multiple sets are similar in efficacy for athletes beginning a strength program; thus one set may be adequate at this stage. As muscular fitness improves, the training volume needs to increase to two to four sets. Within the sets and over the training cycle there should be systematic variation. The number of sets used for each exercise will vary depending on the muscular needs of the sport and athlete. Increases in volume need to be gradual to avoid overtraining.

The rest between sets is also important. Longer rest periods have been shown to be beneficial to the development of strength. As intensity increases, rest between sets should also increase. Athletes who fail to rest adequately between sets will see decreased gains and possibly increases in muscle soreness. Table 6.3 summarizes resistance training for strength development.

Strength training has obvious benefits for short-duration events. Not surprisingly, a solid strength foundation has also been shown to improve endurance performance, even without increases in aerobic power. These aerobic improvements are possibly due to improvements in efficiency of movement along with increased muscle mass recruitment and increased power endurance. Whatever the sport, strength training lays the foundation for power, a necessity in all sports.

Developing Power

Power is the rate of doing work. Almost all sport requires increases in power to improve performance. Traditional strength training methods tend to increase maximal dynamic strength at slow speeds, not at the high speeds generally required for athletic performance. Recent

TABLE 6.3

Resistance Training for Strength

	ATHLETE DEVELOPMENT IN STRENGTH TRAINING		
Training variable	**Beginner**	**Intermediate**	**Advanced**
# Sets	**1-2**	**2-3**	**2-4**
Resistance	50-60% 1RM	60-70% 1RM; vary intensity of each set	80-100% 1RM; vary intensity of each set
Repetitions	10-15	8-12	1-6
Rest	1.5 to 2 min	2 to 4 min	3+ min
Movement speed	Moderate	Moderate to moderately fast	Depends on resistance; use maximal effort
Frequency Sessions per week	1-2	2-3	3-5 per muscle group; split sessions
Equipment	Machines	Machines and free weights	Mostly free weights with machines for sport-specific work

Note: Muscle action should include both concentric and eccentric movements. Isotonic (DCER) training is generally recommended.
DCER = dynamic constant external resistance

research has shown that **power training**, using light to moderate loads at high velocities, most effectively increases sport-specific power and the maximal rate of power development.

It is important to understand that power is the result of applying force (strength) quickly. It is necessary to have adequate strength before one can develop power. Maintaining a reduced volume of strength training during periods of power training seems to enhance the development of power. Some studies have shown that long periods of heavy resistance training for strength will decrease explosive power; thus it is necessary also to maintain some power training during periods of strength training.

A difficulty in developing power is the ballistic nature of free weights and machines during the concentric contraction. The necessary deceleration period is exaggerated during movements with explosive power. Some new machines are being designed to allow for explosive movements, while some more traditional exercises, such as loaded jump squats are possible with varying amounts of resistance and explosive power. Coaches must be innovative in adapting exercises and machines to explosive movements.

Multiple-joint exercises, especially those duplicating movements specific to your sport, are recommended for power training. Rest periods are similar to those for strength training. Adequate rest is needed between sets to allow for high quality on all repetitions. Table 6.4 gives guidelines for power training incorporated along with strength maintenance to enhance the development of power.

Table 6.4 is a general prescription for power training. Well-informed coaches adjust the strength and power program to best fit the needs of their athletes. In the normal progression of resistance training, strength is developed first, then power, and finally power endurance. Athletes who compete in short-duration and sprint sports may not need the power endurance phase. However, for events lasting more than 15 to 20 seconds, coaches and most researchers would suggest that some power endurance training is beneficial. For endurance events, power endurance training may be the major resistance training component.

Developing Power Endurance

It is often not possible to categorize the wide variety of sport activities as strength, power, or endurance events. Most activities include components of all aspects of muscular and energy fitness in varying degrees. In reality, there is a continuum of training needs, from pure strength to multihour endurance. A pure strength event might be slow moving yet require extremely high muscular strength, as in the "World's Strongest Man" contest in which a contestant might attempt to pull an 18-wheeler. An ultramarathon or a 16-hour workday for a wildland firefighter would exemplify the other end of the endurance spectrum. All these events require strength, power, and endurance, though in different ratios. Events that require some aspect of power endurance range broadly from those lasting less than a minute to those lasting several hours.

To describe power endurance, we divide it into three categories as shown in table 6.5. Of most interest to the athlete in terms of resistance training is short-term power endurance. Intermediate- and long-term power endurance are generally trained through practicing sport-specific drills with overload, such as uphill running. For power endurance it is optimal to duplicate, as much as possible, the speed and motion used in the sport.

As athletes fatigue during sets of power endurance training, speed of motion may decline. Some coaches and researchers suggest stopping the activity when the movement speed declines, while others have shown good results even when the speed of movement is declining during 10- to 25RM training. Note that the repetitions in tables 6.5 and 6.6 are shown as RM values. Thus, should the coach decide to have athletes complete sets to failure, the number of repetitions should fall approximately within the range shown.

Strength and power training will independently improve power endurance to some extent. However, greater repetitions and specificity of training produce the greatest gains. Thus it is necessary to evaluate the nature of the sport and to define methods to overload the specific movements in a repeated and sport-specific manner. Basketball players might consider

TABLE 6.4

Resistance Training for Power

	ATHLETE DEVELOPMENT IN STRENGTH TRAINING		
Training variable	**Beginner**	**Intermediate**	**Advanced**
Frequency	**1-2 per week**	**2-3 per week**	**3-5 per week**
Set #1			
Goal	Strength	Strength	Strength
Resistance	60% 1RM	70% 1RM	80% 1RM
Repetitions	10-15	8-12	5-6
Rest	1.5 to 2 min	2 to 4 min	3+ min
Movement speed	Moderate	Moderate	Moderately fast
Equipment	Machines	Machines or free weights	Free weights
Set #2			
Goal	Power	Power	Strength
Resistance	40% 1RM	60% 1RM	90% 1RM
Repetitions	5-6	3-4	3-4
Rest	2-2.5 min	3+ min	3+ min
Movement speed	Fast	Fast	Moderate
Equipment	Machines	Machines or free weights	Free weights
Set #3			
Goal	NA	Power	Power
Resistance	NA	50% 1RM	70% 1RM
Repetitions	NA	5-6	3-4
Rest	NA	3+ min	3+ min
Movement speed	NA	Fast	Fast
Equipment	NA	Machines or free weights	Machines or free weights
Sets #4 & 5			
Goal	NA	NA	Power
Resistance	NA	NA	60%, then 50% 1RM
Repetitions	NA	NA	4-8
Rest	NA	NA	3+ min
Movement speed	NA	NA	Fast
Equipment	NA	NA	Machines or free weights

This power training program includes a maintenance strength program to enhance the development of power.
Power training exercises need to allow for deceleration and are generally multijoint, sport specific, or both.

TABLE 6.5

Power Endurance Training

	Repetitions	Sets	Recovery	Freq/Week	Method
Short-term endurance	10-25RM	3	30-60 sec	3	Circuit or set-rep
Intermediate endurance	30-50RM	1-3	1-2 min	3	Circuit
Long-term endurance	100RM+	1-2	1-2 min	3	Circuit

These exercises are generally sport specific. Short-term endurance is often done with weights, machines, and body weight resistance or with sport-specific exercises. Intermediate- and long-term power endurance are generally done in a sport-specific manner. If possible, the speed of movement should closely imitate that required by the sport.

Adapted, by permission, from B.J. Sharkey, 1986, Coaches guide to sport physiology (Champaign, IL: Human Kinetics), 69.

TABLE 6.6

Resistance Training to Develop Power Endurance

	ATHLETE DEVELOPMENT IN STRENGTH TRAINING		
Training variable	**Beginner**	**Intermediate**	**Advanced**
Sessions per week	**2-3**	**3-4**	**3-5**
Set #1 (Vary between A, B, and C to maintain strength, power, and short-term muscular endurance. These should be done as a circuit of sport-appropriate exercises.)			
A			
Goal	Maintain strength	Maintain strength	Maintain strength
Resistance	60% 1RM	70% 1RM	80% 1RM
Repetitions	10-15	8-12	5-6
Rest	1.5-2 min	2-4 min	3+ min
Movement speed	Moderate	Moderate	Moderately fast
Equipment	Machines	Machines or FW	FW
B			
Goal	Maintain power	Maintain power	Maintain power
Resistance	40% 1RM	60% 1RM	70% 1RM
Repetitions	5-6	3-4	3-4
Rest	2-2.5 min	3+ min	3+ min
Movement speed	Fast	Fast	Fast
Equipment	Machines	Machines or FW	Machines or FW

(continued)

TABLE 6.6

(continued)

	ATHLETE DEVELOPMENT IN STRENGTH TRAINING		
Training variable	**Beginner**	**Intermediate**	**Advanced**
Sessions per week	**2-3**	**3-4**	**3-5**
Set #1 (Vary between A, B, and C to maintain strength, power, and short-term muscular endurance. These should be done as a circuit of sport-appropriate exercises.)			
C: Do two sets of the following in a circuit format.			
Goal	Short-term PE	Short-term PE	Short-term PE
Resistance	20RM	20RM	15RM
Repetitions	15	15-20	10-15
Rest	30-60 sec	30-60 sec	30-60 sec
Movement speed	Intentionally slow	Intentionally slow	Intentionally slow
Equipment	BW or machines	BW, FW, or machines	BW, FW, or machines
Sets #2-4 (The resistance can be varied if desired. Do one to three sets using a circuit format.)			
Goal	Intermediate PE	Intermediate PE	Intermediate PE
Sets	1	2	3
Resistance	25-50RM	25-50RM	25-50RM
Repetitions	Until speed declines	Until speed declines	To fatigue
Rest	2 min	2 min	2 min
Movement speed	Sport imitation	Sport imitation	Sport imitation
Equipment	Sport specific	Sport specific	Sport specific

This program focuses on intermediate power endurance. To adjust for short- or long-term power endurance, substitute the appropriate values from table 6.5 in sets 2 and 3. These exercises are generally sport specific. Short-term endurance work is often done with weights, machines, and body weight resistance or with sport-specific exercises. Intermediate- and long-term power endurance work is generally done in a sport-specific manner. If possible, the speed of movement should closely imitate that required by the sport.

PE = power endurance; BW = body weight; FW = free weights.

multiple vertical jumps wearing weight vests; sprinters may use small parachutes to increase resistance; distance runners may do repeats of 100+ strides uphill; and swimmers may do tethered swimming.

For sports requiring power endurance, the general progression is to first develop strength, next add power, and then focus on power endurance during the precompetition and early competition periods. Endurance athletes may further periodize their training by moving from a focus on power to short-term, then intermediate-term, and finally long-term power endurance. Table 6.6 summarizes a possible training plan for endurance athletes during a period focusing on intermediate-term power endurance. Note that maintenance of strength and power, as well as short-term power endurance, is accomplished by adding one to two sets of each per week.

Many endurance sport coaches incorporate intermediate- and long-term power endurance training into distance training days, especially during the months prior to competition. Starting with a warm-up, they may then incorporate

With innovation or the use of commercially available products, you can design sport-specific resistance exercises.

one to three circuits of sport-specific exercises for power endurance spaced through the workout. The first circuit is generally early in the workout when the athletes are fresh. Later circuits will be completed as athletes tire. This helps train the maintenance of muscular fitness even in the fatigued state.

Now that we have covered how to train for strength, power, and power endurance, it is time to discuss how they all go together into a training program. The next sections will help you plan and periodize training with a systematic and organized approach to resistance training.

Progression and Periodization

The terms progression and periodization are often used interchangeably, but they have different meanings. Progression implies that as athletes get stronger, more weight needs to be added to each set to maintain overload. Progression needs to happen gradually. Generally, progression is applied as maximal strength testing indicates. When testing is not possible, small increments of 2% to 10% 1RM may be added as athletes are able to handle the additional resistance.

Periodization systematically varies the training program to meet specific goals and to maintain interest while avoiding overtraining. Resistance programs are periodized into preparatory, strength, power, and power endurance periods (see figure 6.1). Within each of these periods, training should be varied. Most common is a planned three- to four-week cycle of increasing intensity or volume (or both) followed by a reduced-load week to begin the next cycle (see figure 6.2). Variation of exercises over time is also necessary. In the next sections we cover the periodization of a training cycle and present ideas on how to apply progression and provide variation in the periods of your training program.

Daily periodization within a week, and variation in the sets each day, will help to increase strength and power gains while reducing overtraining. Figure 6.3 shows a theoretical model of how two weeks (with three days of training each week) might be varied during a period of power development. Note that the overall training load is varied daily and that within each day there is variety in the sets.

Preparatory Resistance Training

When starting a resistance training program, it is important to take the time to help each athlete learn proper lifting techniques and prepare the muscles and connective tissues to handle the strain of strength training. This is a great time to decide on appropriate specialty exercises for each athlete. As mentioned earlier, you will probably have a core set of exercises for all athletes

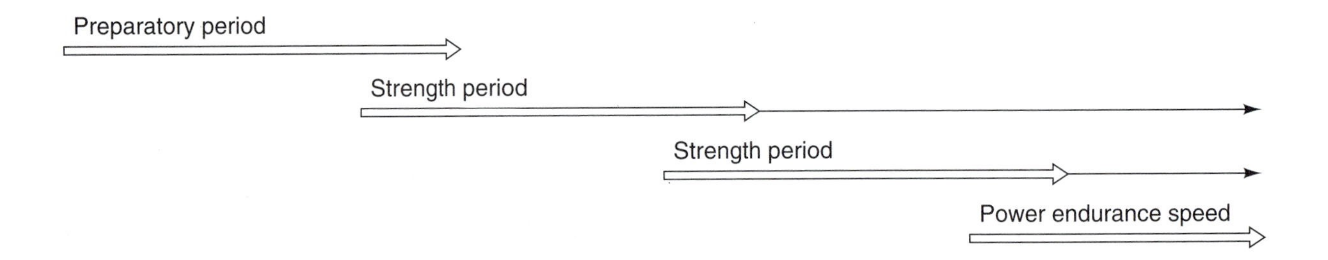

FIGURE 6.1 Season periodization of resistance training. A general progression is necessary to develop muscular fitness, starting with a preparatory period and proceeding to strength, followed by power, and finally power endurance. Each period of training has appropriate resistance, speed of movement, rest periods, and maintenance of previously developed fitness. Note the overlap in the periods showing that there is a transition from one type of training to the next. The arrows extending the strength and power periods indicate that maintenance training is continued for strength and power through the subsequent periods. This figure gives a general overview of the progression of muscular fitness training.

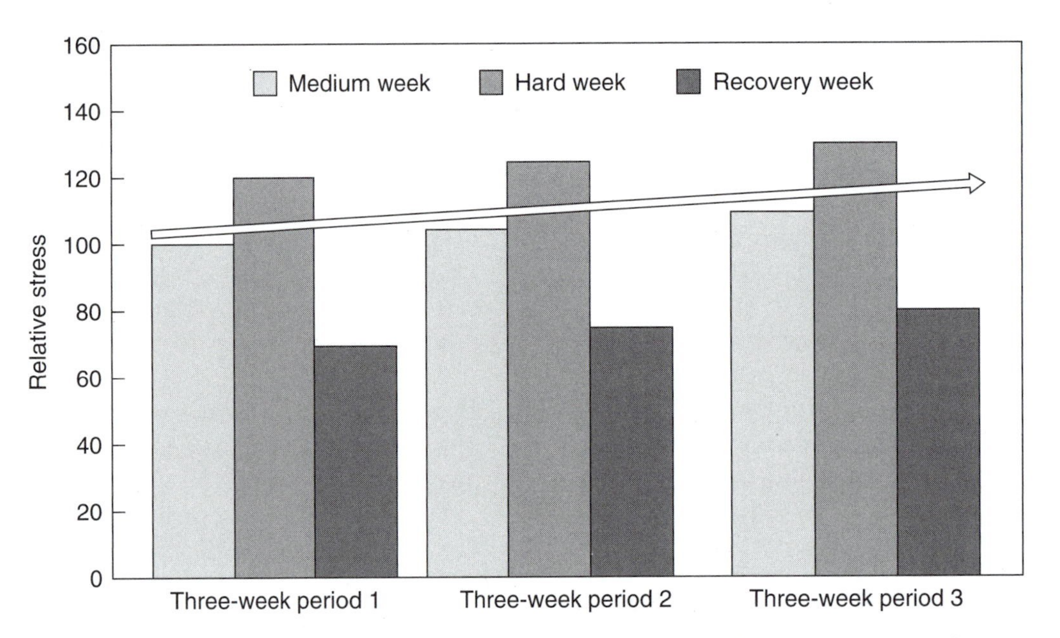

FIGURE 6.2 Weekly periodization of resistance training—a theoretical sample showing how weeks of resistance training stress might be varied. The relative stress is shown based on the first medium week. The weekly cycle progresses from a moderate week to a hard week, followed by an easy week (recovery week). Each following three-week period then demonstrates progression (increase in overall training load). Note that the total stress or load varies dramatically from week to week.

and a few specific to each individual. When you start your program, you may not yet have assessed each individual, but within two to three weeks you should have completed your testing and have individualized several exercises for each athlete. During the first two to three weeks, train athletes in proper lifting techniques. Start with light weights that each individual can lift for 10 to 15 reps. During the second week, you should complete your assessment of each individual using submaximal weights to find a resistance that each athlete can lift 4 to 10 reps to failure (4-10RM). Then use table 5.2 (p. 67) to estimate your 1RM resistance. After determining 1RM values for each individual for each exercise, decide on the appropriate preparatory training resistance. By the end of the third week of preparatory training, your athletes should have progressed to the resistance that you prescribe for strength training, based on table 6.3 (p. 100). Be sure that athletes increase in small

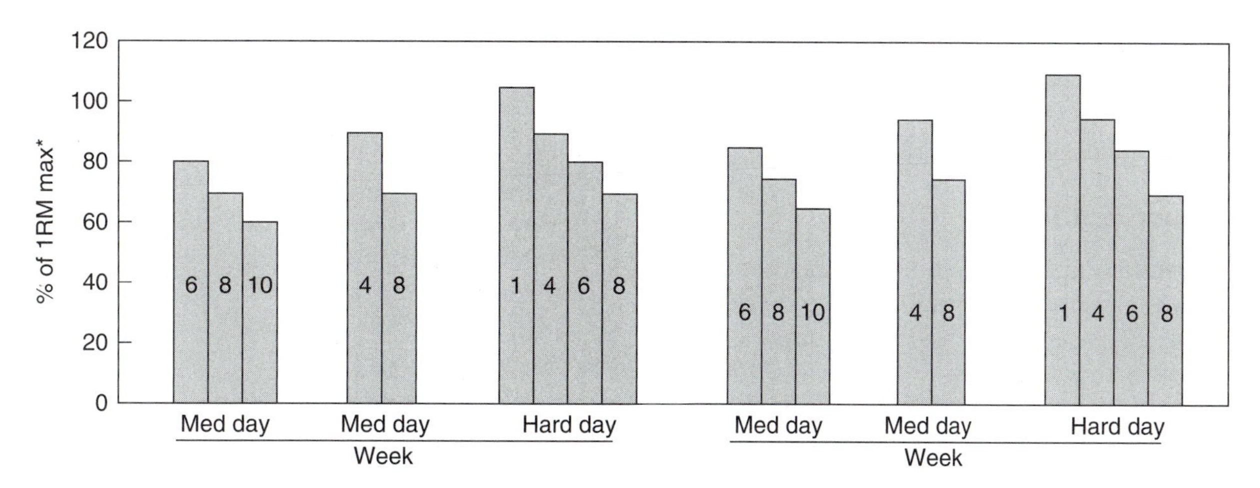

FIGURE 6.3 Periodization of daily strength training. A sample two-week program is shown for an exercise such as the bench press. Each set of bars represents one day. Training days for a muscle group should be separated by at least one day of rest. Each bar represents one set. The height of the bar represents the percentage of the athlete's most recent 1RM (1-repetition) max. The number in the bar represents the number of planned repetitions for that set. Note that on hard days, the effort starts off with a maximal 1RM slightly above the previous 1RM max.

increments. Resistance training should never be a competitive activity. Training should not be rushed if optimal gains are desired.

Strength Training

The length of time required to develop strength depends on the needs of each individual sport. When strength is a high priority, strength development may require 16 to 24 weeks or longer. During the first 6 to 8 weeks there will be little muscle hypertrophy (muscle enlargement). Most of the changes during this time will be metabolic (energy fitness) and improved muscle fiber recruitment. After 6 to 8 weeks, increases in strength will result primarily from muscle hypertrophy along with continued metabolic improvements. Endurance athletes, who desire to increase strength but not muscle mass, should limit their strength period to about 8 to 10 weeks to avoid muscle hypertrophy.

Guidelines for strength training are presented in table 6.3 (p. 100). The values in the table are guidelines only, and coaches must consider individual and sport needs. Varying the sessions within a week will help to keep athletes motivated and healthy. There is no proven best method to vary the daily intensity. Coaches can use both volume and intensity such that each week will have easier and harder days as shown in figure 6.3.

It is useful to test athletes for 1RM (or estimate 1RM from a 3-10RM set) every three to four weeks, generally at the end of a periodized cycle. Adjust individuals' 1RMs and resistances for strength training at these times. The testing also gives athletes motivation and feedback about the effectiveness of their programs. It is normal for athletes to progress differently. If you wish, you may vary the program of those who are not responding to have them lift slightly heavier weights and reduce the volume (total reps).

Varying the intensity and volume of overall weeks during strength training is also useful. A simple method is to start with a moderately easy week with resistances about 5 percent lighter than the guidelines. The second week, increase the average intensity to match the guidelines. The third week, increase the resistance to about 5 percent more. Within each week, adjust the volume so that you have easier and harder days as shown in figure 6.3. Do assessment testing during the third week and reset the 1RM values for each athlete. When you start the fourth week, again start about 5 percent below the guidelines using the new 1RM values. Continue with three weeks of increasing intensity, assess, then start the cycle over with an easier week.

Toward the end of your strength training, begin to add a set of power training (see guidelines in table 6.4, p. 102) in preparation for starting the power period. Many coaches have also found it useful to give their athletes one to two weeks of reduced stress by decreasing volume and resistance during the transition between the strength and power periods.

Power Training

Power training typically follows strength training. As with strength, the length of time to develop power depends on the needs of each sport. When power is a high priority, 7 to 12 weeks of periodized training may be required, followed by a period of 2 to 3 weeks of tapering before major competitions. During power training, strength should be maintained with one to two strength sets each week. It is also possible to dedicate one training session per week to strength training, though some research indicates that it is best to mix the session. Power training results in limited muscle hypertrophy, and most of the changes will be metabolic changes to improve speed of contraction and coordination of muscle recruitment at higher speeds.

Table 6.4 (p. 102) presents guidelines for power training. As with strength training, these guidelines should be considered averages. Intensity and volume, both daily and weekly, should be varied as discussed in the previous section on training for strength. Assessment is important every three or four weeks.

As you near the end of your power training, if necessary for your sport, begin to add sets of short-term power endurance training (see guidelines in tables 6.5 and 6.6, p. 103). If your sport does not require power endurance, you may continue power training to within one to two weeks of competition before beginning a taper and maintenance training. If you are transitioning to a power endurance program, it is useful to give your athletes one to two weeks of reduced volume and intensity to allow for recovery before beginning power endurance training.

Power Endurance Training

For any sport that requires more than a few repetitions, including short sprints, power endurance is important. Through practicing their sport, athletes are training sport-specific power endurance along the short- to long-term continuum. The idea of specific power endurance training is that there will be a period of time during which the athlete focuses more on repeated overloading of specific muscles and movements while maintaining movement speeds similar to those used in competition. Intervals are a form of power endurance training. Power endurance training builds on the increased strength and power from prior training.

As you resistance train your athletes, move from more general exercises during strength training to increasingly specific exercises during power training. Power endurance completes the cycle with exercises as specific as possible in speed and movement patterns. The timing and duration of the power endurance training are dictated by sport-specific needs. Single-repetition sports do not require endurance, and athletes in these sports continue with power training until shortly before the competitive season. Sprints of 40 to 200 meters (44 to 219 yards) require only short-term muscular endurance, and coaches may schedule power endurance training for the period of one to three weeks prior to the main competitions. When power endurance is a high priority, this training may last three to seven weeks prior to competitions.

During power endurance training, strength and power should be maintained with one to two sets each week. Power endurance training results in little, if any, muscle hypertrophy in trained athletes. Most of the changes are increases in the ability of individual muscle fibers to maintain the required energy system for longer periods of time. For short-term power endurance, improvements will be in the adenosine triphosphate-phosphocreatine (ATP-PCr) and anaerobic capacities of recruited fibers. Of particular importance to intermediate- and long-term power endurance is improvement in the oxidative capacity of the fast-oxidative-glycolytic fibers.

Tables 6.5 and 6.6 offer guidelines for power endurance training. As with strength and power, these guidelines should be considered averages only. The training program should be varied weekly and within the period over about a three-week cycle as discussed in the earlier section on strength training. Assessment is important every three or four weeks.

The end of focused power endurance training should be about one to two weeks prior to major competitions. At this time the total training load, for both resistance and energy fitness (chapters 7-9), is reduced to taper for the upcoming competitions. Resistance training enters into a maintenance period so that athletes can benefit from the hard work that they have put into improving their muscular fitness.

Maintenance of Muscular Strength, Power, and Endurance

Individual fitness level and the duration of the competitive season will dictate the volume of resistance training necessary to maintain preseason gains. In general, the higher the fitness, the more training required to maintain that fitness level. This is a paradox for athletes competing at the highest levels. To compete effectively requires rest and a large reduction in training volume. The taper allows the athlete to be adequately rested in order to perform optimally. For a short time following a taper, performance will be enhanced before the lowered training volume begins to result in decreased fitness. The educated coach must therefore help athletes decide which competitions are important enough to warrant reducing training volumes below maintenance levels, and which ones are less important so that it is preferable to sacrifice performance somewhat in order to better maintain long-term fitness.

About two resistance training sessions each week are generally adequate to maintain preseason fitness gains. During the maintenance training sessions, each athlete must decide on the number of sets to dedicate to strength, power, and power endurance. Your sport will dictate the needs. Any form of training helps to maintain some fitness in the other areas, but the proper emphasis will best maintain competitive form. Thus, an endurance athlete might elect to do resistance training twice a week during the competition period with a focus on power endurance training each session and less emphasis, if any, on strength and power.

Muscle Balance

Muscle balance refers to the relative strength, power, and endurance of opposing or bilateral muscles or muscle groups. Opposing muscle balance refers to the balance between primary movers and the antagonist muscles, those that cause the opposite movement. Bilateral balance refers to similar development on the two sides of the body. Although power and endurance are both important to muscular balance, balance in terms of strength is generally the easiest to evaluate. Testing for muscular balance in power and endurance requires more complex laboratory equipment. In this section we address a few areas of muscle balance that are of greatest concern to athletes.

Muscle balance can be difficult to measure, as it varies across different sports. Most of the testing is done measuring strength at slow speeds, while most sports are performed at high speeds requiring high power output. In addition, the free weights or machines used during testing are often not sport specific. Nonetheless, maintaining approximate muscle balance is critical to the athlete. Most training activities, especially resistance training, stimulate increases in only the

Muscle imbalance may occur when sports require development primarily in one limb, such as the dominant arm in tennis. A resistance training program can help to develop muscular balance and improve performance.

trained muscle. If training programs are developed that train only the primary movers, the antagonist muscles are placed under increasing stress to maintain joint stability. Compounding the muscle balance problem is that athletes prefer to train their strengths and ignore their weaknesses. Good strength programs include both the prime movers and the antagonistic muscles. Coaches need to identify weakness in both the primary muscles required for the sport and the opposing muscles that should also be trained. Training for weak muscle groups should take place early in training sessions to achieve the greatest benefit, and maintenance of strengths should be relegated to later sets.

Table 6.7 presents a few commonly accepted muscular strength balance values for major joints. These ratios are guidelines only and are based on slow-speed concentric contractions. Most of this research was done using isokinetic instrumentation that is often unavailable to coaches. Using machines that isolate the specified joint will help you to identify large deficiencies. It should be noted that if isokinetic machines are available, during high-speed movements of 250 to 300 degrees per second the ratio of balance in most of these joints should approach 1.0.

Muscle balance should also be evaluated bilaterally in sports with unilateral use such as throwing and hitting sports. Attention to muscle balance can help athletes avoid debilitating muscle injury such as hamstring pulls and shoulder injuries. Attention to muscle balance in the weight room may keep the athlete from losing many hours on the playing field.

TABLE 6.7

Opposing Muscle Group Strength Ratios for Selected Joints*

Joint	Movement	Strength ratio
Knee	Extension:flexion	3:2
Hip	Extension:flexion	1:1
Shoulder	Horizontal flexion:extension	2:3
Elbow	Flexion:extension	1:1

*Measurement at slow-speed joint movement.

Adapted, by permission, from NSCA, 1994, *Essentials of strength training and conditioning* (Champaign, IL: Human Kinetics), 425.

We hope that this section has helped you to understand the benefits and methods for resistance training. Each coach must use this knowledge to decide how to design program specifics, how long to continue in each period, and how to individualize each athlete's program. You should now understand how resistance exercises move from general core strength toward sport-specific power and endurance. You should have a picture of what is necessary for, and how to move though, the training periods. Coaches and researchers have found many successful models for resistance training. We are continuing to learn more about periodization and the need to maintain strength, power, and endurance through the competitive season.

The complex relationships between strength and endurance never fail to surprise us. The ability of coaches to apply the basic principles and guidelines to sport-specific training continues to help athletes set new records, and will always inspire researchers to evaluate why a program works. Coaches are the innovators, and researchers are the investigators.

Improving Speed

Speed implies acceleration from a starting position. Acceleration implies power, and maintenance of speed requires muscular endurance. Athletes thus require power to accelerate to the speeds necessary for success, whether the speed is that of a ball being released, vertical velocity during the high jump, or running speed during a sprint. For many sports, power is required to accelerate to the competitive speed, and then power endurance is required to maintain that speed. As events become longer in duration, different energy systems are required (chapter 7), and athletes will need longer-term power endurance and a greater reliance on aerobic versus anaerobic power. As this happens, speeds will decrease due to the increased metabolic complexities and requirements for oxygen delivery. Even the 100-meter sprint requires short-term power endurance. Sprint athletes utilize their stored ATP and phosphocreatine within the first 50 to 60 meters (55 to 66 yards) and then struggle to maintain their speed as they produce energy through slower anaerobic pathways. Carl Lewis, one of the great 100-meter sprinters, often accelerated slightly slower than other sprinters from the start to the 50-meter mark, but his greater short-term power endurance allowed him to continue to accelerate through 70 meters (77 yards) while his competitors reached their peak speeds at 50 to 60 meters. Lewis often came from behind to pass his competition during the final 20 meters (22 yards). Intermittent sport and endurance athletes also require speed and acceleration during their events.

Though genetics may play some role in the determination of potential for speed, all athletes can improve their acceleration, maximal, and sustainable speeds. The ratio of fast- and slow-twitch fibers may not change much during training and maturation; but, depending on the training stimulus, the physiology of the cells improves, allowing for greater endurance, power, and strength. Fast-twitch fibers can be trained to improve their endurance qualities and can hypertrophy for increased strength and power. Slow-twitch fibers can improve in oxidative ability and also increase force production through hypertrophy. During acceleration and maintenance of speed, all types of muscle fibers may be recruited. Proper training will help athletes approach their potential.

Power and power endurance are of primary importance in the development and maintenance of speed. In addition to changes in the muscle, the coordinated recruitment of muscle fibers is also critical. Technique and economy of movement are determined by how well our central nervous system tells muscles to contract in a coordinated manner. Throughout this book, we have stressed the importance of specificity. We will continue to do so here and provide you with ideas for applying the overload principle to areas of speed and speed training.

Speed Training

Brad had coached swimming at a small midwestern college for many years with modest success. In 1992, he attended a series of educational programs for college endurance sport coaches. One of the speakers talked about an approach to speed training that made sense to Brad, characterizing improving speed as simply applying the overload principle to speed training. The speaker's concept was that athletes needed to practice at speeds slightly faster than they could currently race until they were comfortable and efficient at the new speed and could then race slightly faster.

Carl Lewis, one of the great 100-meter sprinters, often accelerated slightly slower than other sprinters from the start to the 50-meter mark, but his greater short-term power endurance allowed him to continue to accelerate through 70 meters while his competitors reached their peak speeds at 50 to 60 meters.

For the distance events, this was easy—swimmers would simply swim shorter intervals at slightly faster than race pace, rest, and repeat as long as they could maintain the higher pace. For the sprint races, Brad worked with the engineering department to design a pulley system that would allow for tethered swimming, whereby the swimmers were harnessed to a cable that could assist them in moving through the water about 2 to 5 percent faster than their sprint speeds. Athletes were then able to develop the neural coordination necessary to move faster and to stress their physiological systems with the higher workloads. Within the first year of applying the new speed training system, Brad's swimmers set new school records in nearly every event and distance.

This true story was related to one of the authors (SEG) a few years after he had presented a lecture titled "Training for Speed" to college coaches in the Midwest. There are many aspects to speed training, but the following statement may help coaches focus ideas for their sports: Athletes are most likely to perform at the speeds that they practice most during training. A way to restate this is to say that the athlete who is able to spend the most time practicing at the speeds necessary for success is most likely to be able to perform at that speed during competition.

This concept of speed training has a number of implications:

- Development of speed requires time and patience.
- Most athletes do intervals at a near-maximal speed. This is not productive. Have them keep their interval speeds to only 1 to 5 percent faster than current race pace! They need to slow down to learn to go faster.
- Between intervals, allow athletes to recover for the length of time they personally feel is necessary in order to be able to maintain the required speed and do multiple intervals. During a training session, athletes should be encouraged to do as many intervals as possible as long as the necessary speed is maintained.
- Athletes need to be rested prior to these sessions. Speed training sessions should thus be done following easy days of training.
- During periods of peaking, athletes may do short, maximal intervals on a limited basis.

The following sections will give you more ideas on how to address specific problems and needs of your athletes. Coaches are encouraged to be innovative and to find ways to apply the combined principles of speed, specificity, and muscular and energy fitness to their training programs. Always be quietly critical of what is being done. Ask yourself if the training that you are doing, or that others propose, satisfies the general principles of training. Chapter 10 gives guidelines for using this basic concept of speed training as part of an overall training program.

Power Versus Neuromuscular Recruitment

As previously mentioned, speed requires power for acceleration and muscular endurance for maintenance. Technique is the neural coordination necessary for proper execution of a sport to allow efficiency and economy of movement. During all speed training sessions, coaches must watch athletes carefully to ensure that proper technique is maintained. The adage "Practice makes perfect" should really be stated as "Practice makes permanent." As we repeat a movement multiple times, it becomes automatic and we become able do it without conscious thought. For this reason, improper technique or so-called bad habits are difficult to change. Coaches who continually discuss and demonstrate proper technique help athletes develop the correct images and automated movements. The adage should probably be further modified: "Practice makes permanent, perfect practice makes perfect." Thus, as you develop power, power endurance, and speed training programs for your athletes, focus on specificity and proper technique. Be patient and your athletes will achieve great results.

Plyometrics

A popular method to develop power utilizing both neural and metabolic systems is to train the preload and elastic recoil present in many sport skills. In running, for example, the strong thigh muscles of the leg are slightly stretched

Speed-Assisted Training

Over the past decade the idea of training specificity has led to the understanding that **neuromuscular recruitment** and power are both required for optimal sprinting speed. A difficulty for sprinters in improving neuromuscular recruitment is the inability to move faster than maximal speed. Innovative coaches are learning how to overcome this problem. Examples include sprints on slight downgrades to increase running speed, thus helping the athlete to develop new muscle recruitment patterns. Another example is "bungee running," whereby a fast runner, attached by a long flexible bungee to a second runner, is the "puller." Once the cord is stretched, the second runner will take off. The assist by the bungee cord allows for higher speeds for the second runner for a short sprint. The partners can then switch positions. Hockey skaters in safety harnesses on special treadmills can learn to skate at higher than normal speeds. Overspeed treadmill running is also effective for sprint training. All of these techniques are known as **speed-assisted training**. Any tool that a coach can safely use to help an athlete practice speeds slightly faster than current race speed will help to increase neuromuscular coordination and help athletes more effectively use their muscular power.

or preloaded as they strike the ground. This preload causes a brief lengthening (eccentric lengthening) of the contracting muscle prior to the concentric contraction that provides the forward movement. The eccentric stretch stores up energy that is quickly released during the contraction, in addition to activating the spindle fiber stretch reflex. Similarly, as a tennis player begins the backswing for a forehand, the triceps muscle on the back of the upper arm contracts and the biceps muscle on the front of the arm stretches. When the player begins to swing forward, the biceps contracts powerfully in response to its rapid stretching. This extra power does not take much extra energy, so it provides more power without any added cost—a true example of efficiency. The same principle can be applied in such movements as throwing the javelin or discus and in jumping, as well as in other sports in which a brief stretch can be quickly followed by a contraction. Groups of training methods that apply this principle to develop power and quickness are known as **plyometrics.**

A key feature of plyometric training is the conditioning of the neuromuscular system to permit faster and more powerful changes of direction, such as moving from down to up in jumping, switching leg positions in running, and changing direction in court or field sports. Reducing the time needed for this change in direction or force development increases power, speed, and quickness.

Ski jumpers, cross-country and downhill skiers, basketball and volleyball players, sprinters, high jumpers, and other athletes who require power can profit from plyometric training. In addition to being effective, plyometric exercises

Plyometric training, using the stretch reflex, improves muscular power. Hurdle jumps are among the more popular and effective plyometric exercises for developing explosive leg power.

are relatively easy to teach and learn. Plyometric exercises utilize the following movements: bounds, hops, jumps, leaps, skips, ricochets, swings, and twists. Coaches are continually developing new movements specific to the needs of their sports.

A sample program of plyometric exercises for different muscle groups is shown in table 6.8. A typical program has 6 to 10 reps, three to five sets, and a rest interval of 2 minutes between each set. Intensity refers to the effort that athletes should put into each repetition. Some exercises require high intensity simply to be completed, while for others the effort is self-selected. Coaches need to adapt appropriate exercises to train the muscles required for their sport skills. A description of these exercises is included in appendix C.

If power, speed, and quickness are required by your sport, consider introducing plyometrics into your training sessions. Chapters 10 and 11 will help you with overall plans for selected sports. With a little thought, you may be able to design your own sport-specific plyometrics program to give your athletes an edge over the competition. For example, if you want to improve leg power, you could have your athletes start with one- and two-leg jumps, progressing to explosive jumps up a hill or jumps over low hurdles. As your athletes improve to the high-intensity period, you can adapt the exercises to increase the eccentric loading by using depth jumps. Start gently with low boxes and gradually increase. Athletes who develop soreness or pain around the joints used during plyometric training should avoid or reduce the intensity of the offending exercises.

Improving Flexibility

Next we examine the reasons for including flexibility exercises in muscular fitness training. We also describe several methods for improving flexibility. Specific flexibility exercises can be found in appendix A.

Flexibility Training

Flexibility exercises should be included after a general warm-up that begins every muscular fitness training session. Additionally, research has shown that the most effective time to stretch, in order to avoid injury and maintain long-term flexibility, is after training sessions. Some coaches now schedule flexibility training only at the end of practice to save time. The research remains unclear on the value of stretching prior to exercise. We still believe that a few minutes spent stretching after an adequate warm-up is time well spent, and we further believe (now that age is resulting in more aches and pains) that there is great value in postexercise stretching to maintain flexibility and feel better the next time out.

Connective tissue that surrounds entire muscles **(epimysium)**, bundles of muscle fibers **(perimysium)**, and each individual muscle fiber **(endomysium)** is continuous with the tendons connecting the muscle to bone. Connective tissue limits muscle length and thus limits flexibility. Other causes of reduced flexibility are joint design, lack of joint mobility and joint calcification, injury, and muscle imbalance.

Connective tissue has both elastic (rebounds when stretched) and plastic (remains elongated if adequately stretched) properties. The elastic component of connective tissue is quite limited. The focus of stretching to increase flexibility is to cause a permanent or "plastic" stretch in the connective tissue. Connective tissue is more plastic when warmed by exercise—thus the recommendation to stretch when one is warmed up or after exercise. Fatigued muscle shortens, resulting in a shorter resting length for some period after exercise. Connective tissue and muscle allowed to remain in this shortened state reduce long-term flexibility. Stretching after training appears to increase resting muscle length and causes plastic stretching in the warmed muscle; this results in less discomfort after training and better flexibility at the start of the next practice, even without stretching.

After the warm-up period, and again postexercise, have your athletes concentrate on stretching the muscle groups used in their sport. Five to 10 selected stretching exercises are usually sufficient. One way to decide what needs attention is to have athletes stretch those muscles that are stiff and sore during and after practice. As a minimum, both you and your athletes should be sure to stretch the hamstrings and lower back to avoid low back

TABLE 6.8

Sample Plyometric Training Program

Intensity	Exercise	*Sets × reps*	*Rest (min)*	Progression
Low (4 weeks*)	Double-leg bound	*3 × 6-10*	*1-2*	Add one rep each workout until 10 reps is reached.
	Skipping for height	*3 × 8-10*		
	Lateral hop	*3 × 6-10*		
	Medicine ball chest pass	*3 × 6-10*		
Medium (4 weeks*)	Alternate leg bound	*3 × 8-10*	2	Add one rep each workout.
	Double-leg hop over hurdles**	*3 × 8-10*		
	Split jump	*2 × 8-10*		
	Lateral hop	*3 × 8-10*		
	Drop and catch push-up	*3 × 8-10*		
	Medicine ball twist/toss	*3 × 8-10*		
	Dynamic wall push-off	*3 × 8-12*		
High (balance of season)	Squat jump	*2 × 8-12*	2	Build to 12 reps by adding 1 rep per week. During tapers, reduce reps to 6-8 with maximal effort.
	Hops over high hurdles**	*2 × 8-12*		
	Box jump	*2 × 8-12*		
	Double-leg speed hop	*2 × 8-12*		
	Depth jump	*2 × 8-12*		
	Belly blaster***	*2 × 8-12*		
	Dynamic wall push-off	*2 × 10-16*		
	Heavy bag thrust	*2 × 8-12*		

* During the second 2 weeks of the low-intensity and medium-intensity period, start gradual increases in effort to prepare for the next period. Start light and give athletes time for their muscular systems to adapt to the heavy eccentric loads caused by these exercises.

**These are the same as double-leg bounds, but the emphasis is on height. Start with low hurdles that are easy to clear and progressively add height (1 in./week) as appropriate.

***These are described in appendix B. As athletes develop, these can be done with increasing assistance from the holder, who is standing. The athlete attempts to keep the legs from touching the floor and rapidly rebounds, bringing the feet up toward the holder's stomach.

problems. Regularly remind your athletes of the importance of maintaining flexibility.

Stretching Methods

Two recommended training methods are described for developing flexibility. **Static stretching** is the most accepted form of stretching for teams. In this form of stretching, slow movements are used to the point of a moderate stretch. The position is held for 5 to 10 seconds, and then the athlete relaxes. Research has shown that static stretching, especially when done postexercise, is more

© Photodisc

Research has shown that the most effective time to stretch, in order to avoid injury and maintain long-term flexibility, is after training sessions. A brief period of gentle dynamic stretching following the warm-up is also recommended by many successful coaches and provides a period of relaxation prior to the training session.

effective than dynamic stretching. Static stretching increases connective tissue plasticity and is less likely to cause injury. Dynamic, or ballistic/bounding stretching, is not held long enough to cause plastic stretching. However, sports are dynamic, and some easy dynamic movements may help athletes feel more ready to practice and compete. If any athlete is suffering from muscle soreness, static stretching is a great way to bring relief. Have athletes start on a rug or mat, stretching first the hamstrings and then the lower back. Then move on to the other muscles, using the contract–hold–relax technique. Illustrations and descriptions of static stretching exercises can be found in appendix A.

In **PNF (proprioceptive neuromuscular facilitation) stretching,** the athlete first stretches a muscle or muscle group, then contracts that muscle before again stretching. For many stretches, this may require a partner to assist in the stretch and to hold a position while the athlete contracts the muscle. The proprioceptive feedback from the spindle fibers (organs in the muscle that sense tension) allows the muscles to relax following contractions. A greater stretch to the connective tissue is thus possible. After the usual static stretch, have athletes briefly contract the stretched muscle, then relax and do the static stretch again. Athletes should feel the stretch in the tendon after the muscle is relaxed. Work systematically around the body, contracting and relaxing the major muscle groups. See the illustration of an athlete and partner using this technique to stretch the hamstring (posterior thigh) muscle.

It is wise to stretch daily. Some researchers and coaches recommend stretching twice daily to improve flexibility. For athletes who lack adequate flexibility, it may be necessary to devote several sessions a week to general and sport-specific stretching in order to improve. Flexibility is lost rather slowly at first and is not hard to regain. It is wise to advise your athletes to maintain flexibility during the off-

Proprioceptive neuromuscular facilitation (PNF) stretching uses properties of the spindle fiber stretch sensory organs to allow for greater stretch to the connective tissue by relaxing the muscle fibers.

season; this will make the return to training less stressful.

Is it possible to become too flexible? Adequate flexibility for a sport requires that an athlete have free mobility throughout a slightly greater range of motion than that necessary for the given sport skills. Hamstring flexibility is necessary for all athletes to avoid lower back problems. Excessive flexibility is associated with reduced joint stability. Baseball pitchers require free range of movement through the pitching motion, but overly flexible shoulders may result in shoulder dislocations and injury. A practical note to coaches: Inflexible athletes need encouragement to improve flexibility (athletes don't like to work on areas of weakness), and flexible athletes may need attention to ensure that they do not overstretch.

Improving Agility and Balance

As a final note to this chapter, it is appropriate to make some comments on agility and balance. These attributes are sport specific and are learned through sport-specific drills. While some athletes may seem to have more innate agility or balance than others, all athletes can improve. Scandinavian researchers and coaches have stressed the need for young athletes to participate in multiple sports, believing that this will improve agility and balance in other sports as they mature. Though little research supports this concept, there is lots of anecdotal evidence to support the idea of many activities for youth with specialization in the midteen years.

Agility and balance both require strength, power, and endurance. As athletes develop speed and power they will become more agile. Dynamic balance also seems to be enhanced with increased power and quickness. Additionally, agility and balance both require game knowledge and the ability to anticipate the need for changes of direction. Thus, strength is the platform from which we build power, muscular endurance, speed, balance, and agility, while experience and anticipation give athletes the grace and fluidity that define great athletes. Perceptive coaches learn to recognize specific muscular fitness strengths and weaknesses in each of their athletes and help each one develop a long-term program to improve and maintain as needed.

Agility and balance are sport specific and are learned through sport-specific drills. While some athletes may seem to have more innate agility or balance than others, all athletes can improve with regular practice.

SUMMARY

This chapter discussed the types of training that are necessary to develop strength, power, power endurance, speed, flexibility, agility, and balance. Resistance training requires following a simple set of basic principles, including these:

- Overload and recovery: In order to stimulate physiological changes in muscles, you must overload them more than they are accustomed to, then allow them time to recover and respond.

- Specificity principle: Recruited muscles will respond to the specific overload applied.
- Individuality principle: Each individual will respond differently to your training program.
- Maintenance and reversibility principles: Once a new fitness level is achieved, it will take less training to maintain the fitness than it did to attain that level of fitness.
- Progression, variation, and periodization: As athletes get stronger they need to increase the training load. The progression needs to be done using variation and periods of easier training interspersed into the harder weeks.

The chapter also covered terminology and types of strength training and presented information on organizing a strength training program through understanding the individual goals and performing needs assessments. Training typically starts with core stability and then moves to specific exercises for the sport. Programs also tend to move from multi-joint lifts to more sport-specific training as the competitive season approaches. Finally, a strength program generally begins with light weights (for learning the exercises and preparing muscles to lift heavy weights for strength [1-6RM]), then moves more to power (6-12RM), then to speed or power endurance as needed. The chapter also covered training for flexibility, speed, and agility.

KEY TERMS

assistant exercises (p. 89)
circuit training (p. 88)
core stability (p. 88)
dynamic constant external resistance (DCER) (p. 85)
endomysium (p. 114)
epimysium (p. 114)
intensity (of resistance training) (p. 88)
isokinetics (p. 86)
isometrics/static contractions (p. 85)
isotonics (p. 85)
major exercises (p. 89)
muscle balance (p. 109)
neuromuscular recruitment (p. 113)
percent maximum strength (% max) (p. 84)
perimysium (p. 114)
plyometrics (p. 113)
PNF (proprioceptive neuromuscular facilitation) stretching (p. 116)
power training (p. 101)
rep (repetition) (p. 87)
repetition maximum (RM) (p. 84)
rest period (for strength training) (p. 88)
set (p. 87)
set-rep training (p. 88)
specialty exercises (p. 89)
speed-assisted training (p. 113)
static stretching (p. 115)
supplementary exercises (p. 89)
volume of training (p. 83)

REVIEW QUESTIONS

Answer the following questions by filling in the blank(s) with the appropriate word or words.

1. The FIT acronym commonly associated with the overload principle in resistance training stands for __________ , __________, and __________.
2. _________ are muscles that are primary movers while __________ are the muscles opposing the primary movers.
3. A __________ is the number of times that an athlete can lift a load before exhaustion.
4. __________ are contractions against an immovable object.
5. __________ refers to resistance training using either free weights or weight machines and means "same weight."
6. In __________ resistance training, the speed and movement are controlled and variable resistance by the athlete is allowed for.
7. When organizing a resistance program, it is important to consider these four types of exercises: __________ , __________ , __________ , and __________.
8. __________ is a popular method of developing power that utilizes both neural and metabolic systems by training the preload and elastic recoil phases.
9. In __________ stretching, slow movements are used to the point of a moderate stretch so as to increase connective tissue plasticity and decrease risk of injury.
10. In __________ stretching, the athlete first stretches a muscle or muscle group, then contracts that muscle before stretching again.

Determine if the following statements are true or false and then circle the correct answer.

11. T/F Isokinetic contractions are generally eccentric and often produce muscle soreness.
12. T/F The order of exercises does not have any impact on the efficacy of strength, power, or power endurance development.
13. T/F During a workout it is important to train large muscle groups before small groups, perform multiple-joint exercises before single-joint exercises, and perform high-intensity exercises before lower-intensity exercises.
14. T/F The higher the fitness level, the less training required to maintain fitness.
15. T/F Athletes are most likely to perform at the speeds that they most practice during training.

PRACTICAL ACTIVITIES

1. At the beginning of this chapter we introduced you to John, a young football lineman whose father decided to have him get professional help in designing a training program to improve muscle mass, power, and speed. Write out a possible time line for a periodized strength and power program that John might have been prescribed. Include the type of training during each period of the year for one year. You don't have to define specific exercises.

2. You coach a group of young swimmers who must be proficient in all of the strokes. Design a general flexibility program for your team, noting when (relative to the practice) it will be done, what stretches will be done, and how they will be done (dynamic, static, PNF, etc.).
3. Bob is a 6-foot 3-inch (190.5-centimeter) high school soccer player who is excellent at heading the ball except that his vertical jump could be better. He often misses scoring opportunities when opposing players are able to jump higher and deflect balls that John might otherwise be able to head into the goal. Design a power program, including plyometrics, that might improve both John's vertical jump and his quickness in heading the ball in the air.

PART III

Energy Fitness Training

Muscles require a continuous supply of energy to fuel contractions. This energy comes from food that is consumed, digested, stored, and eventually converted for use as fuel in metabolic pathways located in the muscle fibers. This part of the book describes the energy systems used in sport, provides ways to assess the capacity of these systems, and demonstrates means of improving each system with training.

7

Anaerobic and Aerobic Energy Systems

This chapter will help you

- understand how energy systems power the muscular contractions used in sport,
- differentiate between anaerobic and aerobic energy systems,
- identify the components of anaerobic energy,
- determine the components of aerobic energy and how they relate to performance,
- determine the sports and events best suited to each energy system,
- understand how anaerobic and aerobic training affect energy sources, and
- recognize the role of specific training in the development of energy fitness.

Jerry had always enjoyed running, so when he started high school he decided to go out for track. After several weeks of preseason practices, the coach decided to try him in the 400-meter event. In the first few dual meets our young runner was unable to do better than third, behind two fine upper-class athletes from his own school. He didn't feel that the short intense race was best for him, so one day at practice he asked the coach if he could work out with the longer-distance runners. The task that day was a time trial to select athletes for the 1,600-meter run in the next meet. When the trial started, Jerry found himself running comfortably behind the school's best distance runner. In laps 2 and 3 he and the senior pulled away from the pack. As the time trial entered the last lap, Jerry decided he could pick up the pace, and he passed the surprised senior, finishing comfortably in front. The coach had found an athlete who, by virtue of his natural ability and dedication to hard work, would dominate the longer runs, compete in 4 × 400 relays, and captain the team in his senior year.

It's clear that Jerry had some physical inclination toward distance running. The authors of a classic study (Costill et al. 1976) investigated the muscles of some of the world's finest distance runners. Muscle biopsies were taken to determine fiber type and enzyme activity. The results indicated a high percentage of slow-twitch (slow-oxidative) fibers and an elevated level of aerobic enzyme activity. Not surprisingly, the athletes also demonstrated a superior aerobic capacity (maximal oxygen intake) and world-class running performances. The results of this study have been confirmed in subsequent studies, with elite men averaging 79 percent slow-twitch fibers (women averaged 69 percent) and with peak slow-twitch values ranging over 90 percent for some male and female athletes.

This type of study does not differentiate between the effects of heredity and training. However, animal and some human studies clearly show that slow and fast fibers can improve their capacity for aerobic (oxidative) metabolism and endurance, thus proving the saying, "Fitness can neither be bought nor bestowed; like honor, it must be earned" (Anonymous). But coaches don't need muscle biopsies, fiber types, and enzyme activities to select athletes. In chapter 8 we will show you how to determine an athlete's capabilities, and in chapters 9 and 10 we'll lay out steps to maximize potential in anaerobic and aerobic events.

Let's observe the contractions of a muscle fiber during a run. Early on, the fiber gets its energy from two stored compounds, **adenosine triphosphate (ATP)** and **phosphocreatine (PCr)** (figure 7.1). Since both are in limited supply, the muscle will soon run out of energy unless something else occurs. After about 10 seconds of exercise, just as ATP and PCr are running low, the muscle starts to use stored carbohydrate **(glycogen)** to produce more ATP. Soon thereafter a by-product of glycogen breakdown, lactic acid, begins to appear in the muscle. Unfortunately, acid by-products can inhibit energy and force production, so another source of energy better step up or fatigue will ensue. After a couple of minutes of exercise, the muscle gets its second wind as the cardiorespiratory systems gear up to provide an adequate supply of oxygen that allows the efficient formation of ATP from the oxidation of carbohydrate and fat.

When our muscle is called on to increase the pace, it will recruit fast-twitch fibers that rely on muscle glycogen to produce ATP. But muscle glycogen stores are also limited, and won't last more than 1 or 2 hours. Eventually, the body has to call on another source of energy, blood glucose supplied from glycogen stored in the liver. When muscle and liver glycogen is depleted, the muscle is forced to rely on fat for energy. When this happens to a runner, the pace slows dramatically. Why? Energy metabolism is less efficient when we run out of carbohydrate since more oxygen is required to burn fat. The fat molecules carry less oxygen than carbohydrate does, so more must be supplied during the effort.

The energy for muscular contractions comes from two energy systems, the anaerobic and

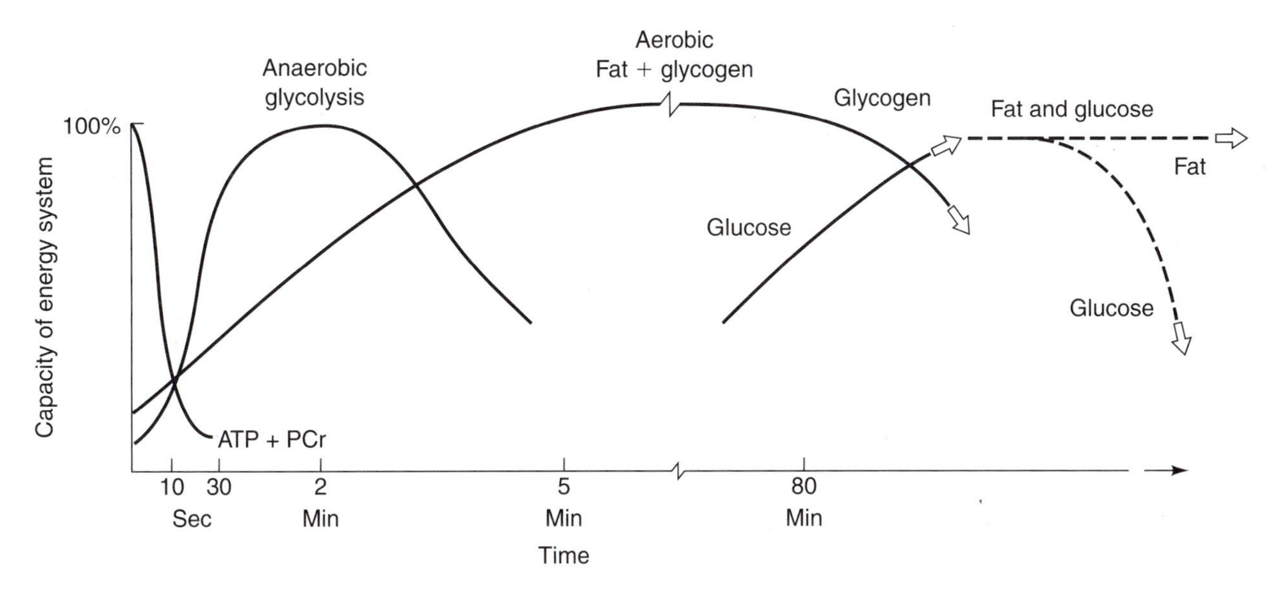

FIGURE 7.1 Patterns of energy production. Duration of the athletic event determines the primary energy production pathway. For long events, as muscle glycogen is used up, blood glucose temporarily fills the demand for carbohydrate.
Reprinted, by permission, from B.J. Sharkey, 1986, *Coaches guide to sport physiology* (Champaign, IL: Human Kinetics), 83.

the aerobic. **Aerobic** means "with oxygen"; the aerobic energy system uses oxygen to produce energy from the metabolic breakdown of carbohydrate and fat, while the **anaerobic** system derives energy from carbohydrate without the need for oxygen. The anaerobic system provides the energy we use at the start of exercise, when we increase the pace, and when we engage in high-intensity effort such as a sprint. Otherwise, the aerobic system is the primary source of energy. The anaerobic system is limited and inefficient; it wastes energy. The aerobic system uses abundant supplies of fat and carbohydrate, and it uses them efficiently. One of the major goals of training is to enhance aerobic fitness in order to conserve limited anaerobic energy sources.

Anaerobic Energy System

The nonoxidative or anaerobic system relies on the breakdown of stored compounds to produce energy. Since the supply of these high-energy compounds, ATP and PCr, is quite limited, anaerobic effort is brief though intense. The anaerobic breakdown of stored muscle

Lactic Acid

Lactic acid produced during anaerobic effort, or when the oxygen supply is limited, can accumulate in the muscle and affect its ability to perform. The acid changes the acid–base balance in the muscle fiber, thereby reducing enzyme activity and ATP production in aerobic energy pathways. The lactic acid can also reduce the contractile force of the fiber. Lactic acid is probably responsible for the burning sensation caused by intense effort, and it is associated with physical and mental fatigue. After exercise the lactic acid accumulation diminishes rapidly, especially if you engage in active recovery. Some of the lactic acid may be taken up by rested fibers and used as a source of energy. Most of the lactic acid is removed from circulation within an hour, so it cannot be blamed for muscle soreness. Training reduces production of lactic acid and improves the body's ability to remove it from the muscle.

Incidentally, in this book we use the terms **lactic acid** and **lactate** interchangeably. Anaerobic glycolysis forms lactic acid, but it rapidly loses its hydrogen ion (H^+) and combines with sodium (Na^+) or potassium (K^+) to form lactate, a salt of lactic acid.

glycogen produces a small amount of ATP energy along with a metabolic by-product, lactic acid. So this energy system is important at the onset of exercise and for events of short duration. The anaerobic system also fuels the sprint or kick at the end of a longer event.

Aerobic Energy System

The aerobic energy system, located in the mitochondria of the muscle fiber, breaks down or oxidizes carbohydrate and fat to produce energy in the form of ATP. The anaerobic breakdown of one molecule of glucose yields 3 units of ATP, while the yield is 36 units when accomplished aerobically, so the aerobic system is many times more efficient. Efficiency can be very important when energy supplies are limited. Table 7.1 indicates the approximate amount of energy available from the energy sources.

As you can see, ATP and PCr do not last long without backup. Even carbohydrate energy is limited. At 100 kilocalories per mile, you couldn't make the 20-mile (32-kilometer) mark of a marathon (26.2 miles or 42 kilometers) on carbohydrate alone. (Of course you could carbo load in the days before the event or consume solid and liquid carbohydrate supplements during the race.) But fat is clearly our most abundant source of energy; the example in table 7.1 has enough fat to fuel a run of 300 to 700 miles (483 to 1,127 kilometers)! Better yet, training improves the muscle's ability to use fat as a fuel, thereby conserving the glycogen needed for high-intensity effort.

What about the protein that is sold in energy bars and drinks? Is it needed? Protein is essential to build, maintain, and repair tissues, as well as to add muscle and enzymes stimulated by training. But protein is generally a minor source of energy for muscles. An exception is the situation in which glycogen and glucose stores are low, as when one is on a starvation diet. Then protein from muscle is broken down and converted to the glucose required by the nervous system and for other energy needs. A high-protein, low-carbohydrate diet is not recommended for active individuals, especially athletes. Athletes need adequate protein, but they also need sufficient carbohydrates to ensure the energy required to train and compete. Without adequate energy, an athlete cannot achieve the benefits of training.

TABLE 7.1

Available Energy Sources

Source	Supply	Energy (kcal)	Miles*
ATP and PCr	Small quantities in muscles	4-5 kcal	0.045
CARBOHYDRATE			
Muscle glycogen	20 g/kg of muscle	1,600 kcal	16
Liver glycogen	80 g	320 kcal	3.2
Blood glucose	4 g	16 kcal	0.16
FAT			
Muscle	Limited; varies with training	1,500	15
Adipose tissue	Variable**	30,000-70,000 kcal	300-700

*Assumes 100 kcal/mile and all energy in working muscles.

**Depends on body weight and percent body fat: 10% fat × 150 lb = 15 lb fat × 3,500 kcal/lb = 52,500 kcal.

PCr = phosphocreatine

Adapted, by permission, from B.J. Sharkey, 1990, *Training for cross-country skiing racing* (Champaign, IL: Human Kinetics), 425. Adaptations from Sharkey 1990.

Protein

Protein needs range from 0.8 grams per kilogram of body weight for sedentary individuals to 1.2 to 1.4 grams per kilogram for endurance athletes and 1.4 to 1.8 grams per kilogram for athletes engaged in serious weight training (1 kilogram = 2.2 pounds, so a 70-kilogram [154-pound] strength/endurance athlete may need 1.4 grams × 70 = 98 grams of protein daily, vs. 56 grams for a sedentary counterpart). Fortunately, these needs are usually supplied in a well-balanced diet. Since protein yields about 4.3 kilocalories per gram, our example would require 420 kilocalories of protein daily. In general we recommend that athletes try to get 15 percent of their calories from good-quality protein. Amounts of protein in excess of these recommendations have not been shown to enhance training benefits or to improve performance. In fact, part of the excess protein contributes to weight gain, while the rest of the amino acid molecule (ammonia) is excreted in the urine.

Anaerobic Power

The anaerobic energy system is composed of two parts: short-term energy (ATP and PCr) and the breakdown of glycogen to lactic acid (called anaerobic glycolysis). The maximal all-out effort that this energy system can sustain for several seconds is called **anaerobic power.**

- ATP-PCr energy. Adenosine triphosphate and PCr are high-energy compounds stored in muscles and other cells. When the muscle is signaled by the nervous system to contract, the ATP is split (from adenosine triphosphate to adenosine diphosphate or ADP), releasing the energy needed to make the muscle contract). The role of PCr is to back up the ATP, to restore it to full power, and keep the muscle contracting. So, with the aid of an enzyme called creatine kinase, it splits to provide energy to restore ATP. During intense activity, such as a sprint, the PCr level rapidly declines. While ATP also declines, some is preserved since it is needed for other cellular functions. As ATP levels continue to drop, the muscle becomes less able to contract and coordinated movement becomes difficult (picture a 400-meter runner who ran too fast in the first half of the race, as he struggles down the stretch). Indeed, at critically low levels of ATP, muscular contractions are inhibited. While ATP stores do not change much with training, there is some evidence that PCr storage can be increased with training. However, since ATP and PCr together provide for only 3 to 15 seconds of all-out effort, it makes sense to look to anaerobic glycolysis for additional help.

- **Anaerobic glycolysis** is the term used for the nonoxidative breakdown of glycogen. Glycogen is a clump of glucose molecules stored in the muscle. This pathway restores a small amount of ATP (three units for each molecule of glucose) while producing the by-product, lactic acid. The six-carbon glucose molecule is broken down in a series of enzymatic steps to a three-carbon lactic acid molecule. The energy released is used to restore ATP. The lactic acid, which can inhibit enzyme activity and reduce muscle force, also serves to transport unused energy to other cells via the circulation. When sufficient oxygen is available, glucose can be processed via the aerobic pathway to yield 36 units of ATP. This aerobic process does not lead to the accumulation of lactic acid. The availability of oxygen at the cellular level, along with the concentration of aerobic enzymes, dictates which pathway, anaerobic or aerobic, is employed, while exercise intensity dictates the demand for oxygen. When you exercise at a high intensity, lactic acid begins to accumulate in the muscle.

In running events such as 400 to 800 meters, lactic acid levels can exceed 20 mmol/L. The goal in these events is to run out of anaerobic energy as you cross the finish line. In longer events and sports with continuous motion (soccer, basketball), the athlete alternates between anaerobic and aerobic energy pathways to provide the energy for contractions. One of the great benefits of aerobic fitness is that it improves the ability to recover from anaerobic bursts of effort, such as a fast break in basketball.

Lactate Threshold

The **lactate threshold (LT)** defines an increased level of lactic acid accumulation during exercise. From a blood level of about 1 millimole per liter (mmol/L) at rest, levels during easy exercise will stay low (1-2 mmol/L) until you reach a level of exercise that feels moderate. The point where lactate rises above 2 mmol/L, called the first LT (or sustainable threshold), approximates the level of effort you can sustain for a number of hours. While this point has been called a threshold, it is better understood as a transitional zone that signals a shift from predominantly slow-oxidative muscle fibers to fast-oxidative-glycolytic fibers. Later in this book (chapter 10) we will refer to training in the easy training zone (EZ) representing the intensity for long slow distance and recovery workouts. The EZ is based on the first LT.

As intensity increases, blood lactate continues to rise. At about 4 mmol/L you reach the second LT. Swimming coaches have long used this threshold to guide interval training and to determine the optimal racing pace. The second LT indicates a transition zone above which increasing numbers of fast-glycolytic fibers are recruited and lactate accumulates rapidly, inhibiting aerobic metabolism and reducing the force of contractions (figure 7.2). Depending on the length of an event, your athletes may perform below, at, or slightly above the second LT. The intensity that your athletes will use for competition is called the performance threshold (PT). The PT depends on the race duration and athlete fitness. The small range of speed slightly above the PT (from PT to 5 percent above PT) is called the performance training zone (PZ), because that is where you train to achieve gains in performance.

We will provide simple ways to estimate EZ and PZ training intensities in chapter 9.

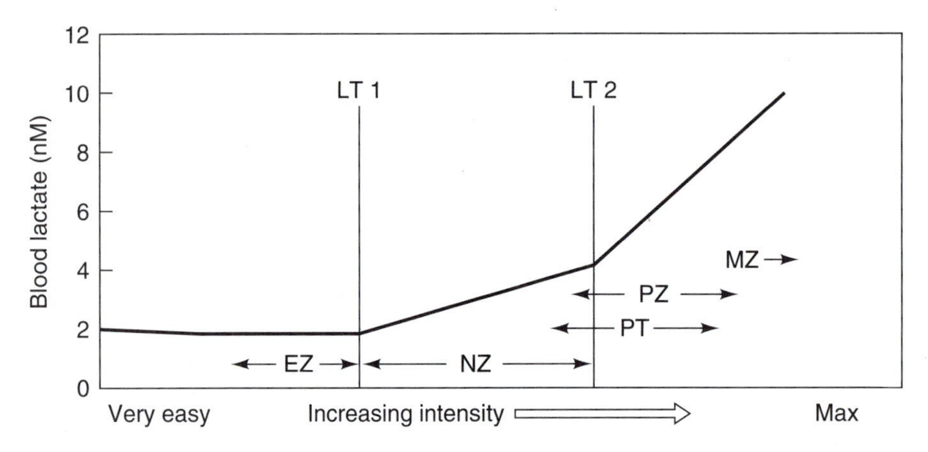

FIGURE 7.2 Blood lactate and training zones. The lactate thresholds are shown at the points where blood lactate increases. The first lactate threshold (LT 1) is associated with the top of the easy training zone (EZ). The no-training zone (NZ) starts at LT 1 and extends up to the performance threshold (PT) defined as current race speed. The width of NZ is determined by the location of PT, the normal range of which is shown by the arrows on either side of PT. For events longer than 60 minutes, PT generally occurs below the second lactate threshold (LT 2), while for shorter races, PT will occur above LT 2. The performance zone (PZ) will also depend on PT and extend from PT to 5 percent above PT. Speed training and races lasting less than 5 minutes are typically in the maximal zone (MZ).

Aerobic Energy

Aerobic energy pathways utilize abundant energy sources, carbohydrate and fat, to provide ATP for muscular contractions. These pathways are more efficient than the anaerobic pathway; they get more energy from each molecule of glucose. Here we explore the capacity, power, and endurance of aerobic energy pathways.

Inefficient anaerobic energy must be used when the supply of oxygen reaching the muscle is inadequate to meet the demand for aerobic metabolism, or when the muscles lack the enzymes to utilize available oxygen. But when sufficient oxygen and enzymes are available, three carbon fragments from glycogen breakdown move into the mitochondria, where aerobic enzymes and pathways are located. Mitochondria are cellular energy factories that are designed to make metabolism efficient. Carbohydrate and fat fragments move along an assembly line of enzymes that systematically liberate the energy they contain. This energy is ultimately used to form additional ATP. Carbohydrate fragments are processed in two major pathways, the citric acid cycle and the electron transport system. The final step in the process involves the union of oxygen with carbon and hydrogen to form carbon dioxide, water, and energy.

$$C_6H_{12}O_6 + 6O_2 = 6CO_2 + 6H_2O + \text{energy (36 ATP)}$$

$$\text{glucose + oxygen = carbon dioxide + water + energy}$$

Fat stored in muscle or adipose tissue is oxidized in a pathway called beta-oxidation. Stored fat (triglyceride) is composed of three fatty acids and glycerol. The fatty acids are separated from glycerol, and the fatty acid fragments are systematically cleaved into two carbon segments that enter the aerobic pathways in the mitochondria. It has been said that fat burns in the flame of carbohydrate. Indeed, when carbohydrate stores are depleted, fat metabolism becomes less efficient, and the pace slows dramatically. Endurance-trained muscle contains more intramuscular fat, and highly trained endurance muscles are better able to use fat, even during moderately intense activity. As intramuscular supplies of fat are used up, additional fat is released from adipose tissue cells and transported via the circulation to the working muscle. Figure 7.3 summarizes the steps in energy production.

Fat is a dense source of energy, providing over 9 kilocalories per gram versus 4 kilocalories per gram for carbohydrate. Unfortunately, since fat requires more oxygen in order to burn, this abundant fuel cannot be a major source of energy during high-intensity effort, when oxygen supply becomes a limiting factor. Now let's look at specific components of aerobic fitness.

Aerobic Capacity

A progressive treadmill or bicycle ergometer test is usually used to determine the ability to take in, transport, and utilize oxygen. Exercise intensity is increased until the athlete reaches a plateau in oxygen intake, or until the athlete cannot continue. The result, in liters of oxygen utilized per minute (L/min), defines the capacity of the aerobic system, including the respiratory system (taking in), the cardiovascular system (transport), and the muscles (utilization). A large **aerobic capacity** (also called maximal oxygen intake or $\dot{V}O_2max$) defines the size of the aerobic engine, just as a car engine is rated for size (e.g., 3 liters). The aerobic capacity is an excellent predictor of performance in non-weight-bearing sports such as bicycle racing, swimming, and

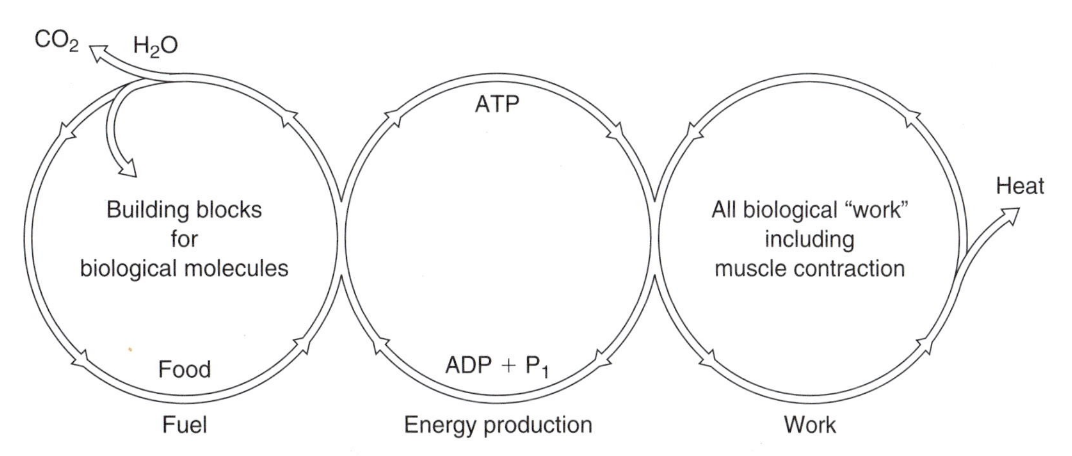

FIGURE 7.3 The production of energy. The energy stored in food (fuel), primarily carbohydrates and fats (and a little protein), is used to convert adenosine diphosphate (ADP) into adenosine triphosphate (ATP; the energy currency of life), which is then used to do all biological work, including muscle contraction.

rowing. Values range from 3 to over 6 L/min, and higher scores are correlated with superior performances.

Aerobic Power

When the aerobic capacity score in liters of oxygen per minute is divided by the body weight in kilograms, you get the measure called aerobic power, which is measured in milliliters (ml) of oxygen per kilogram of body weight per minute ($ml \cdot kg^{-1} \cdot min^{-1}$). It defines one's ability to use oxygen per unit of body weight. For example, for a 60-kilogram (132-pound) athlete with a 4 L/min aerobic capacity, the aerobic power becomes:

The bicycle ergometer test is one method of determining aerobic capacity.

4 L/min / 60 kilograms =
66.7 ml of O_2 per kilogram of body weight per minute (or $ml \cdot kg^{-1} \cdot min^{-1}$)

This measure is a predictor of performance in weight-bearing activities such as running, cross-country skiing, and speed skating. It is highly correlated to performances that take 5 to 15 minutes (e.g., 1,500- to 5,000-meter runs). Values range from below 40 to over 80 $ml \cdot kg^{-1} \cdot min^{-1}$. Our example (66.7 $ml \cdot kg^{-1} \cdot min^{-1}$) could be a moderately successful high school runner. Elite male runners can exceed 80, and we have tested elite females with a score of 70 $ml \cdot kg^{-1} \cdot min^{-1}$. Of course this measure is relevant to many sports, including soccer, lacrosse, basketball, and even football. If you want your team to avoid fatigue in the fourth quarter, make sure your training program includes aerobic power.

Lactate Threshold

Once called the anaerobic threshold, the LT is now known to represent a transition zone that involves increasing dependence on anaerobic energy pathways. It defines the upper limit of your ability to clear lactic acid from muscle and blood, and the transition zone that involves increasing lactate production resulting from the recruitment of fast-glycolytic muscle fibers. Threshold measures are usually determined during a progressive treadmill test. Blood samples are taken during each stage of the test, and the lactic acid level is analyzed. The first LT defines the top of the easy training zone (EZ) and is defined as the point where the lactate begins to rise, usually around a level of 2 mmol/L (see figure 7.4). The second LT is the point where the lactate level rises dramatically, usually around 4 mmol/L. You can go for several hours at the first threshold, but exceeding the second threshold leads to rapid fatigue. Fortunately, it is possible to increase both thresholds with training. The second LT is highly related to success in events lasting 30 minutes or more, such as longer running, swimming, or cycling races, and is important to success in all sports requiring endurance.

As you remember, we have three main types of muscle fibers; slow-twitch (slow-oxidative or SO) fibers that are efficient in the use of oxygen; a faster-contracting fast-twitch type that can work with oxygen or without (fast-oxidative-glycolytic or FOG); and a fast-twitch fiber that uses muscle glycogen for short, intense contractions (fast-glycolytic or FG). As we go from a walk to a jog to a run to a sprint, we recruit SO, FOG, and then FG fibers to help us go faster. Recruit too many FG fibers and the effort becomes anaerobic; the fibers produce lactic acid, and we are forced to

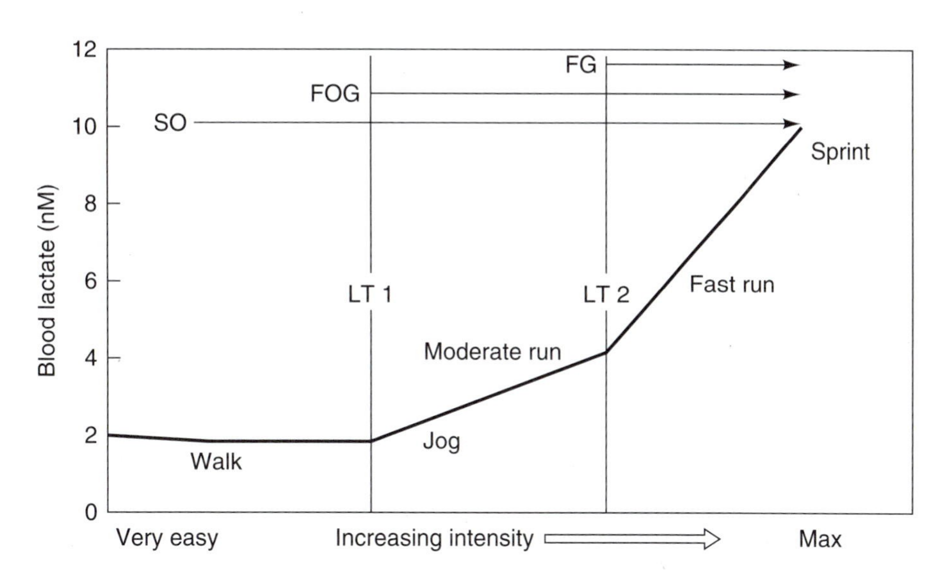

FIGURE 7.4 Lactate thresholds. As exercise intensity increases, we recruit first slow-oxidative (SO) fibers, then fast-oxidative-glycolytic (FOG) fibers, and finally fast-glycolytic (FG) fibers. More blood lactate accumulates above the first lactate threshold (LT 1) because we begin to produce more lactic acid than we can clear from the blood. Above the second lactate threshold (LT 2), the accumulation of blood lactate becomes more rapid because of the increased recruitment of FG fibers that produce more lactate and because more muscle fibers are active and therefore unable to remove (take up) lactate.

slow down or stop. Figure 7.4 illustrates the transition from slow to fast fibers, the accumulation of lactate, and the LTs.

Exercise at EZ, below the first LT, is accomplished primarily with SO fibers. Above EZ is an area of increasing recruitment of FOG fibers, and the second LT marks the transition from SO and FOG fibers to more FG fibers, increased anaerobic metabolism, and lactate production. Studies show that well-designed training programs can improve the oxidative abilities of all three fiber types. In effect, training shifts the lactate curve to the right, allowing a higher intensity or velocity at all lactate values. By improving the oxidative ability of fast-twitch fibers, you can run faster without accumulating excess lactic acid.

An intensity associated with the second LT is called the performance threshold or PT. The PT refers to the intensity that can be maintained by an individual for the duration of the given event. Events longer than 45 to 60 minutes will have a PT below the second LT, while shorter events will have their PT above that LT. We will show you how to use the PT to determine training intensities for intervals.

There are multiple dimensions of aerobic fitness, most of which can be determined in a single treadmill test. The PT is generally determined during a time trial similar to the event.

Sport-Specific Training

It is important to understand the energy demands of the sport you coach or intend to coach. You need to know how to determine the energy capabilities of your athletes, and then you need to train them specifically for their event or

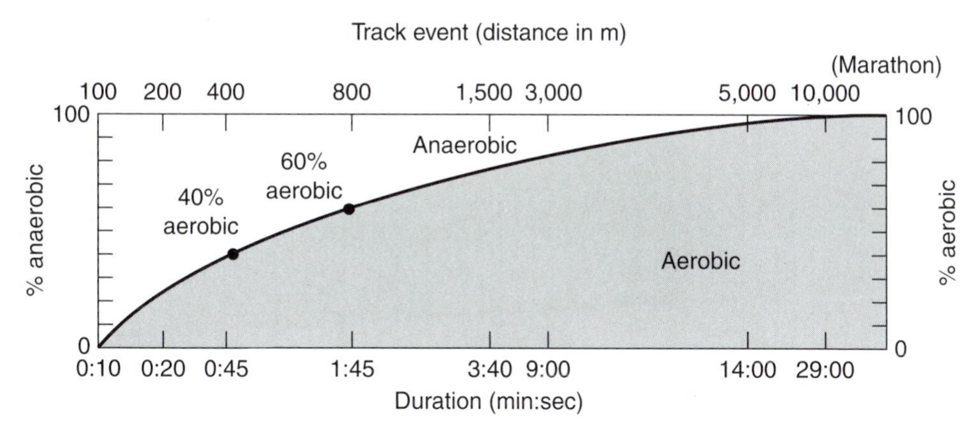

FIGURE 7.5 Anaerobic and aerobic energy sources in relation to distance and duration of events. Shorter events like the 400-meter run are primarily anaerobic (60 percent), while the 800-meter run is 60 percent aerobic. For distances greater than 1,500 meters (longer than 4 minutes), training should concentrate on aerobic fitness.

Adapted, by permission, from B.J. Sharkey, 2002, *Fitness & health*, 5th ed. (Champaign, IL: Human Kinetics), 372.

position. If all this sounds daunting, relax; we will do our best to make it understandable.

Energy Demands

Studies have determined the relative energy requirements of track events ranging from 100 meters to the marathon (42 kilometers), or for events lasting from 10 seconds to over 2 hours (see figure 7.5).

You can see that the energy for the 100-meter run is 100 percent anaerobic and for the marathon (42 kilometers) is virtually 100 percent aerobic. You can use the time scale to estimate energy requirements for other sports. Understanding energy demands helps to focus on the training needed for success. Understanding each athlete's strengths and weaknesses helps to focus on areas most in need of improvement (see chapter 8).

Energy Training

Since 1967, when Holloszy first reported the cellular effects of training in muscle (rat), studies have demonstrated the specific effects of training on animal and then human muscle fibers. This section summarizes the documented effects of training on anaerobic and aerobic energy systems.

Anaerobic Energy Training

In sports requiring maximal or near-maximal force production (e.g., heavy lifts, sprints), most of the energy needs are met with stored ATP and PCr and with the nonoxidative (anaerobic) breakdown of muscle glycogen. Short intense contractions lead to improvements in muscular strength and may increase enzymes in the ATP-PCr system. Recent studies have demonstrated the short-term effects of creatine supplementation—the ability to do more sets and reps during weight training. Unfortunately, the long-term health effects of this supplement have not received adequate study.

Anaerobic glycolysis is essential in events lasting up to 2 minutes. Efforts to train this system with 30-second bouts of intense effort have yielded limited (10-25 percent) improvements in the activity of glycolytic enzymes. However, the relationship of the training to improvements in performance is not overwhelming. It is possible that the improvements, when seen, are related to neuromuscular improvements, including increases in strength. So we can conclude that the benefits of anaerobic energy training are relatively subtle. Because of this and because high-intensity effort can cause injuries, we recommend moderation in the use of intense interval training, especially for younger athletes. Sprint and power athletes need to do sufficient anaerobic training to form a foundation for performance, to overcome soreness, and to allow efficiency at high intensities of effort. Just don't overdo it. We'll provide training guidelines in chapter 9.

Aerobic Energy Training

Unlike those for anaerobic training, the effects of aerobic training are clear, dramatic, and specific. Aerobic training has been shown to increase the volume of mitochondria in muscle fibers, to double the enzyme activity in aerobic pathways, to double the oxygen intake of mitochondria, and to dramatically increase the fiber's ability to generate ATP aerobically (figure 7.6). Aerobic training also increases the number of capillaries serving each muscle fiber, the capillary/fiber ratio, providing more fuel and oxygen to the working muscles. High-intensity training is necessary for trained athletes to achieve improvements in aerobic capacity and power. This training is also associated with improvements in cardiovascular and respiratory systems.

In addition to improvements in aerobic pathways, endurance training can increase the amount of fat (triglyceride) stored in muscles. Training improves the ability to burn fat and opens up access to the abundant supply of energy stored in adipose tissue. Training improves the ability to mobilize fat (fatty acids) from adipose tissue, to transport the fat via the circulation, and to metabolize the fat in the working muscle. When we use fat, we conserve limited stores of muscle and liver glycogen. All this has enormous implications for endurance performances, as well as for health.

The ability of the trained muscle to utilize fat may be the most important health benefit of aerobic fitness. Aerobically fit individuals also

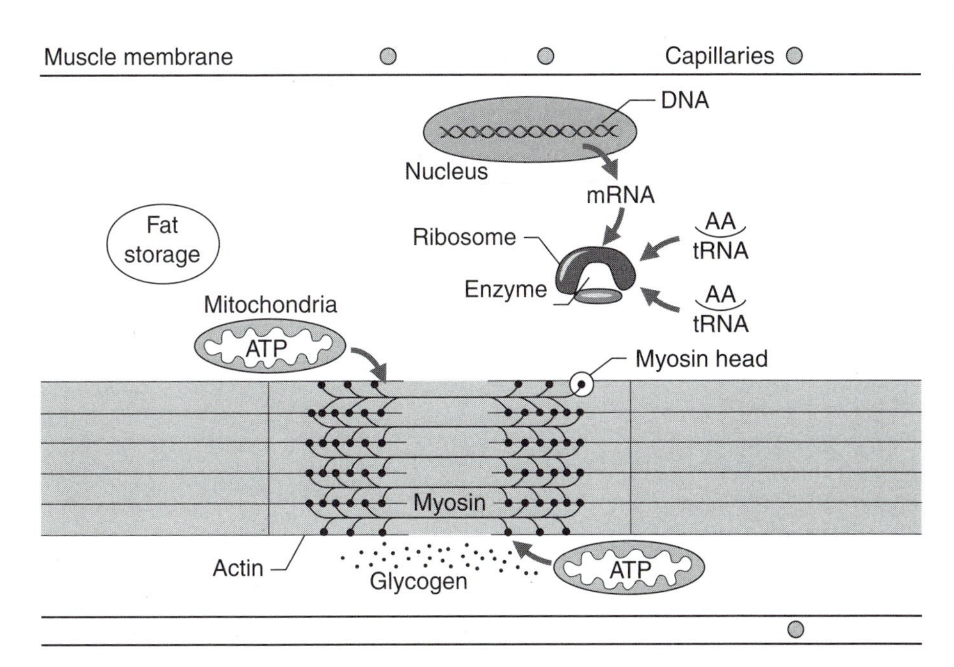

FIGURE 7.6 Cellular effects of training. Training has a number of direct effects on muscle cells. Increased capillaries deliver more oxygen-carrying blood. Increased fat and glycogen stores increase endurance capacity. Increased numbers and size of mitochondria allow for more rapid production of adenosine triphosphate (ATP) aerobically. Increases in both anaerobic and aerobic enzymes increase the overall speed of ATP production. With muscular fitness training, there will be increases in the muscle proteins (actin and myosin), increasing strength and power.

Adapted, by permission, from B.J. Sharkey, 2002, *Fitness & health*, 5th ed. (Champaign, IL: Human Kinetics), 87

use fat during rest and recovery, allowing the muscle to store more glycogen. And by burning fat you minimize the likelihood that it will be deposited in adipose tissue or in the coronary arteries, leading to the number-one killer, heart disease.

All these improvements take place in the muscles and fibers used in training. Thus it is important that training activities be specific to the sport and event. Swim training doesn't improve running performances; training the arms won't improve the legs; and anaerobic training does not improve aerobic capabilities, or vice versa. Experienced coaches design training programs that will specifically improve the muscles and energy pathways used in the sport.

Other Effects of Training

Not all the effects of endurance training are quite so specific. Training leads to general effects in the cardiovascular and respiratory systems. We call them general because most of the improvements gained in one activity, such as running, transfer to other activities, like swimming and cycling.

Blood Volume

Aerobic endurance training can increase **blood volume** 10 to 15 percent. An individual with a blood volume of 5 liters can increase to 5.5 or even 5.75 liters of blood. This increase is important for a number of reasons. With an increased volume you are able to return more blood to the chambers of the heart during rest and exercise. Consequently the heart is able to pump more blood per beat, and when it does that it doesn't need to beat as often. Increased blood volume means slower resting and exercise heart rates. It also leads to an increase in the **stroke volume,** the volume of blood pumped with each beat of the heart. Maintaining blood volume is essential to the maintenance of exercise capacity. You can see why it is so important to maintain hydration during vigorous exercise, since dehydration eventually leads to a loss of blood volume. And when an athlete stops training, blood volume begins to decline, the heart rate rises, and the stroke volume declines.

Cardiac Output

Years ago, before we understood the cellular effects of training, or the effects of training on blood volume, we thought that the most important effects of training were on the heart. It is true that **cardiac output,** the volume of blood pumped per minute, is related to aerobic capacity:

$$\text{cardiac output} = \text{heart rate} \times \text{stroke volume}$$

It is also related to performance in some sports. But the effects of training on the heart, already the body's finest endurance muscle, are subtle. The dramatic effects of training on heart rate and stroke volume (figure 7.7) are related to changes in blood volume as much as to changes in heart muscle.

The heart does undergo some changes with endurance training, including subtle increases in wall thickness and muscle mass. But you should not focus training on the heart. Focus on sport-specific muscles, and the heart will adjust to the demands.

Respiration

Like the heart, the respiratory system becomes more efficient with training, providing more air with fewer breaths. Studies show subtle improvements in the respiratory capacity of breathing muscles, like the diaphragm. But the major feature of the trained respiratory apparatus is a greater tidal volume, allowing a lower frequency of respiration to deliver a given level of air to the lungs (ventilation).

$$\text{ventilation} = \text{tidal volume} \times \text{frequency}$$

Since respiration is driven by the need to exhale CO_2, as well as the need to inhale O_2, it is an essential component of one's exercise capacity. Fortunately, you don't need to pay special attention to heart or respiratory dynamics during training; the adjustments are automatic.

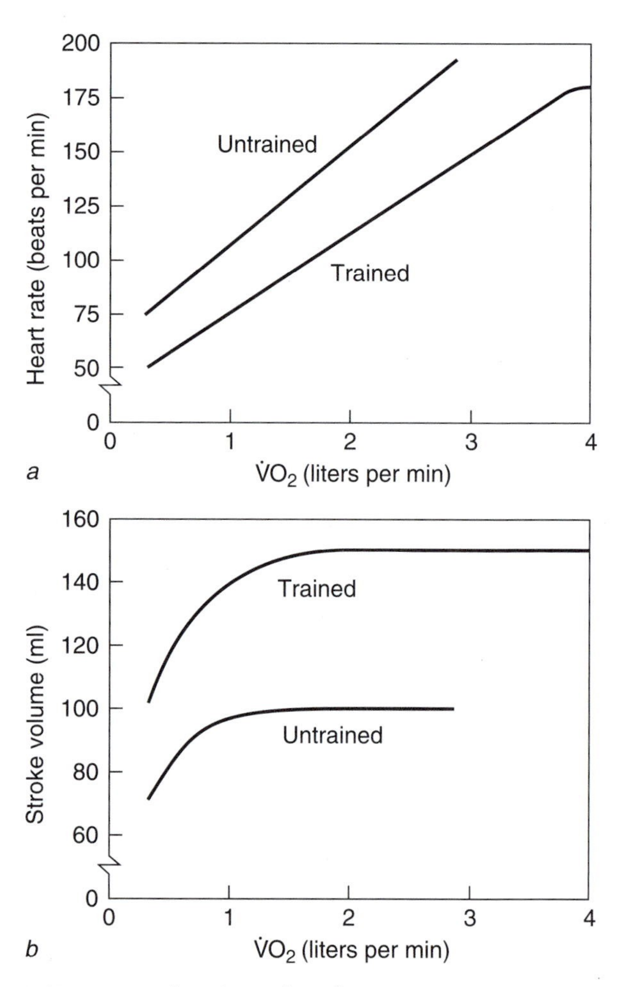

FIGURE 7.7 The relationship of exercise intensity (oxygen intake or $\dot{V}O_2$) to *(a)* heart rate and *(b)* stroke volume, before and after training.

Reprinted, by permission, from B.J. Sharkey, 1984, *Training for cross-country skiing racing* (Champaign, IL: Human Kinetics), 34.

SUMMARY

This chapter began with an illustration of the pattern of energy use during exercise, defining the anaerobic and aerobic energy systems. We described the short-term sources of energy for muscular contractions, ATP and PCr, and the carbohydrate and fat utilized in aerobic exercise and to replenish the short-term sources. We defined the components of aerobic energy, the lactate thresholds (first and second), the aerobic capacity, aerobic power, and we suggested ways in which each can be trained. We finished the chapter with a discussion of the effects of training, emphasizing the need for specificity. The most important features of energy training take place in skeletal muscle fibers—not in the heart or the respiratory system, but in the muscle fibers used in the sport. This accounts for the need for specific training. We closed with other important effects of aerobic training, including significant increases in blood volume that permit the heart to pump more blood with each beat, allowing lower resting and exercise heart rates.

KEY TERMS

adenosine triphosphate (ATP) (p. 124)
aerobic (p. 125)
aerobic capacity (p. 129)
anaerobic (p. 125)
anaerobic glycolysis (p. 127)
anaerobic power (p. 127)
blood volume (p. 133)
cardiac output (p. 133)
glycogen (p. 124)
lactate threshold (LT) (p. 128)
lactic acid (lactate) (p. 125)
phosphocreatine (PCr) (p. 124)
stroke volume (p. 133)

REVIEW QUESTIONS

Answer the following questions by filling in the blank(s) with the appropriate word or words to complete the statements.

1. In the first 10 seconds of exercise, muscle fibers get their energy from the two stored compounds known as _________ and _________.
2. _________ and _________ are the two energy systems that provide energy for muscular contraction.
3. The _________ energy system uses oxygen to produce energy from the metabolic breakdown of carbohydrate and fat to produce ATP.
4. The _________ energy system derives energy from carbohydrates without the need for oxygen by relying on the breakdown of stored compounds to produce energy.
5. The _________ defines the upper limit of the ability to clear lactic acid from muscle and blood and is also the transition zone that involves increased lactate production resulting from the recruitment of fast-glycolytic muscle fibers.
6. The _________ refers to the intensity that can be maintained by an individual for the duration of a given event.
7. Fat provides _________ kilocalories of energy per gram.
8. Carbohydrates provide _________ kilocalories of energy per gram.

Match the type of athlete on the left with the amount of protein needs on the right.

9. Nonathlete
10. Endurance athlete
11. Strength athlete

a. 1.2 to 1.4 grams per kilogram body weight
b. 0.8 grams per kilogram body weight
c. 1.4 to 1.8 grams per kilogram body weight

Determine if the following statements are true or false and then circle the correct answer.

12. T/F Energy metabolism is less efficient when we run out of carbohydrate because it requires more oxygen to burn fat.
13. T/F In short-duration events that require maximal or near-maximal force production, most of the energy needs are met with stored ATP and PCr along with the aerobic breakdown of muscle glycogen.
14. T/F High-intensity training is necessary for trained athletes to achieve improvements in aerobic capacity and power as well as improvements in the cardiovascular and respiratory systems.

15. T/F Aerobic training can increase blood volume 10 to 15 percent.
16. T/F There is enough energy in the average body to fuel a run of 500 miles.
17. T/F Protein needs range from 2 to 4 grams per kilogram of body weight.
18. T/F The first lactate threshold defines a level of exertion that can be sustained for a number of hours.
19. T/F Aerobic capacity tells more about performance in running than does the aerobic power score.
20. T/F The second lactate threshold represents a transition zone between predominantly aerobic effort and increasing anaerobic effort.

Select the correct response (more than one may be correct).

21. The primary source of energy for muscular contractions is
    a. ATP
    b. glycogen
    c. PCr
    d. glucose
    e. fat
22. When the immediate sources of energy are running low, we use one or more of the following to replenish the supply (include the two most likely to be used):
    a. fat
    b. protein
    c. carbohydrate
23. In prolonged exertion, when muscle and liver glycogen become depleted, we are forced to rely on _________ for energy.
    a. protein
    b. fat
    c. energy drinks
24. Produced in muscle fibers when the effort exceeds the oxygen supply:
    a. anaerobic glycolysis
    b. lactic acid
    c. amino acids
25. Your most abundant source of energy is
    a. carbohydrate
    b. protein
    c. fat
    d. McDonald's

PRACTICAL ACTIVITIES

1. You are the coach of the high school track team. Indicate methods you can use to help athletes find their best event, the one in which they are most likely to succeed.

2. You are seeking an athlete to do the 5,000-meter run. Which component of aerobic fitness best describes that ability? How would you test to determine that ability? What level of aerobic power would be likely to achieve success?
3. Explain to an athlete why training needs to be specific to the sport—why swimming, for example, will not do much to improve the athlete's ability to run. Where do the major effects of training take place?

8

Assessing Energy Fitness

This chapter will help you

- determine the energy systems used in your sport;
- understand how to evaluate an athlete's energy capabilities;
- identify ways to assess anaerobic and aerobic energy;
- develop field tests to assess athletes' capabilities as well as the effects of training; and
- understand the need to utilize sport-specific assessment techniques, so you know where you are going and don't end up someplace else.

Casey had been involved in soccer since grade school and had advanced to the U12 squad in middle school. She played on her high school team and on the U17 team that traveled to regional tournaments. Over the years she played several positions, but seemed best suited to midfield. She was delighted when she was offered the chance to play for the university team. When she arrived for fall camp, she was surprised to find that the coach required the players to complete a battery of tests. One of the tests, called the maximal oxygen intake test ($\dot{V}O_2$max), was used to evaluate her endurance. The test was administered in the university's human performance laboratory.

After a warm-up and practice on the treadmill, Casey was fitted with electrodes and a breathing valve. The test started at a slow jog and increased in speed until she was at a full run. Her heart rate was recorded, a computer calculated her oxygen intake, and blood lactate measures were taken at the end of each stage of the test. When she was unable to keep up with the increasing speed, the test was terminated and she was allowed to cool down with easy jogging. Days later the coach showed her the results of the tests, pointing out strengths and weaknesses and describing how the test results would be used to guide her training. She was surprised to find that she had a relatively high level of aerobic power, a definite advantage for a midfielder.

The testing Casey's college coaches performed on their athletes is supported by research. In one study (Kemi et al. 2003), the authors attempted to develop a soccer-specific test of maximal oxygen intake. Ten male soccer players participated in a treadmill test and a soccer-specific field test consisting of dribbling, repetitive jumping (heading), accelerations, decelerations, turning, and backward running. The maximal oxygen intake values were similar in the two tests, as were maximal heart rate and respiration. The authors concluded that the soccer test yielded results similar to those for the lab test and was therefore a valid test of endurance (maximal oxygen intake) in soccer players. So, you don't need a lab and expensive equipment to assess your athletes! But remember the words of Malcolm Stevenson Forbes: "To measure the man, measure his heart."

As a coach you need to know the relative importance of the energy systems required for success at various positions within the sport. While soccer midfielders need a high level of aerobic power, the goalkeeper does not. She needs explosive power to intercept shots, and a strong leg to clear the ball. Figure 8.1 summarizes the anaerobic and aerobic energy requirements of different sports. It was developed using data collected on athletes engaged in each sport. Use it to determine the energy demands of your sport.

Once you identify the energy requirements of the sport, you can design training programs for the athletes by utilizing approaches proven to develop the energy system required for success. And you can individualize the program by knowing each athlete's strengths and weaknesses. For example, our soccer player, Casey, had excellent endurance (aerobic power) but was somewhat slow to the ball and tended to tire during a series of fast runs. So her coach had her do additional sprints to focus on speed and anaerobic glycolysis. In chapter 9 we will show the contributions of different training methods to the development of aerobic and anaerobic capabilities.

Anaerobic Power

You can assess your athletes' capabilities using established tests or sport-specific methods that you design. There are two components of anaerobic power, the adenosine triphosphate-phosphocreatine (ATP-PCr) system and anaerobic glycolysis.

Adenosine Triphosphate-Phosphocreatine

Adenosine triphosphate-phosphocreatine is tested in an all-out effort lasting up to 5 seconds, such as a short sprint. We've used a stair run test that was first suggested by an Italian physiologist

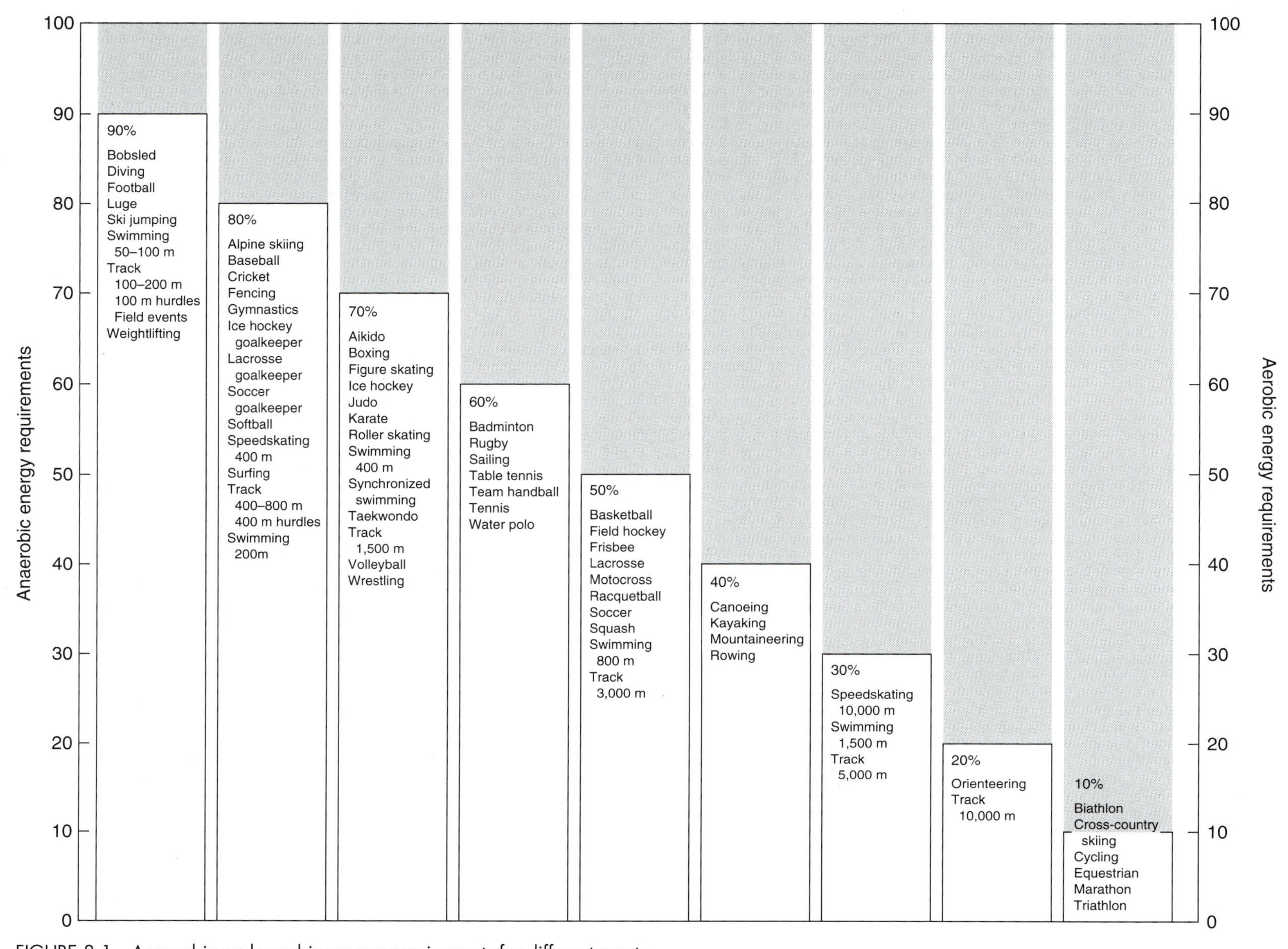

FIGURE 8.1 Anaerobic and aerobic energy requirements for different sports.

Reprinted, by permission, from B.J. Sharkey, 1986, *Coaches guide to sport physiology* (Champaign, IL: Human Kinetics), 100.

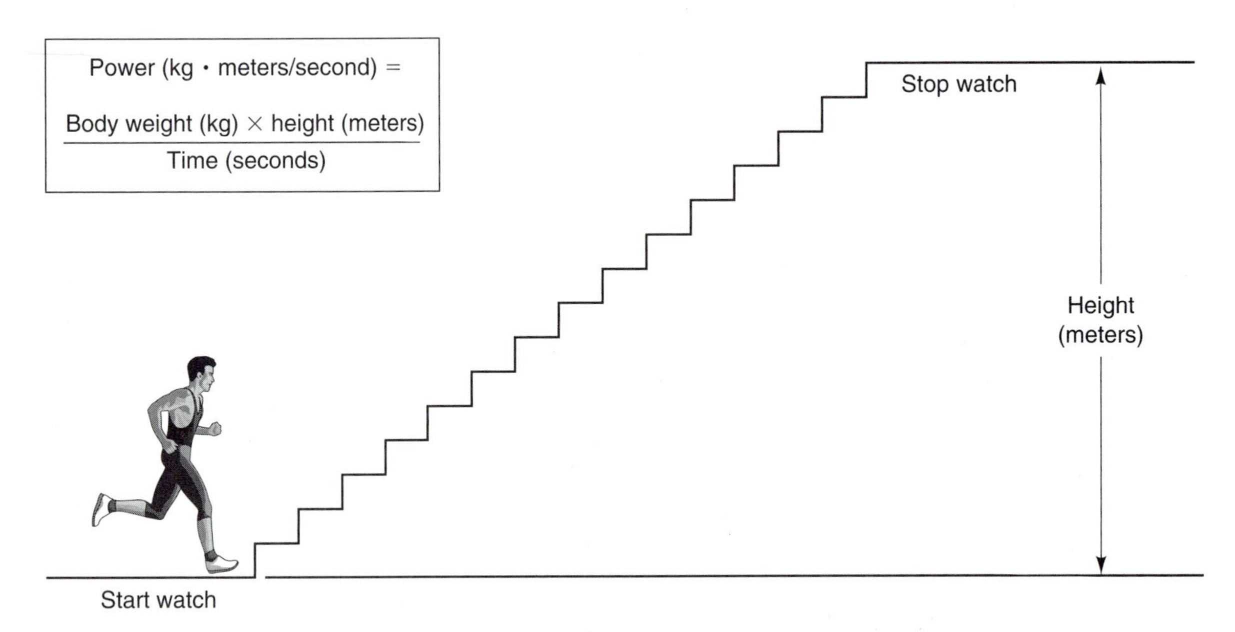

FIGURE 8.2 Stair run power test.

(Margaria) to measure this aspect of anaerobic power (see figure 8.2).

You don't need an electronic timing device; just start a stopwatch when the athlete's foot strikes the starting step, and stop the watch when the foot strikes the final step. Athletes should not stop on the final step but should keep going another few steps. Select any vertical distance that can be completed in less than 5 seconds (use efforts approaching 5 seconds when hand timing). To score the test, multiply the athlete's weight in kilograms times the vertical distance in meters and divide the total by the time in seconds to arrive at power in kilogram-meters per second. For example:

[60 kilograms × 2.84 meters] / 1.7 sec = 100.2 kilogram-meters per second

where

power = [force × distance] / time,
body weight = 60 kilograms (132 pounds),
vertical distance = 2.84 meters (16 steps, 7 inches [18 centimeters] each, taken two at a time), and
time = 1.7 seconds.

Be sure to let the athletes warm up and practice the stair run before the test. Time three trials, allowing at least 1 minute between trials. An uphill dash or a short sprint (e.g., 40 meters [44 yards]) is another way to assess anaerobic power. The stair run, hill run, and sprint scores are correlated to each other, as are short all-out jumping tests and other anaerobic efforts using the same muscles. Select the test most specific to your sport.

Anaerobic Glycolysis

Anaerobic glycolysis can be measured in an all-out run lasting 30 to 60 seconds. Cyclists use the Wingate test, a 30-second sprint against high resistance on a stationary bicycle, to measure sport-specific anaerobic power (figure 8.3). Soccer coaches use a 30-second uphill sprint on a grassy slope or stadium steps to assess this component of energy fitness. If anaerobic glycolysis is important in your sport, devise a sport-specific way to assess your athletes' capacity and the effects of training.

Aerobic Energy Fitness

There are several dimensions of aerobic fitness. Aerobic capacity (maximal oxygen intake in liters per minute), a measure of intensity, is best suited for non-weight-bearing sports, such as road cycling, swimming, and rowing. Aerobic power (maximal oxygen intake in milliliters per kilogram of body weight per minute) is an intensity measurement appropriate for use in weight-

Football Power Test

One of our graduate students developed an anaerobic power test specifically for interior linemen in football. Using a one-man blocking sled, the athlete pushes the sled in four 4-second bouts spaced 30 seconds apart (simulating a series of downs). Power is calculated using the weight of the sled times the total distance covered, divided by total time, 16 seconds. The test, called the football power test, was popular among the players, who competed to have the best time. The grad student found that the test scores were highly correlated to the coaches' ratings of the players' abilities.

A common football practice activity, using the blocking sled, can be slightly altered to serve as a power test.

© James D. Smith/Icon SMI

bearing activities lasting 5 to 15 minutes (e.g., running, cross-country skiing, cycling uphill). The first lactate threshold (LT 1) is a measure of exercise duration related to performance in very long-duration events (over 3 hours). It is most often used to establish the top of long-duration (EZ) training activities. In the lab, the second lactate threshold (LT 2) is the best indicator of performance for events lasting 30 minutes to 2 hours, while the related performance zone is often used to guide the intensity of interval and race-pace training. Use the measures most appropriate and specific for your sport.

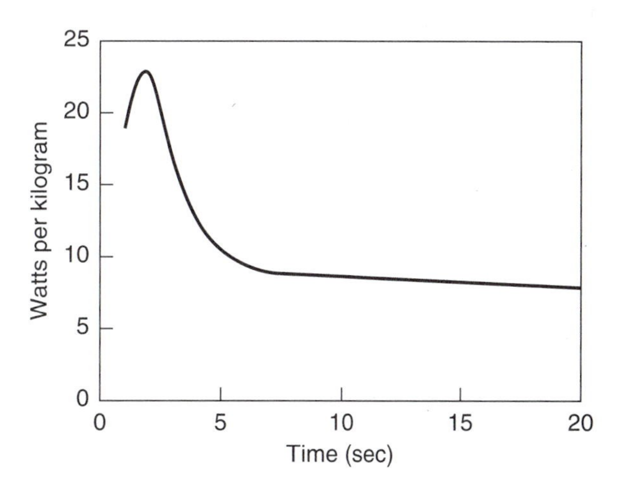

FIGURE 8.3 Wingate power test. Note the initial rise, then rapid drop in power as muscular adenosine triphosphate and phosphocreatine are depleted in the first 3 to 4 seconds. The more sustained power output is anaerobic glycolysis that will decline near the end of the 30-second test.

Aerobic Capacity

The maximal oxygen intake in liters defines the size of the engine in a human being, much as in a sports car. So a 4-liter engine is more powerful than a 3-liter engine, all things being equal.

Economy and Efficiency

Cars differ in engine size, gross vehicle weight, fuel economy, and performance. Cars that get 30 or more miles per gallon can go farther on one tank of gas than those that get 20 miles per gallon or less. Athletes become more fuel efficient with training. For example, as a runner eliminates extraneous movements, he or she can go farther or faster on a given amount of energy. Modest weight loss contributes to fuel economy. And resistance or power training may further enhance economy of movement. Elite runners expend up to 10 percent less energy at a given running speed than others. Studies have shown that excellent economy can make up for a minor deficit in aerobic power. While economy can be measured in the laboratory, it is not easy to estimate in the field. We mention it to emphasize its importance, so you will be aware of its contribution to performance.

But things are seldom equal, and other factors such as vehicle weight influence performance. For example, if two vehicles weigh the same, the larger engine should prevail. But larger engines usually mean larger weights, and the same is true in athletes. Since aerobic capacity is strongly influenced by heredity, there are limits to its development via training. The coach's role is to improve an athlete's aerobic capacity, efficiency, and skill to allow the best possible performance. As we've said, the aerobic capacity is best related to performance in events lasting 5 to 15 minutes, in sports in which the body weight is supported, such as cycling, swimming, and rowing. Improving maximal oxygen intake will improve performance, as will improvements in skill and efficiency (the ability to use energy efficiently; i.e., without wasting energy on movements that do not contribute to performance).

Aerobic capacity should be evaluated in a sport-specific test: stationary bicycle for cyclists, tethered swim for swimmers, rowing ergometer for rowers, and so on. While it is desirable to conduct the test with metabolic measurement apparatus, similar information can be gleaned from a sport-specific test. For example, Swedish researchers have shown that aerobic capacity can be estimated from a bicycle ergometer test, using heart rate measurements at each stage of the test. Our lab developed a road cycling test that predicted aerobic capacity and power. You can best assess an athlete's capacity for a sport in a test that closely simulates the movements and intensity of the event.

Aerobic Power

Aerobic power indicates the oxygen intake per unit of body weight and is related to performance in weight-bearing sports such as running and cross-country skiing. Aerobic power is best suited to intense activities that last 5 to 15 minutes. It is also an excellent measure of endurance fitness in popular sports such as basketball, soccer, and even football. Competitive runners soon learn the penalty of excess body weight. One study showed that an extra 5 kilograms, carried in a vest, added 5 percent to a runner's 10-kilometer time. The extra 5 kilograms (11 pounds) added 1.8 minutes to a 36-minute 10K, enough to turn a competitor into an "also ran."

Aerobic power should always be evaluated in a sport-specific test, either in the lab or in the field. An excellent alternative to the laboratory treadmill test for runners is the 1.5-mile (2.4-kilometer) run. Physiologists Bruno Balke and Jack Daniels validated the prediction of aerobic power from a running test in separate groups. We have merged the data into a graph that allows prediction of aerobic power in a wide range of athletes (see figure 8.4). We added altitude adjustments so the test could be used at sea level, in Denver, in Laramie, or wherever you live.

Use an actual performance, lab tests, validated field tests, or tests that you design to assess an athlete's ability. Then use the information to design training programs and to place athletes in events appropriate to their abilities (e.g., 400-meter vs. 3,200-meter running events). Repeat the tests periodically to gauge progress toward

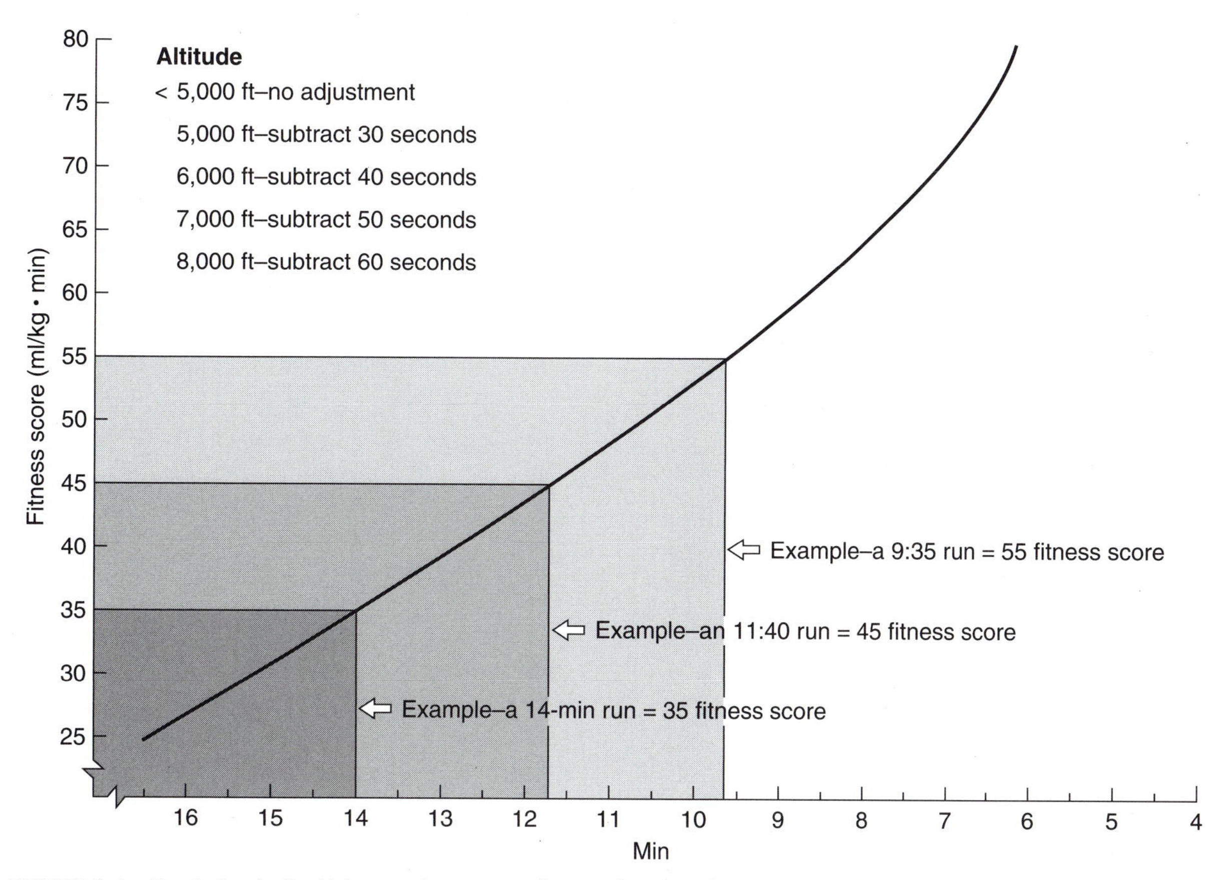

FIGURE 8.4 The 1.5-mile (2.4-kilometer) run test. Subtract altitude adjustment from 1.5-mile run time, then use the graph to find the score.

Reprinted, by permission, from B.J. Sharkey, 1986, *Coaches guide to sport physiology* (Champaign, IL: Human Kinetics), 141. Data modified from Balke 1963; Cooper 1970; Sharkey 1977.

training goals and to adjust training accordingly.

Lactate Threshold Measures

Aerobic capacity and power are measures that reflect the maximal intensity one can attain in an all-out effort. Threshold measures, on the other hand, provide information about the duration of time one can persist at a given velocity or level of effort.

Once called the aerobic threshold, the first LT defines long-term (over 3 hours) endurance capacity. It is an excellent guide to the level of effort required in long slow distance training. The threshold defines a training zone that we call the **easy training zone (EZ).**

The second break in the lactate curve is called the second LT. This intensity is closely related to performance in events lasting 30 minutes to 2 hours. Both threshold measures can be defined as a level of lactic acid (2 or 4 millimoles per liter), as the heart rate at that level of lactic acid, or as the velocity at that level of lactic acid (vLT). Since this threshold requires lab testing, we rely on the athlete's current ability or **performance threshold (PT),** which within a sport will be well related to vLT. The PT can then be used to accurately define the **performance training zone (PZ).**

For everyday use, we recommend the rating of perceived exertion as a way to estimate the thresholds and **training zones** (see table 8.1). The use of perceived exertion forces the athlete to focus on the sensations associated with a level of effort. Once aware of the signals, athletes can use them in competition to guide their efforts and to avoid fatigue.

Studies indicate that a rating of perceived exertion (RPE) of 12 is closely related to the LT 1. Train at or below RPE 12 (between light and somewhat hard) and you will be in the EZ train-

Max?

Sport physiologist Carl Foster has had a long relationship with the sport of speed skating. While he used a variety of lab and field tests to guide the training of national-caliber skaters, he recognized that the best measure of an athlete's ability is his or her performance in an actual event. Foster conducted experiments during time trials in order to see how close to their "maximal" oxygen intake the athletes were able to perform. To his surprise, some athletes actually exceeded the maximal oxygen intake values that had been determined in a controlled lab test. In effect, the challenge of the time trial evoked a greater effort than a sterile laboratory test. Others have found that athletes sometimes exceed their "maximal" heart rates during intense interval training. Whenever possible, utilize the actual event as an essential measure of performance.

Testing during competition rather than in the lab can produce more realistic results.

ing zone, the ideal level for long slow distance training or recovery days.

The second LT defines the transition from predominantly oxidative fibers (slow-oxidative and fast-oxidative-glycolytic) to the recruitment of fast-glycolytic fibers. Above that point the effort involves a greater proportion of inefficient anaerobic metabolism and a rapid accumulation of lactic acid. This LT is highly related to performance in events lasting from 30 minutes to 2 hours. More importantly, it provides a level of training intensity that has been associated with improvements in the LT and performance.

The LT 2 is closely related to a measure called the respiratory compensation threshold, which defines the point where breathing rate and depth increase dramatically. This so-called breakaway ventilation is a signal that you are exceeding the abilities of oxidative muscle fibers, recruiting fast-glycolytic fibers, and blowing off large quantities of carbon dioxide. Athletes can be taught to recognize breakaway ventilation and can use it as an indicator of LT. Another perceptual clue is the RPE. Studies show that an RPE of 15 to 16 is associated with the LT 2 and the respiratory compensation threshold. Train at or slightly above 15 (hard or heavy) and you will be in or near the PZ. The vLT indicates the speed one can achieve without exceeding the second LT. Training in the PZ helps move the lactate curve to the right, allowing one to run a given pace with less lactic acid accumulation, or to run at a faster pace with the same level of lactate without exceeding the threshold.

Threshold and training zone measures are invaluable aids to training and performance. Coaches can use lactate measurements or heart rate monitors to help athletes train at the desired intensity. However, since the RPE accurately reflects the essential features of the thresholds, we recommend using this simple, inexpensive approach to guide training intensity. Coaches can determine PZ most easily by recording an athlete's velocity (distance/time) during a race or time trial. Of course, threshold and training zone data must always be determined and utilized in a sport-specific manner.

This chapter has focused on two main points: the energy demands of your sport and ways to assess athletes' energy capabilities. Knowing the

TABLE 8.1

Training Zones and Ratings of Perceived Exertion

Borg scale (Descriptor)	Borg rating	Training zone	Lactate thresholds	Heart rate
No exertion at all	6			
Extremely light	7			
	8	EZ		Race HR – 30%
Very light	9	EZ		
	10	EZ		
Light	11	EZ		
	12		LT 1	Race HR – 20%
Somewhat hard	13	NZ		
	14	NZ	↑	
Hard (heavy)	15	NZ	LT 2 (PT ↕)	Somewhat hard
	16	PZ	↓	Avg. race HR
Very hard	17	PZ	↓	
	18	PZ	↓	Race HR + 5%
Extremely hard	19	MZ	↓	
Maximal exertion	20	MZ	↓	

The Borg scale is a 6- to 20-point scale with descriptive terms. These terms can be related to the training zones. The top of the no-training zone (NZ) and bottom of the performance zone (PZ) depend on the performance threshold (PT). For races longer than 60 min, the PT will be below or near the second lactate threshold (LT 2), while for shorter races the PT will be above LT 2. The performance zone extends from PT to 5% faster.

Performance Threshold

The PT can be defined as the lactate, heart rate, perceived exertion, or velocity at which each athlete can currently race. To improve that pace we train in the PZ, generally using intervals slightly faster than race pace.

For races shorter than 2 hours, performance is related to the LT 2. Training for a marathon may involve intervals just below LT 2. For shorter events, training slightly above the LT 2 (or vLT) is an effective way to improve racing speed and shift the lactate curve to the right. So for all but extremely long events, effective, quality training is conducted in the athlete's race- and pace-specific PZ. See figure 8.5 for an overview of the training zones in relation to race speed.

(continued)

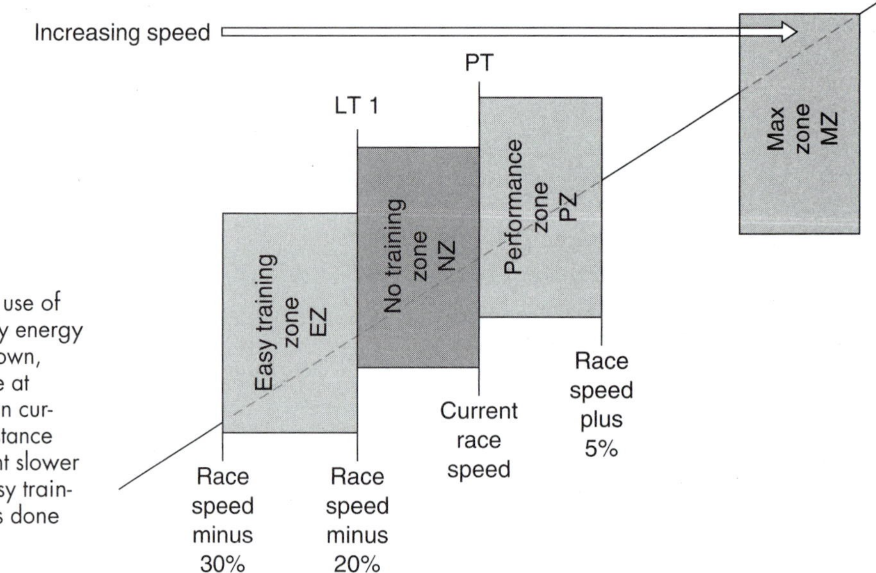

FIGURE 8.5 Training zones. The use of the four training zones can simplify energy training. Once racing speed is known, performance zone training is done at intervals up to 5 percent faster than current race speed. Recovery and distance training is done at least 20 percent slower than current race speed, in the easy training zone (EZ). Short speed work is done maximally (MZ).

energy demands helps you design appropriate training programs. Knowing an athlete's capabilities helps you place the athlete in the right position or event, and provides a way to track progress toward training goals. Throughout the discussion we have emphasized the need to utilize sport-specific measures to assess performance. A treadmill test may provide a measure of aerobic capacity, but it will not relate to performance in swimming. The muscle fibers used in running are not the same as those used in swimming, so the treadmill test tells you little about an athlete's ability to perform in the pool. We will provide more examples of sport-specific testing and training in later chapters.

SUMMARY

Determining the energy requirements of the sport you coach is not a difficult task. Figure 8.1 (p. 141) provides guidance for most sports. If yours is not listed, use similar ones or insights you have gained from personal experience. Then you are ready to assess your athletes' capabilities using time trials, lab tests, established field tests, or sport-specific methods of your own design. We've provided some useful tests for your consideration. Regular use of field tests motivates athletes to engage in the right kind of training for the sport. Knowing the demands of the sport and each athlete's capability, you are ready to begin planning seasonal and year-round training programs. You'll note that we recommend the rating of perceived exertion as the way to estimate exercise intensity and training zones.

KEY TERMS

easy training zone (EZ) (p. 145)
performance threshold (p. 145)
performance training zone (PZ) (p. 145)
training zones (p. 145)

REVIEW QUESTIONS

Answer the following questions by filling in the blank(s) with the appropriate word or words to complete the statement.

1. The _________ system and _________ are the two components of anaerobic power.
2. Aerobic capacity and power are measures that reflect the maximal _________ one can attain in an all-out effort.
3. Threshold measures provide information about the _________ one can persist at a given velocity or level of effort.

Match the terms in the left column with the correct description in the right column.

4. Aerobic capacity
5. Aerobic power
6. First lactate threshold
7. Second lactate threshold

a. Measure of duration related to performance in very long-duration events (over 3 hours); defined long-term endurance capacity
b. Measure of intensity, best suited for non-weight-bearing sports
c. Best indicator of performance for events lasting 30 minutes to 2 hours; determines transition from predominantly slow-oxidative and fast-oxidative-glycolytic fibers to fast-glycolytic fibers
d. Intensity measurement appropriate for use in weight-bearing activities lasting 5 to 15 minutes

Determine if the following statements are true or false and then circle the correct answer.

8. T/F Whenever possible, do not utilize the actual event as an essential measure of performance.
9. T/F The easy training zone (EZ) includes all intensities of RPE 12 and above.
10. T/F For races shorter than 2 hours, performance is related to the first lactate threshold.
11. T/F Aerobic capacity should be evaluated in a sport-specific test; for example, using a stationary bicycle for cyclists or a rowing ergometer for rowers.
12. T/F The principle of specificity demands that you use a sport-specific test to assess progress toward training goals.
13. T/F Determine the performance zone (PZ) by recording an athlete's velocity (distance/time) during a race or time trial.
14. T/F Training below the PZ is the best way to improve racing speed.
15. T/F Appropriate training shifts the lactate curve to the right.
16. T/F The second LT defines the transition from predominantly oxidative fibers (slow-oxidative and fast-oxidative-glycolytic) to the recruitment of fast-glycolytic fibers (FG).

Select the correct response (more than one may be correct).

17. The EZ training zone is defined by
    a. second LT
    b. first LT
    c. aerobic power
    d. aerobic capacity
18. The performance training zone (PZ) is based on
    a. first LT
    b. second LT
    c. anaerobic glycolysis
19. The no-training zone (NZ) is between
    a. first LT and aerobic power
    b. second LT and aerobic capacity
    c. first LT and second LT
20. To guide training intensity you can use
    a. heart rate monitor
    b. lactate measurements
    c. rating of perceived exertion
    d. all of the above
21. A runner can compensate for a minor deficit in aerobic power by improving
    a. stride length
    b. economy/efficiency of running
    c. sport-specific weight training
    d. b and c

PRACTICAL ACTIVITIES

For the sport you coach (or hope to coach):

1. Determine the energy systems that are essential for success in the sport.
2. Identify laboratory methods for the evaluation of your athletes' energy capabilities: Indicate the energy system(s) and how you will test in the lab.
3. Develop a field test that mimics the lab test with sport-specific activities. If your sport is track or swimming, indicate the event that is related to the energy system.

9

Developing Energy Fitness

This chapter will help you

- understand how to develop short-term anaerobic systems,
- determine the types of training used to develop long-term aerobic systems,
- be aware of the cellular changes that occur with different intensities and durations of training, and
- appreciate the changes in the cardiovascular system that affect aerobic fitness.

Mary grew up playing tennis. By the time she went to high school she was playing tennis at least 3 hours every day. Her hard work paid off when she made the high school team and played doubles. Mary, and most of the team members, had never done any fitness training for tennis beyond actual tennis playing. She didn't understand the need when her coach initiated both running and weight training for the team. Initially, she and the other players didn't like the training and felt that it was taking time away from tennis. However, after a few weeks, the training became easier and the girls starting winning more of their matches. Before the conference meet, the coach reduced their physical training and allowed more rest time. Mary and her partner played well at conference and qualified, along with several other team members, for the state meet. The players appreciated the extra fitness training that had helped them to improve their game and fitness to finish strong.

Like Mary and her coach, many tennis players have learned the importance of improving muscular and energy fitness, as well as skill, to improve their competitiveness. A number of researchers have evaluated successful tennis players and shown that they have high levels of aerobic fitness. It is now common practice for tennis players to include running and even interval work as part of their training. While there is a large body of anecdotal evidence supporting improved energy fitness for intermittent sports (e.g., tennis, soccer, basketball, volleyball, lacrosse), few experimental studies have shown improved competitive performance related to energy fitness. We expect that the number of such studies will increase.

One study (Groppel and Roetert 1992) evaluating fitness and performance in tennis players showed that energy fitness related to tournament play. The researchers observed positive correlations between tournament play and a number of fitness parameters. The findings showed that fitness development in tennis players should incorporate flexibility, strength, and endurance training to enhance performance and also maintain muscular balance and reduce the incidence of injuries. Tennis was found to have aerobic and anaerobic components that relate to tournament play, with the predominant energy supply coming from the phosphagen energy system. Other results showed that elite tennis players had high levels of aerobic fitness. These findings suggest that training programs should be designed that are specific to the actual energy and muscular demands of the game. The need for further research still exists in order to provide a better understanding of the game and to identify assessment measures that are sport specific. Longitudinal studies will yield greater insight into the relationship of physiological capabilities and success.

Most sports require a balance of anaerobic and aerobic energy systems (for example, the need for power as well as endurance in tennis players). Training needs to be specific to the energy needs of each individual.

Although we often think of training in terms of aerobic and anaerobic systems, it is difficult to separate the two systems, as they always occur together, albeit in different proportions. The dependence on either energy system has to do with the sport, the time during the event, fatigue, and individual genetics. This continuum between energy systems must be evaluated for all sports and considered during the planning of training.

Energy fitness is our ability to produce adenosine triphosphate (ATP), the high-energy molecule that is the energy currency of life. In the simplest sense, physical training increases our ability to generate ATP and make it available for muscular contraction. As we overload the neuromuscular and cardiovascular systems during training, they respond with specific changes that allow for more efficient and rapid production of ATP. In chapter 6 on muscular fitness training, we emphasized the principles of overload, recovery, adaptation, progression, and periodization. These principles are equally applicable when one is training energy fitness.

Overload

"If you always do what you always did, you will always get what you always got" can sum up the principle of overload. In order to increase the capacity of an energy system, that system must be specifically overloaded; it will respond with improved capacity. One can control overload of energy systems by attention to the same three training variables: frequency, intensity, and time (FIT). Of the three variables, intensity is the most difficult to define and control. Duration and intensity are inversely related; as one increases, the other must decrease.

Intensity

Three methods are used to control energy fitness training intensity: speed, perceived exertion, and heart rate (HR). In spite of the widespread use of HR, the most effective methods to control intensity are to use the speed related to your athlete's current competitive performance or perceived exertion. As Finnish national ski coach Antti Lepavouri said, "No race was ever won or judged by HR; speed is the only criteria. Heart rate monitors are good to help athletes go slow, but to go fast you need to monitor speed." The information that follows will teach you how to use these three methods of controlling intensity. Coaches should be familiar with all three of these methods.

As intensity of effort increases, the speed, perceived exertion, and HR increase while the duration over which the activity can be sustained decreases. Additionally, as intensity increases

In order to increase the capacity of an energy system, that system must be specifically overloaded, and it will respond with improved load capacity. Overload of energy systems can be controlled by attention to the three training variables: frequency, intensity, and time (FIT). Here, rugby players train using a sport-specific drill.

Heart rate monitors are good to help athletes go slow, but to go fast it is necessary to monitor speed.

from easy to hard, there is a gradual transition from aerobic (oxidative) to anaerobic (nonoxidative) energy pathways. During easy activity, the main fuel source is fat. As we increase to more intense aerobic activity, we rely more on carbohydrates. At high intensities that cannot be maintained aerobically, we utilize carbohydrates anaerobically with the resultant buildup of lactic acid. Finally, when our energy utilization rate exceeds what we can produce by anaerobic glycolysis, we turn to our final short-term energy sources, ATP and phosphocreatine—sources that last only a few seconds (figure 9.1). In the following sections, we explore methods to control the intensity of training to effectively target sport-specific energy fitness. You can decide which intensity methods work best for you and your athletes.

Speed to Control Training

Using speed to control training intensity requires knowledge of the current race speed of each athlete. Training intensity is then controlled through monitoring of the training speed relative to current performance speed. Speed-based training is especially appropriate for timed events (swimming, running, cycling, etc.) but also works well for other aspects of energy fitness training.

Performance speed is the average speed that your athlete can maintain during the given event. You can determine this speed using a time trial or an actual event. To convert the times and distances to average speed, use the information in the sidebar "Calculating Performance Speed" on page 161. These calculations will give you the average speed in units that can easily be converted into interval and training speeds. This book uses metric units. We will show you how to convert times and distances.

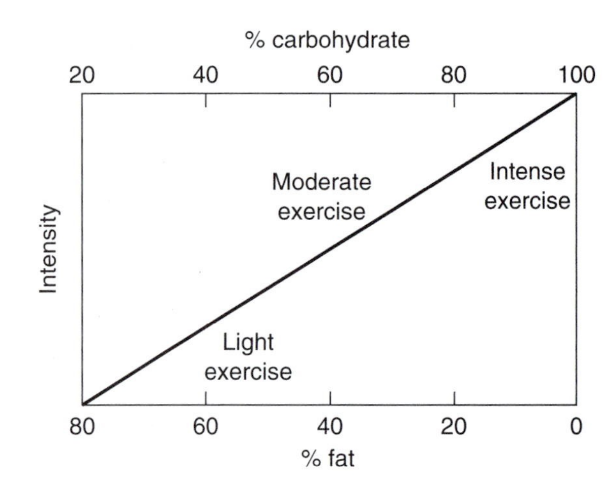

FIGURE 9.1 Energy utilization during exercise. During light exercise, most energy is from the metabolism of fat, while during intense exercise, carbohydrates are the primary source of energy. Protein is not a major source of energy during exercise.

Reprinted, by permission, from B.J. Sharkey, 1986, *Coaches guide to sport physiology* (Champaign, IL: Human Kinetics), 84.

The intensity of most intervals should be controlled so that they are only slightly faster than the athlete's current race pace. Intervals that are performed "as hard as possible" are of limited usefulness and break a basic rule of training—that of specificity. Here, a coach monitors pace with the poolside clock.

Heart Rate to Control Training

Using HR to control training intensity requires that you monitor the average HR during an actual event or time trial. The average race HR is called the performance heart rate and is specific to each athlete. Training intensity is then controlled relative to performance HR. The use of performance HR agrees with the specificity and overload principles of training. Heart rate training works well for monitoring the intensity of aerobic foundation training and recovery activities in the easy training zone (EZ) for all energy fitness training.

For determination of the average performance HR, the athlete wears a HR monitor, preferably one with a memory, during an aerobic competition or time trial; the HR values are then averaged over the entire activity. We recommend that you do not use the HR values during the first and final minutes of the event, as these may be too low or too high. If the HR monitor does not have a memory function, have the athlete glance at the monitor frequently during the competition and report the average HR. If you determine the performance HR from a time trial, be sure that the athlete has given a competitive effort.

Perceived Exertion Training

A large body of research has shown that athletes are able to use sensory feedback to gauge training intensity. In some sports, or with young athletes, the use of HR or speed is inappropriate for setting training intensity. For many team sports, ratings of **perceived exertion** (RPE) such as easy, moderate, or hard are more meaningful and easier to use. If using HR or competitive speed excessively complicates your athletes' training, or if they would rather be freed from the bother of frequent HR checks, consider using RPE. The RPE scale and descriptors relating perceived effort to breathing and the activity duration are shown in table 9.1.

During the early stages of training, coaches should have athletes use all possible methods to measure intensity: HR, speed, and RPE. Athletes will soon develop a feeling for intensity and for how to interpret their sensations. Coaches working with endurance athletes have found that they quickly learn to judge their effort related to HR and how fast they were going, even with changing conditions and varying terrain.

After athletes have had adequate time to recognize the feelings associated with different HRs, speeds, and levels of perceived exertion, they should be able to tell you their perceived exertion, their approximate HR, and how fast they are going at any time during training. Most importantly, the descriptive, breathing/talking, and duration scales provide your athletes clear and simple training intensity guides.

Past Use of Heart Rates

Coaches and athletes have historically used percentages of the **maximal heart rate** (HR; maximum possible beats per minute) to control energy training. The classic chart used to determine these ranges is shown in figure 9.2. The drawback to percentages is that each athlete is different. Each percentage must be calculated individually; these values cannot be generalized.

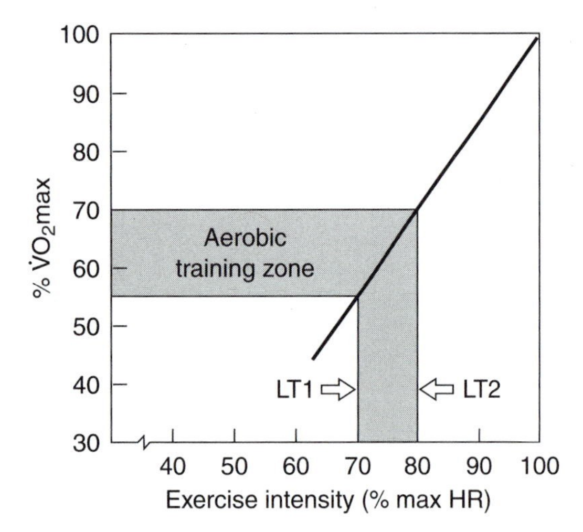

FIGURE 9.2 Classic aerobic training zone. General aerobic fitness improves during exercise in the "aerobic training zone." This principle applies to general fitness but is inadequate for athletes in aerobic sports. The authors consider this zone the "no-training zone," with easy and distance training at an easier intensity and performance training performed just above current race pace.

Adapted, by permission, from B.J. Sharkey, 2002, *Fitness & health*, 5th ed. (Champaign, IL: Human Kinetics), 102.

The aerobic training zone recommended by the American College of Sports Medicine and American Medical Association is defined as 50 to 85 percent of heart rate reserve (HRR). For coaches to calculate this range, they need to know both resting heart rate (HRrest) and maximal HR (HRmax) values. The athlete's aerobic training zone of 50 to 85 percent HR reserve is calculated as the range between

$$50\%\ \text{HRR} = ([\text{HRmax} - \text{HRrest}] \times 0.50) + \text{HRrest}$$

and

$$85\%\ \text{HRR} = ([\text{HRmax} - \text{HRrest}] \times 0.85) + \text{HRrest}$$

This training zone is adequate for health and fitness purposes, but does little to help a coach define the proper training zones for athletes. The difficulties include the following:

- Using an estimate of maximal HR (220 – age) may introduce a large error, up to ±12 beats per minute for many individuals and as large as 36 beats for a few. This means that the HR of a 20-year-old athlete with an estimated maximal HR of 200 (220 – 20) could be as low as 164 or as high as 236. Figure 9.3 shows the classic age-adjusted aerobic training zone.
- As an athlete improves in fitness training, HRs change.
- The range of 50 to 85 percent HRR is large. At 50 percent, the athlete might find that the training is extremely easy, while at 85 percent the athlete might find it hard to maintain the intensity for more than a few minutes. Thus, the optimal training intensity will likely be somewhere within this HR zone. Training HRs for athletes need to be more precise, hence our recommendation to use performance speed or a measured performance HR to set training intensity.

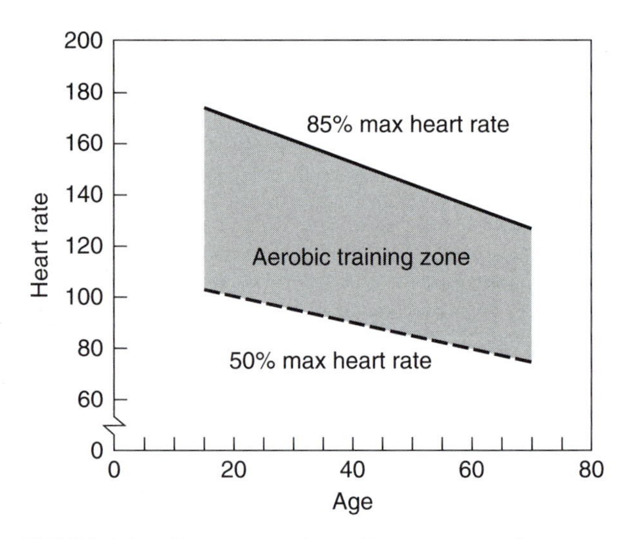

FIGURE 9.3 Classic age-adjusted heart rate aerobic training zone. This figure shows the commonly accepted heart rate zones for aerobic fitness. This method is appropriate for general fitness, but is too broad and nonspecific for athletes.

TABLE 9.1

The Rating of Perceived Exertion (RPE) Scale

Borg scale descriptor	Borg scale	Speaking and thinking scale	Duration*	Training zone
No exertion at all	6			
Extremely light	7	Can sing in full stanzas	All day	EZ**
	8			EZ**
Very light	9	Can sing a few lines	All day with breaks	EZ**
	10		5-10 hr	EZ**
Light	11	Can talk in full sentences	4-5 hr	EZ**
	12		2.5-4 hr	NZ
Somewhat hard	13	Can talk—a few sentences at a time	1.5-2.5 hr	NZ
	14	Can talk—a few words at a time	50-90 min	NZ-PZ***
Hard (heavy)	15		25-50 min	NZ-PZ***
	16	Can't talk, but can think clearly	15-25 min	PZ***
Very hard	17	Need to focus to maintain intensity	7-15 min	PZ***
	18	Struggle to maintain intensity	3-7 min	PZ***
Extremely hard	19	"Brain dead"—really hurting	30 sec-3 min	PZ-MZ****
Maximal exertion	20	"End is imminent"	<30 sec	PZ-MZ****

The Borg Scale is a 6- to 20-point scale with descriptive terms. Work with athletes has shown that other descriptors, especially the speaking and thinking and duration scales, may be more understandable, especially for younger athletes.

*Duration = time during which the intensity can be maintained *comfortably*.

**The easy training zone (EZ) includes all intensities of RPE 11 and below. Recovery training should be at lower RPE values, while distance and aerobic base training intensity will depend on the duration and purpose. See the sections on training zones.

***The transition from the no-training zone (NZ) to the performance training zone (PZ) will depend on the event duration. Check the duration of your event in the duration column and find the next higher RPE to determine the appropriate RPE for your PZ.

****The maximal zone (MZ) is used for limited, very short speed intervals. For anaerobic events, the PZ may overlap. See the section on training zones.

For a description of the training zones, please see page 158.

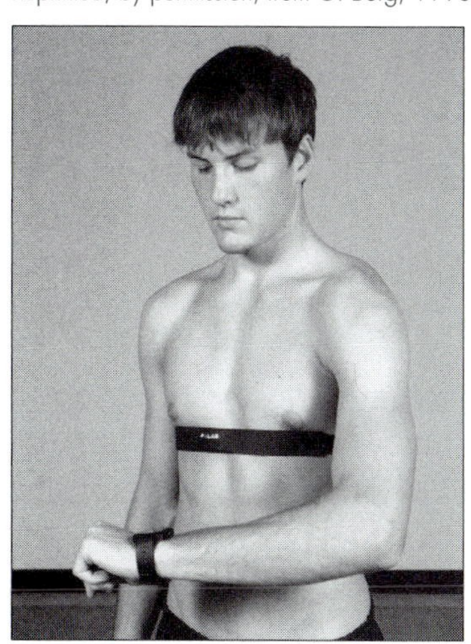

Heart rate monitors have become a popular tool for monitoring training intensity. Basing training on current performance heart rate may be useful, but using standard percentages of maximal heart rate has drawbacks and should be avoided by athletes.

Understanding ratings of perceived exertion relative to heart rate and speed is an important first step for all athletes as they learn about training zones and how to control the intensity of their training.

Training Zones

We divide training into four zones: the easy (EZ), **no-training (NZ),** performance (PZ), and **maximal training (MZ) zones.** We discussed the EZ and PZ in the previous chapter, but will do so in more depth here. Limits for each of these zones can be set using speed, HR, or RPE. The training zones are depicted in figure 9.4.

Easy Training Zone (EZ)

The training in the EZ is low intensity and is used for easy distance and recovery sessions. The HR should be maintained at about 20 to 30 percent below the average performance HR. The RPE should be 11 or easier (fairly light, allowing talking in full sentences, and an intensity that could be maintained comfortably for 4+ hours). Most athletes train too hard to recover on their "easy" days. By monitoring speed, HR, or RPE and staying within this easy training zone, athletes will be able to recover while they build their aerobic foundation. Athletes using perceived exertion should occasionally check speed and HR to ensure that their perceptions are correct. This low-intensity training also helps athletes avoid the "no-training zone" (discussed next). For endurance athletes we recommend that about 70 to 90 percent of training time be spent in this training zone, with some as short recovery training and the remainder as easy distance training.

No-Training Zone (NZ)

The intensities in this zone range from 1 to 20 percent below performance speed and HR. Values for RPE range from 12 to 15 (see table 9.1, "The Rating of Perceived Exertion Scale"). Surprisingly, most recreational athletes and many serious athletes tend to do most of their training in this zone. By doing so, they are violating many of the principles of training, including overload (easier than competitive intensity), specificity (becoming efficient at a speed or intensity below that required for performance), and recovery (working too hard to recover). "No-training zone" does not mean that you can never do any training at this intensity. Rather, it means that time spent in this zone will be less productive than EZ or PZ training. For individuals interested in general fitness and health, NZ is an appropriate intensity as it provides general benefits to all energy fitness systems. However, athletes need to be more focused to improve specific energy fitness and should spend little time in this zone.

The Performance Zone (PZ)

The PZ is fundamental to improving performance. The PZ is defined as speeds or HRs slightly higher (1 to 5 percent) than individual race values. To use RPE for this training, find the event duration as shown in table 9.1, then use one RPE higher. We recommend using speed to control intensity in this zone. The PZ provides the intensity for intervals that is needed to improve competitive speed and performance. This intensity provides an overload specific to the energy demands of the activity. Importantly, in that they are just slightly harder than race intensity, performance intervals can be maintained long enough to duplicate the energy systems required for the activity while maintaining technique and form. As the athlete

→ Increasing Intensity →					
	EZ Training Zone Speed Range: 20-30% below racing speed HR Range: 20-30% below racing HR RPE 9-12	**NZ Training Zone** Speed Range: 20% below to current racing speed HR Range: 20% below racing HR to current racing HR RPE 13-15*	**PT** Current Race Speed or Average Race HR RPE 14-17*	**PZ Training Zone** Speed Range: Current racing speed to 5% above HR Range: Current racing HR to 5% above RPE 15-18*	**MZ Training Zone** Range: Maximal efforts RPE 19-20

Example: A runner with a max HR = 195, resting HR = 50, and whose average racing HR = 166 bpm completes a 1,600M race in 6:20 (Average race speed = 1,600 meters/6.333 min = 253 meters/minute = 9.44 mph). His EZ training speed would be 177-202 meters/min or 6.6 to 7.5 miles per hour. His 440-meter intervals would be at 260 meters/minute or 92 seconds per lap.

	EZ Guidelines	**NZ Guidelines**	**PT**	**PZ Guidelines**	**MZ Guidelines**
Using HR	116-133 bpm	134-165 bpm	166 bpm	167-174 bpm	190-195 bpm
Using mph	6.6-7.5 mph	7.6-9.3 mph	9.44 mph	9:45-9.9 mph	Max speed
Using m/min	177-202 m/min	203-252 m/min	253 m/min	254-266 mph	Max speed
Using sec/400m	1.59-2.16 sec/400m	1:58-1:34 sec/400m	1:35 sec/400m	1:30-1:34 sec/400m	Max speed

*Rating of Perceived Exertion—RPE will depend on event duration. RPE values for longer durations will be lower while shorter duration will have higher RPE.

bpm = beats per minute

mph = miles per hour

m/min = meters per minute

sec/400m = time in seconds for a 400-meter lap

FIGURE 9.4 Training zones: heart rate, speed, and rating of perceived exertion. The easy training zone (EZ) is much slower than race speed. The no-training zone (NZ) is an intensity of limited training between the EZ and the performance threshold (PT). The performance threshold is defined as current racing speed and heart rate. The performance training zone (PZ) is the intensity for intervals, extending from current race pace up to 5 percent faster. The maximal training zone (MZ) is for short speed intervals and activities that are to be done at a maximal effort.

repeats multiple intervals within the PZ, the appropriate energy systems are improved. For aerobic fitness, intervals of 4 to 8 minutes with full recovery (3-5 minutes) work well. See table 9.2 for appropriate interval lengths for different event durations. Performance zone intervals should be continued until the target speed can no longer be maintained. We recommend that only about 10 to 30 percent of training time be spent at this intensity (including races and intervals).

Maximal Speed Zone (MZ)

A small amount of training to improve power and maximal speed can utilize very hard, short intervals. Intervals in MZ do not need to be controlled by speed or HR. Rather, they are maximal efforts (RPE 20) of 5 to 15 seconds. After each interval, athletes should be given a long rest period, 2 to 3 minutes. This training is done after both the aerobic and sport-specific foundations are well established. Intervals in the MZ are always performed with maximal effort. Endurance athletes will spend less than 1 percent of total training time in this zone. For sprint athletes, this zone is not different from the PZ and requires speed assistance to perform slightly above race speed. Sprint athletes will include more training time in this zone.

Training Zone Summary

Training is done in three primary zones. Easy zone training accounts for the majority of the training time, develops the aerobic foundation, and allows for recovery from periods of hard training. Most intervals are done as PZ intervals, slightly faster than current race speed. About 1 percent of training time is spent doing MZ intervals. The fourth training zone, no-training zone (NZ), should be avoided by athletes. These training zones are summarized in table 9.4 and shown graphically in figure 9.6.

Factors That Affect Intensity

Several factors complicate the use of the HR, speed, or RPE as an indicator of training intensity. Some of these factors are emotions, illness, heat, altitude, fatigue, type of exercise being used, travel, loss of sleep, and other life stresses.

- Emotions: Heart rate can speed up when people are emotionally involved in a situation.

TABLE 9.2

Duration of Event, Energy Systems, and Interval Duration

Duration of event	Energy system	Optimal interval duration	Main fiber type(s)**
<20 sec	Creatine phosphate-ATP, anaerobic glycolysis	5-20 sec*	Fast-glycolytic
20-60 sec	Anaerobic glycolysis, creatine phosphate-ATP, aerobic energy (CHO***)	10-45 sec*	Fast-glycolytic, fast-oxidative-glycolytic (FOG)
1-5 min	Anaerobic glycolysis, aerobic energy (CHO)	30 sec-2 min	FOG, slow-oxidative
5-15 min	Aerobic energy (CHO), anaerobic glycolysis	2-4 min	Slow-oxidative, FOG
15-60 min	Aerobic energy (CHO)	3-6 min	Slow-oxidative, FOG
>60 min	Aerobic energy (CHO/fat)	4-10 min	Slow-oxidative, FOG

Energy systems are listed in order of importance for each event length.

*For the short sprints, interval training faster than race pace will require a speed assist (see text).

**During sprinting, fast-glycolytic fibers provide explosive power, along with FOG and slow-twitch fibers that are recruited, adding to the force generation.

***CHO = carbohydrate.

Calculating Performance Speed

You will need to know the race lengths and your athletes' performance times to calculate speed. You must then convert these speeds to meters per minute in order to calculate interval times. To determine the speed of each athlete and convert to meters per minute, do the following:

Convert the length of the event into meters if it was not measured in meters:

1 mile = 1,620 meters (multiply miles by 1,620)

1 yard = 0.91 meters (multiply yards by 0.91)

1 kilometer = 1,000 meters

1 foot = 0.305 meters (multiply feet by 0.305)

Example: If an athlete runs a mile race, the distance = 1,620 meters.

Convert your athletes' times into decimal times:

Convert into minutes and seconds (example: 5 minutes and 20 seconds = 5:20).

Convert the seconds into decimal seconds by dividing the seconds by 60 (example: 20 seconds = 20 / 60 = 0.333 minutes).

Write the total times as minutes (example: 5 minutes and 20 seconds = 5.333 minutes).

Calculate the athlete's average speed in meters per minute (example: 1,620 meters / 5.333 minutes = 303.8 meters/minute).

Use table 9.3 to decide on the appropriate times for specific length intervals. Use table 9.2 to determine the proper length interval for the athlete.

Situations creating fear and excitement are two examples of conditions in which the HR is not an accurate estimate of effort. Emotionally arousing situations such as a highly competitive practice can elevate the HR, but the effect wears off as the session continues.

- Illness: A fever, or even the beginnings of a cold or respiratory infection, can elevate the body temperature and the HR. If morning HRs are elevated, this can be a sign of illness that could get worse if the athlete does not reduce training or take the day off. After a period of bed rest, the HR is also unusually high. If an athlete has been sick, have the athlete return to vigorous activity gradually.

- Heat: Exercise in a hot environment raises the HR because the body sends blood to the skin for cooling as well as to the working muscles. This means that less oxygen is delivered to the working muscles, and speed will be reduced. In this case, it is best to use RPE as a training guide. Using HR as an additional guide will ensure that you maintain a safe intensity. If the athlete is allowed to become dehydrated, the HR will also increase. Athletes should be encouraged to drink plenty of fluids during practice and competition.

- Altitude: During the first week at an altitude above 5,000 feet (1,524 meters), the exercise HR could be elevated and performance will be depressed. Above 7,000 to 8,000 feet (2,134 to 2,438 meters), performance will be further depressed; it will rebound somewhat after acclimatization, but never to sea level values. The use of RPE will help athletes maintain a reasonable intensity during the first week until they begin to acclimatize and can set new HR and speed targets.

Using Intervals

Performance zone intervals should be only slightly faster (about 1-5 percent) than current race pace. For aerobic events lasting 4 minutes or longer, intervals should be no longer than 25 percent of the race distance, up to a maximum length of about 10 minutes. For race durations shorter than 4 minutes, see table 9.2 for appropriate interval lengths. Full recovery is necessary between intervals. The recovery duration should be determined by the athletes and should be such that they can maintain their speed on subsequent intervals. Intervals may be repeated until the athlete is unable to maintain the target time. About 10 to 30 percent of all training time may safely be done as competitions, time trials, or intervals. Events lasting less than 5 to 6 minutes may require about 30 percent of total training time in the PZ, while longer endurance events require only 10 to 15 percent. Use table 9.3 to determine "performance zone" interval times based on current race performance. The method to calculate your athletes' current race speeds is shown on page 161. Figure 9.5 shows how one of the authors used increasingly fast PZ intervals over two years to help athletes achieve high international standards.

TABLE 9.3

Performance Threshold Interval Times

Avg. race speed (m/min)	Avg. race speed (mph)	INTERVAL TIMES FOR DIFFERENT-LENGTH INTERVALS 4% FASTER THAN AVERAGE RACE SPEED									
		100 m (sec)	200 m (min: sec)	400 m (min: sec)	800 m (min: sec)	1 mi (min: sec)	1.5 mi (min: sec)	2 mi (min: sec)	1 km (min: sec)	2 km (min: sec)	3 km (min: sec)
170	6.3	33.9	1:08	2:16	4:31	9:09	13:43	18:18	5:39	11:18	16:56
180	6.7	32.0	1:04	2:08	4:16	8:38	12:58	17:17	5:20	10:40	16:00
190	7.1	30.3	1:01	2:01	4:03	8:11	12:17	16:22	5:03	10:06	15:10
200	7.5	28.8	0:58	1:55	3:39	7:47	11:40	15:33	4:48	9:36	14:24
210	7.8	27.4	0:55	1:50	3:39	7:24	11:06	14:49	4:34	9:09	13:43
220	8.2	26.2	0:52	1:45	3:29	7:04	10:36	14:08	4:22	8:44	13:06
225	8.4	25.6	0:51	1:42	3:25	6:55	10:22	13:50	4:16	8:32	12:48
230	8.6	25.0	0:50	1:40	3:20	6:46	10:09	13:32	4:10	8:21	12:31
235	8.8	24.5	0:49	1:38	3:16	6:37	9:56	13:14	4:05	8:10	12:15
240	9.0	24.0	0:48	1:36	3:12	6:29	9:43	12:58	4:00	8:00	12:00
245	9.1	23.5	0:47	1:34	3:09	6:21	9:31	12:42	3:55	7:50	11:45
250	9.3	23.0	0:46	1:32	3:04	6:13	9:20	12:27	3:50	7:41	11:31
255	9.5	22.6	0:45	1:30	3:01	6:06	9:09	12:12	3:46	7:32	11:05
260	9.7	22.2	0:44	1:29	2:57	5:59	8:58	11:58	3:42	7:23	11:05
265	9.9	21.7	0:43	1:27	2:54	5:52	8:48	11:44	3:37	7:15	10:52
270	10.1	21.3	0:43	1:25	2:51	5:46	8:38	11:31	3:33	7:07	10:40
275	10.3	20.9	0:42	1:24	2:48	5:39	8:29	11:19	3:29	6:59	10:28

Avg. race speed (m/min)	Avg. race speed (mph)	INTERVAL TIMES FOR DIFFERENT-LENGTH INTERVALS 4% FASTER THAN AVERAGE RACE SPEED									
		100 m (sec)	200 m (min: sec)	400 m (min: sec)	800 m (min: sec)	1 mi (min: sec)	1.5 mi (min: sec)	2 mi (min: sec)	1 km (min: sec)	2 km (min: sec)	3 km (min: sec)
280	10.4	20.6	0:41	1:22	2:45	5:33	8:20	11:06	3:26	6:51	10:17
285	10.6	20.2	0:40	1:21	2:42	5:27	8:11	10:55	3:22	6:44	10:06
290	10.8	19.9	0:40	1:19	2:39	5:22	8:03	10:44	3:19	6:37	9:56
295	11.0	19.5	0:39	1:18	2:36	5:16	7:55	10:33	3:15	6:31	9:46
300	11.2	19.2	0:38	1:17	2:34	5:11	7:47	10:22	3:12	6:24	9:36
305	11.4	18.9	0:38	1:16	2:31	5:06	7:39	10:12	3:09	6:18	9:27
310	11.6	18.6	0:37	1:14	2:29	5:01	7:32	10:02	3:06	6:12	9:17
315	11.8	18.3	0:37	1:13	2:26	4:56	7:24	9:53	3:03	6:06	9:09
320	11.9	18.0	0:36	1:12	2:24	4:52	7:17	9:43	3:00	6:00	9:00
325	12.1	17.7	0:35	1:11	2:22	4:47	7:11	9:34	2:57	5:54	8:52
330	12.3	17.5	0:35	1:10	2:20	4:43	7:04	9:26	2:55	5:49	8:44
335	12.5	17.2	0:34	1:09	2:18	4:39	6:58	9:17	2:52	5:44	8:36
340	12.7	16.9	0:34	1:08	2:16	4:34	6:52	9:09	2:49	5:39	8:28
345	12.9	16.7	0:33	1:07	2:14	4:31	6:46	9:01	2:47	5:34	8:21
350	13.1	16.5	0:33	1:06	2:12	4:27	6:40	8:53	2:45	5:29	8:14
355	13.2	16.2	0:32	1:05	2:10	4:23	6:34	8:46	2:42	5:25	8:07
360	13.4	16.0	0:32	1:04	2:08	4:19	6:29	8:38	2:40	5:20	8:00
365	13.6	15.8	0:32	1:03	2:06	4:16	6:24	8:31	2:38	5:16	7:53
370	13.8	15.6	0:31	1:02	2:05	4:12	6:18	8:24	2:36	5:11	7:47
375	14.0	15.4	0:31	1:01	2:03	4:09	6:13	8:18	2:34	5:07	7:41
380	14.2	15.2	0:30	1:01	2:01	4:06	6:08	8:11	2:32	5:03	7:35
385	14.4	15.0	0:30	1:00	2:00	4:03	6:04	8:05	2:30	4:59	7:29
390	14.6	14.8	0:30	0:59	1:58	3:59	5:59	7:59	2:28	4:55	7:23
395	14.7	14.6	0:29	0:58	1:57	3:56	5:54	7:53	2:26	4:52	7:17
400	14.9	14.4	0:29	0:58	1:55	3:53	5:50	7:47	2:24	4:48	7:12
405	15.1	14.2	0:28	0:57	1:54	3:50	5:46	7:41	2:22	4:44	7:07
410	15.3	14.0	0:28	0:56	1:52	3:48	5:41	7:35	2:20	4:41	7:02
415	15.5	13.9	0:28	0:56	1:51	3:45	5:37	7:30	2:19	4:38	6:56

Coaches will first need to know the athlete's current average race speed. Please see page 161.

(continued)

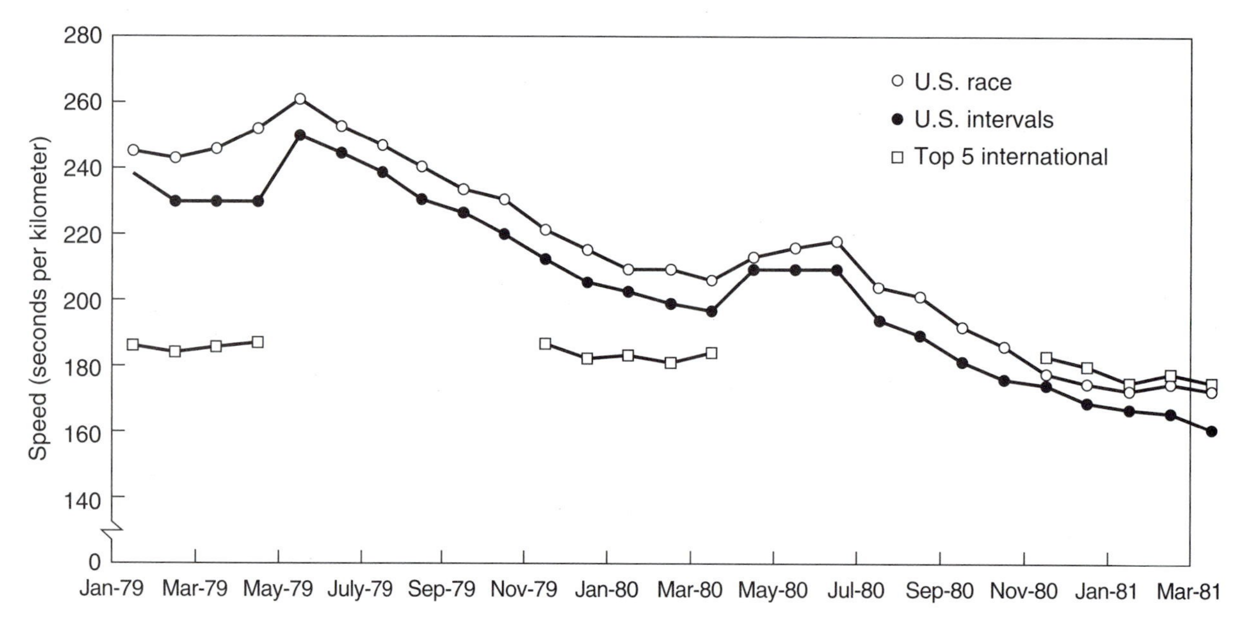

FIGURE 9.5 Performance zone intervals used by Nordic combined skiers. This graph shows how the U.S. Nordic combined team used performance zone intervals for a two-year period. The average speed of the top five international finishers in each race is shown as open squares. The open circles show the average race (or time trial) speed for the U.S. team. The group tested race speed monthly and did two to three performance zone interval sessions each week, year-round, at the speed indicated by the filled circles. By the third winter, the average U.S. race speed was faster than that of the top five international skiers, with Americans winning nearly every race!

For events lasting less than 4 minutes, the same concepts can be applied, but the energy systems will become increasingly more anaerobic. The difficulty with speed-based intervals for sprints is intuitively obvious to most coaches. As race distances become shorter, it becomes difficult to run intervals faster than race speed. For some events, this barrier can be overcome with assisted intervals. Tethered swimming can help a swimmer move slightly faster than current race speed, teach neuromuscular coordination, and overload the energy system. Bungee cord-assisted sprinting can help a sprinter move slightly faster than performance speed for short distances. Downhill running (1-2 percent) can also help increase speed. Coaches need to be inventive to find ways to help their athletes perform short intervals that are slightly faster than their current performance speed.

- Fatigue: Athletes who have not recovered from previous days of hard training may have elevated resting and exercise HRs. The fatigue index (p. 237) will help athletes identify those days when they have not recovered. Back-to-back days of hard training are often part of training plans. On subsequent days of hard training, athletes can expect higher HRs if they maintain similar training speeds.

- HR drift: During long training sessions, HR will slowly increase if the workload remains the same. This drift is common and appears to be due to dehydration and a shift of fluid from blood plasma into interstitial (between cell) spaces. It is common for HR to drift as much as 5 to 10 beats during an hour ride on a cycle ergometer at a fixed workload. Thus, if you are controlling the intensity of a long exercise session with a HR monitor, you may find that you need to either slow down later in the exercise to remain within your easy training zone or allow your HR to exceed your target. There is no scientific evidence that says which is better, slowing down or maintaining your training speed and letting your HR slowly increase but the authors would suggest maintaining your pace, if possible, and allowing your HR to increase. During long workouts, HR may be used as a basic guideline, but the RPE will account for HR drift and other factors affecting intensity.

- Type of exercise: Speed and HR are generally good intensity guides during aerobic energy training, but are seldom used for muscular fitness. During weightlifting, especially isometric

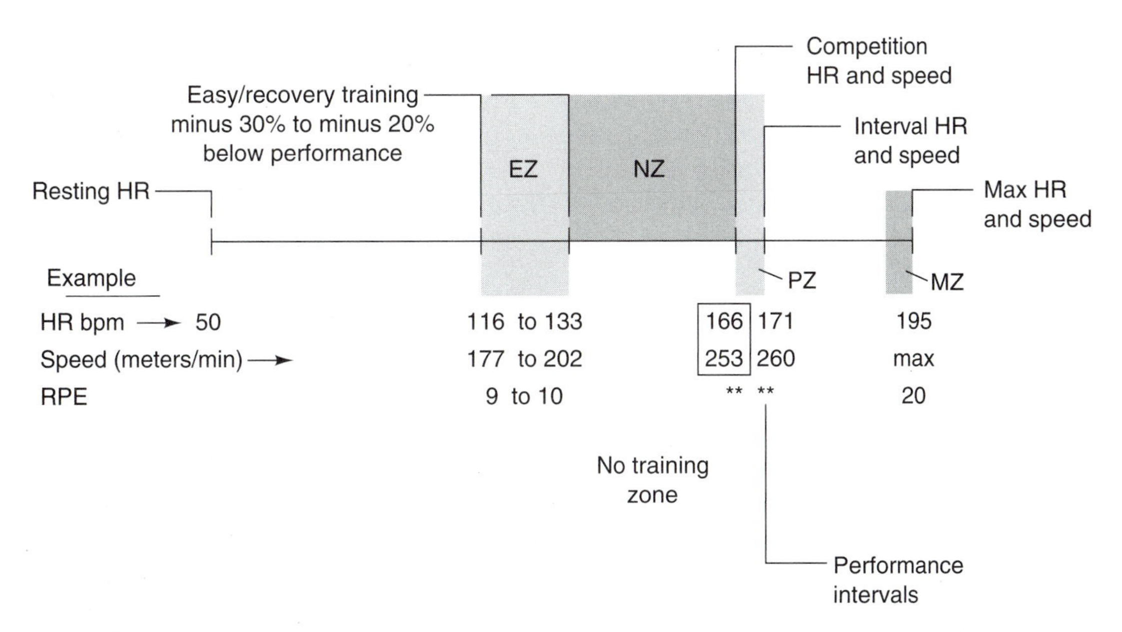

FIGURE 9.6 The energy fitness training zones. Training is done primarily in the easy (EZ) and performance (PZ) zones. The determination of these zones is always based on either the speed or heart rate (HR) during an actual competition or high effort time trial. In the example, an athlete has a HR of 166 beats per minute (bpm) and a speed of 253 meters per minute (9.44 miles per hour). Their EZ training should be done at 20 to 30 percent below race intensity at a speed of 177 to 203 meters per minute or at a HR of 116 to 133 bpm. Performance zone intervals should be done from 1 to 5 percent faster than race speed. In this example, the athlete chose 3 percent faster which is a speed of 260 meters per minute or a HR of 171 bpm. A small amount of training (less than 1 percent) will be done at maximal zone (MZ) intervals of very short maximal sprints with long recovery periods.

TABLE 9.4

Training Zone Summary

Training zone	Speed range	HR range	RPE*	Training example**
Easy training zone (EZ)	20-30% below	20-30% below	≤11	30 min recovery run, HR = 125 bpm; 50 min distance run, speed = 7 mph
No-training zone (NZ)	2-20% below	2-20% below	12-15*	Not recommended
Performance zone (PZ)	1-4% above	1-5% above	*	400 m intervals, HR = 183; 400 m intervals, time = 92 sec; 8-15 intervals with full recovery
Maximal zone (MZ)	Max speed	NA	20	5-10 sec maximal effort, long recovery (2-3 min)

*See table 9.1 to use RPE (rating of perceived exertion) for intensity zones.

**For an athlete who races 1,600 m in 6:20 with an average HR of 178 bpm.

exercises, the HR increases beyond that expected for the work being done. This happens because the contracting muscles restrict blood flow, leading to increased blood pressure, making the heart work harder. Additionally, when small muscles are involved, as in weightlifting with the arms, the HR and blood pressure increase more than would be expected with equal leg work.

- Sleep and travel: Travel can elevate the HR and hamper performance, especially when you traverse several time zones. Air travel increases the risk for upper respiratory tract infections due to the recirculated dry air and the large number

Training Intensities: An Example

Consider John, a high school cross country runner with a personal best of 18:20 for 3 miles. His speed is 6:07/mile or 265 meters/minute, and his average HR is 172 beats per minute. From table 9.3 we find that his interval speed should be about 275 meters/minute or 5:54 per mile. We recommend using speed to control PZ intervals, but if he uses HR to control intensity, the rate should be about 177 beats per minute. Table 9.3 shows that for his event, John should consider intervals of 3 to 6 minutes in length. He decides on 1/2- and 1-mile intervals in 2:57 and 5:54, respectively. His perceived exertion during races is 17 on the Borg scale (very hard), and his speaking scale is "Need to focus to maintain intensity." For distance training, he uses a HR monitor occasionally to help him stay in the 20 to 30 percent below race HR (120 to 138 beats per minute) zone. His RPE during his short and easy or long-distance runs is about 11, at a HR of 130, and he can easily talk in full sentences with running partners. For distance training, he runs for 2 to 3 hours once a week, though he feels like he could easily go farther. John has learned how to control his training intensities using multiple methods. As his race times improve, he will adjust his interval speeds based on his new race speeds.

Sport and physical activities often require training and performing in extreme environments such as *(a)* heat, humidity, and *(b)* altitude. Understanding how those environments affect us, and what precautions to take in those environments, will allow successful coaches to best protect and help their athletes.

of people in a relatively small space. Sleep loss is another factor, like stresses and physical fatigue, that raises the HR. Athletes should strive to get adequate daily sleep and allow extra rest after travel.

- Other stressors: Coaches and athletes often forget that there are many types of stress in our lives. One of the main duties of a coach is to help reduce and control nontraining stress experienced by their athletes. Homework, social problems, family issues, work, tests at school, and many other factors can create stress and decrease performance. Figure 9.7 shows the "stress barrel." This concept suggests that each athlete can handle only a certain amount of stress before performance will become affected. Some athletes can handle only a little stress, others a great deal. The idea of training is to overload (stress) the physiological systems, then allow time for recovery and growth. Some athletes are good at managing stress and can limit most of the stress entering their barrel. Some athletes are able to get rid of stress effectively; nothing seems to bother those with large spigots on their barrels. As a coach, one of your duties is to help your athletes learn to deal with stress. If the level of accumulated stress from all causes becomes too great, athletes are at an increased risk for illness. If the accumulated physical stress remains too high, and the athlete does not get adequate rest, he or she risks both overtraining and illness (see chapter 12).

- Adjusting intensity during periods of stress: Just as the intensity of training and the total load should be reduced following periods of high physical stress, training load and intensity should be reduced during and following periods of nontraining stress. The use of RPE can help athletes to appropriately adjust their EZ training. Athletes will self-regulate PZ intervals. When athletes have difficulty maintaining their target speed, the intervals should be stopped.

Duration of Energy Training

For energy fitness, the duration of training depends on the sport and the primary energy systems utilized. Some sports, such as soccer, demand both explosive speed and endurance, requiring training of multiple systems, while other sports are more specific, such as sprinting or distance running. The duration of training sessions and the total annual hours of training

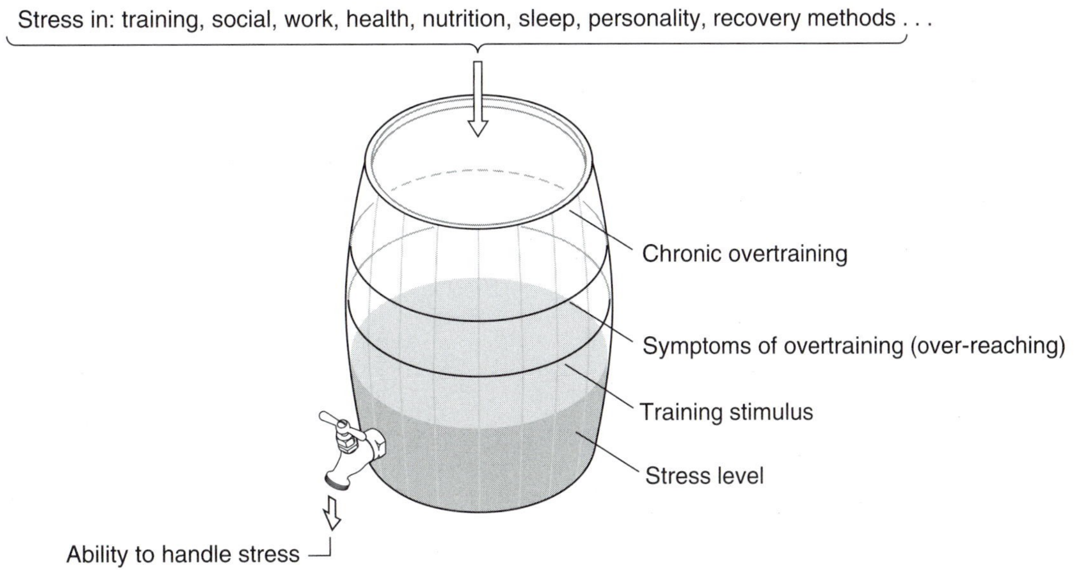

FIGURE 9.7 The "stress barrel." Stress is accumulated by athletes in many forms. Athletes also have differing abilities to reduce stress, as suggested by the spigot. If stresses accumulate more quickly than an athlete can reduce them, the athlete's stress barrel begins to fill. Training is the process of reaching critical levels of physical stress to exceed the training stimulus level. Some experts also believe that stress needs to briefly get to the overreaching level followed by periods of recovery. This model also suggests that if athletes accumulate excessive stress, they may reach a chronic overtrained state from which it is difficult to quickly recover.

depend on the ability of each athlete to handle an overload.

The Training Pyramid

Energy fitness training is based on the training pyramid illustrated in figure 9.8. The emphasis that coaches place on the various energy fitness components will vary depending on the specific fitness requirements of their sport. In endurance events like running or swimming, EZ and PZ training are the major emphases. For shorter, faster events like wrestling or baseball, EZ training is reduced while PZ and MZ training meet the major fitness needs. Let's review the four levels of training in the training pyramid, and the duration of training required to develop an aerobic energy foundation.

EZ (Aerobic) Training

In all sports, athletes should first build a solid aerobic foundation. This foundation of EZ training prepares the respiratory and circulatory systems, toughens tendons and ligaments, provides energy for sustained work, and provides the mechanisms to recover more rapidly from anaerobic work. This period of sustained low-intensity work also seems to be a time when mechanical efficiency is increased and athletes learn how to relax and move fluidly. Easy zone training also develops metabolic efficiency, increasing the stamina and improving the energy systems of the slow-twitch fibers. Fat utilization during low-intensity work and recovery is improved, and increased stores of muscle carbohydrates and aerobic enzymes and mitochondria enhance higher-intensity work. This foundation training also enhances the ability of the fast-oxidative-glycolytic and fast-glycolytic fibers to recover faster. Shown in table 9.5 are general guidelines for aerobic training during the recovery and basic training seasons. To use the table you must estimate the amount of continuous effort your athletes expend in their activity, and then compare this to the basic training season goals. During the competitive season, it is beneficial for power and speed athletes to maintain some aerobic training.

Performance Zone (PZ)

Previous books have dealt with training near the anaerobic or lactate threshold. While use of

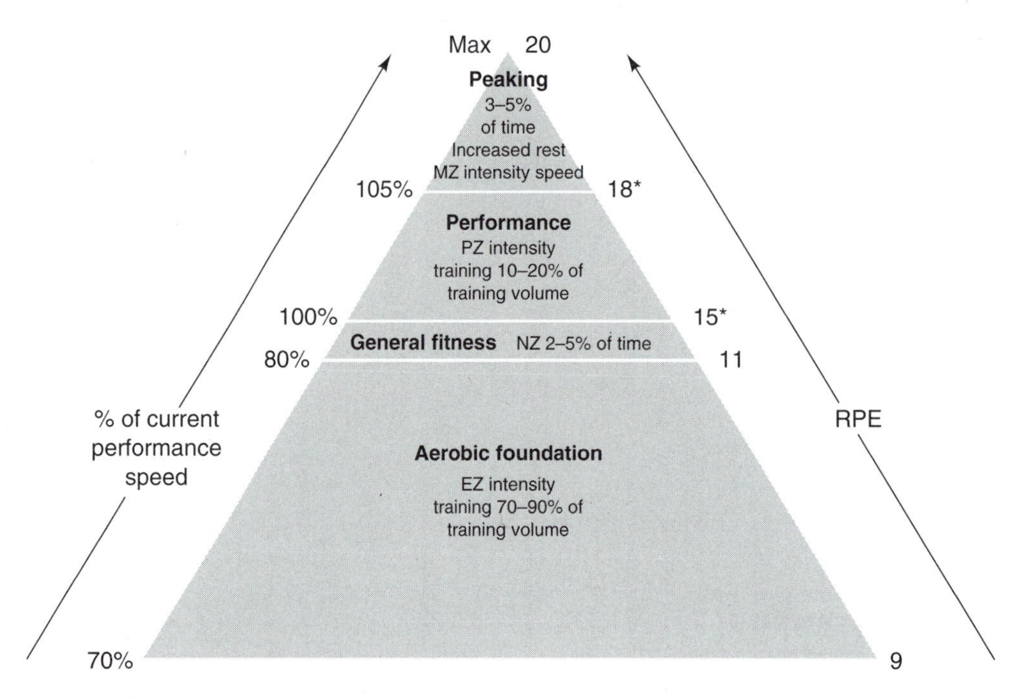

FIGURE 9.8 Energy fitness training pyramid. The area of each portion of the pyramid is roughly equivalent to the relative time spent in each training zone.

*The rating of perceived exertion will depend on the event duration. For performance zone (PZ) training, we suggest using speed relative to current race speed.

metabolic thresholds has proven helpful, races come in a variety of distances requiring a combination of energy systems, and are seldom contested at metabolic threshold intensities. Thus, we have chosen to use the term performance zone as a method to help athletes individualize their programs. Your performance speed can be thought of as the maximum speed (and HR) that you can sustain for an event.

Performance zone training is generally done as interval training based on speeds (table 9.3, p. 162). Table 9.6 shows general guidelines for volume of PZ training during the off- and preseason. To use the table you must estimate the duration of continuous effort your athletes expend in their activity, and then compare this to the off-season and early-preseason goals. The goals are stated in terms of frequency per week. The times shown in the off- and preseason goal columns reflect the total summed interval time for each training session (i.e., 15 minutes would be 10 intervals of 1.5-minute duration). During the competitive season, all athletes, even those in power and speed sports, should maintain some PZ training. The recommended lengths for PZ intervals are shown in table 9.2 (p. 160).

MZ (Anaerobic) Training

Improvements of the anaerobic system should begin when the oxygen-using abilities of the fast-oxidative-glycolytic muscle fibers are well developed through six to eight weeks of PZ training. The anaerobic system will not improve as significantly as the aerobic system, and it is

TABLE 9.5

Easy Zone (EZ) Training Goals

Continuous effort in sport	Off-season goals*	Preseason goals*
Under 10 sec	10-15 mi/wk or 1.5-2 hr/wk	5-10 mi/wk or 45 min-1.5 hr/wk
10 sec-2 min	15-20 mi/wk or 2.5-4 hr/wk	7-15 mi/wk or 1-2 hr/wk
2-15 min	20-30 mi/wk or 3-5 hr/wk	15-20 mi/wk or 2.5-4 hr/wk
15-30 min	30-40 mi/wk or 5-7 hr/wk	25-35 mi/wk or 4-6 hr/wk
Over 30 min	Over 40 mi/wk or >7 hr/wk	Over 35 mi/wk or >6 hr/wk

This training is at an intensity of RPE 9-11, or about 20-30% below competition speed and HR.

*Runners, football and basketball players, and others who play on foot use miles per week. Swimmers, cyclists, skiers, and other nonrunners use hourly goals.

TABLE 9.6

Performance Threshold Interval Training Frequency and Duration

Continuous effort in event	Off-season goals*	Preseason goals*
Under 60 sec	1/week: SAS	3/week: SAS**
1-4 min	1/week: 10-20 min	2/week: 15-20 min
4-15 min	1/week: 20-25 min	2/week: 15-20 min
15-30 min	1/week: 25-35 min	2/week: 20-30 min
Over 30 min	1/week: 35-45 min	2/week: 30-40 min

*Number of sessions per week and total time doing intervals per session.

**SAS = speed-assisted intervals. See text.

rarely necessary to devote more than two to three times per week for six to eight weeks to anaerobic training. This means that for high school and collegiate sports, the preparatory work (EZ and PZ training) needs to be well established at least six to eight weeks before the important competitions begin. This requires that student-athletes do much of the preparatory work on their own. Excessive emphasis on the difficult, high-intensity MZ training often leads to fatigue, illness, and injury. Anaerobic training develops short-term energy sources and pathways while preparing the fast-glycolytic muscle fibers for competition. For endurance athletes, MZ training is often synonymous with speed training and is generally done as short (4-15 seconds) maximal efforts with long recovery periods. For short-event and sprint athletes, these intervals may need to be assisted in order to generate speeds slightly above race speed. Improvements in sprint events will come primarily through muscular power training and improved technique and coordination.

Peaking

Peaking, the top of the pyramid, focuses more on neuromuscular development than on improving energy training systems. Peaking involves increased rest, decreased EZ training, PZ intervals with decreasing rest periods, and more MZ speed training. The high-intensity MZ work is generally performed as short bursts (<15 seconds) of maximal effort, followed by long rest periods of easy activity. Speed training is used to sharpen and hone skills before competition. Endurance athletes will spend less than 1 percent of their training time doing these maximal bursts of speed. For sprinters and power athletes, the final training period leading to major events will include a larger percentage of time working on speed. A final means of peaking for aerobic event athletes that some coaches use,

Speed Play

During the basic training period, it is always good to build some "speed play" into your workouts. This speed play (**fartlek** or natural intervals) refers to pickups during EZ runs. These relaxed speed intervals provide variation to the training and allow athletes to gradually increase their speed work. Athletes take turns leading the intervals and use local features (e.g., telephone poles) to determine their length. Start easily, gain and hold speed, then slow gradually.

Natural intervals or fartlek training requires athletes to pick up their speed into the performance zone for short periods during easy zone training. These relaxed speed intervals provide variation to the training and allow athletes to gradually increase their speed work.

but that can have mixed results, is to incorporate a few sessions of maximal-effort, long (3-5 minutes) intervals into the two weeks prior to a major competition. These sessions are generally limited to four to eight intervals. This training is believed to increase the athlete's anaerobic capacity. Coaches may wish to test these intervals earlier in the season to determine which athletes respond favorably.

The most important aspect of peaking is the taper, a period of 3 to 14 days during which the training load is gradually reduced. For endurance athletes, a longer taper (10-14 days) seems to be optimal, while sprint and power athletes may require only three to four days. Training volumes are typically reduced by 30 to 50 percent during a taper for a major competition.

Energy Fitness Maintenance

Regardless of your sport, the energy fitness training program must be founded on the principle of progression. As your athletes progress through the four training stages, do not ignore previous stages in the training pyramid. Aerobic training should be included at every stage in the training season. To maintain previous gains, continue to schedule EZ and PZ training during the early competition and peak performance (main competition) periods.

Two EZ, at least one PZ, and one MZ session per week will help to maintain the aerobic and anaerobic energy systems. Include maintenance in your plan. During a hectic competitive schedule, time needs to be devoted to maintenance training. Studies have shown that athletes lose energy fitness as the season progresses. Coaches who neglect maintenance often find the end of the season less productive than hoped for. Preoccupation with game plans, strategy, offense, and defense take practice time that was previously available for energy and muscular fitness development. The smart coach ensures that athletes maintain the capabilities they've worked so hard to develop throughout the entire competitive season. In chapter 10 we consider the development of a season plan in more detail.

SUMMARY

This chapter focused on how we proceed to develop anaerobic and aerobic energy systems. We determined the types of training used to develop short-term anaerobic and aerobic systems, and we outlined the specific cellular changes that occur with different intensities and durations of training. To simplify training, we have divided training intensity into zones and described how you can use those zones to achieve specific goals. Need to improve race pace? Train in the performance zone, using pace, perceived exertion, or heart rate as the measure of intensity. If you use heart rate, be certain that you recognize those factors that can affect intensity, including heat, emotions, fatigue, illness, and heart rate drift, among others.

KEY TERMS

fartlek (p. 170)
maximal heart rate (p. 156)
maximal training zone (MZ) (p. 158)
no-training zone (NZ) (p. 158)
peaking (p. 170)
perceived exertion (p. 155)
performance speed (p. 154)

REVIEW QUESTIONS

Fill in the blank(s) with the appropriate word or words to complete the following statements.

1. _________ is our ability to produce adenosine triphosphate (ATP), the high-energy molecule that is the energy currency of life.
2. _________ , _________ , and _________ are the three methods used to control the intensity of a workout.
3. _________ training is especially appropriate for timed events but works well for other aspects of energy fitness training.
4. The most important aspect of peaking is the _________, a period of 3 to 14 days during which the training load is gradually reduced.

Match the training zones in the left column with the appropriate description in the right column.

5. Easy training zone (EZ)
6. No-training zone (NZ)
7. Performance training zone (PZ)
8. Maximal training zone (MZ)

a. Speed range is maximal effort; RPE is 19-20.
b. Speed range is 20 percent below to current racing speed; HR range is 20 percent below racing HR to current racing HR; RPE is 13-15.
c. Speed range is 20 to 30 percent below racing speed; HR range is 20 to 30 percent below racing HR; RPE is 9-12.
d. Speed range is current racing speed to 5 percent above; HR range is current racing HR to 5 percent above; RPE is 15-18.

Determine if the following statements are true or false and then circle the correct answer.

9. T/F In order to increase the capacity of an energy system, the system must be specifically overloaded, and it will respond with improved capacity.
10. T/F During intense exercise, most energy is from the metabolism of fat, while during light exercise, carbohydrates are the primary source or energy.
11. T/F Heart rate training works well for monitoring the intensity of aerobic foundation training and recovery activities in the easy training zone for all energy fitness training.
12. T/F Factors such as emotions, illness, heat, altitude, fatigue, type of exercise, travel, loss of sleep, and life stresses complicate the use of HR, speed, or RPE as indicators of training intensity.
13. T/F Time spent training in the no-training zone (NZ) is more productive than time in the easy training zone (EZ) or performance zone (PZ).
14. T/F A cold or fever can elevate the heart rate and complicate the use of the HR during training.
15. T/F The training zone in which training is least productive is the MZ.
16. T/F PZ intervals should last between 4 and 8 minutes with full recovery between intervals.

17. T/F Most athletes train too hard to recover on their easy days.
18. T/F During the early days of training, coaches should have athletes use all three of the methods used to control intensity in training.

Select the correct response (more than one may be correct).

19. Methods used to control intensity of training include
    a. heart rate
    b. speed
    c. perceived exertion
    d. all of the above
20. Reasons why you shouldn't use an estimated heart rate to control training intensity include
    a. the fact that individuals differ in resting and maximal heart rates
    b. the fact that estimates of the max HR are highly variable
    c. the fact that as an athlete improves the HR must change
    d. all of the above
21. The most important effects of training occur in this zone:
    a. NZ
    b. PZ
    c. EZ
    d. MZ
22. During training in the PZ, recovery between intervals should be
    a. 30 seconds
    b. 60 seconds
    c. 2 minutes
    d. full
23. The heart rate drift that occurs during a long training session is due to
    a. dehydration
    b. body temperature increase
    c. fluid shifts
    d. all of the above

PRACTICAL ACTIVITIES

1. Three methods are used to control intensity in training: speed, heart rate, and perceived exertion. For all three methods, describe how you would determine the appropriate intensity for an athlete who is training to run 5,000 meters (3.1 miles).
2. Training zones: Describe each training zone; indicate what it is used to accomplish and the effects of training in the zone.
3. How does an athlete peak for a major event? Describe the steps and indicate the role of tapering.

PART IV

Systematic Training Program Development

Energy fitness supplies the adenosine triphosphate (ATP) our muscles need in order to contract and create movement. Muscular fitness defines how our muscles respond to specific training so that they have the strength, power, and endurance needed to perform. In this final part, we will teach you a systematic approach to developing energy and muscular fitness in an integrated training plan. We build on a foundation of the training principles and incorporate those principles into lessons learned from science and from top coaches.

10

Developing Training Programs

This chapter will help you

- understand how to apply the principles of training to program design and
- learn a four-step method to help you design a program for your athletes.

Howard was both a wonderful history teacher and a passionate high school football coach. He cared greatly for his athletes both on and off the field. His athletes were expected to be, and were, model students and community leaders. Howard preached and lived the concept of mind and body working in harmony. Other coaches had great respect for Howard's knowledge of skills and tactics, and his coaching seminars were always well attended. Howard's coaching weakness was that his teams started strong in the first half, only to struggle late in the game. Howard attributed this problem to a lack of conditioning. Unfortunately, it seemed that the harder the team trained, the fewer the rewards.

Howard sought the advice of Pete, one of his good friends who was also the tennis coach. Pete had just returned from a coaching seminar where participants had talked about training programs and the need to build an aerobic foundation, even for explosive sports. Howard and Pete talked further and together developed a plan. The football practice field was located a mile from the school, and the team had always been bussed to and from practice. Starting with preseason practice, the athletes were asked to walk to the field, and over the next two weeks, they worked up to jogging the entire way to and from practice. Howard changed the practices so that they became more varied in intensity. Following sessions of high-intensity work, he planned entire workouts with easy drills and technique to allow recovery. In addition to daily variations in training intensity, Howard planned harder and easier training weeks. That year the team won the league championship and outscored their opponents in the second half of every game. Howard was a quick study, learning the fundamentals of energy fitness and the concept of periodization.

All sports require a balance of training for both aerobic and anaerobic energy. Sports such as football, which have long been thought to be solely in the domain of speed and power, have been shown to benefit when modest aerobic training is added to the basic training period.

While the need for strength and power has long been recognized in football linemen, only recently have we recognized the need for modest aerobic fitness to improve their performance. Surprisingly, few muscular fitness studies have been completed to evaluate strength and power in football players. In one such study (Hoffman and Kang 2003), both intensity and volume of resistance training were evaluated during a twice-weekly in-season football program. Fifty-three collegiate athletes were tested on the bench press (1-repetition maximum or 1RM) and squat (1RM) on the first day of training and again during the final week of the regular season. Subjects were required to perform three sets of six to eight repetitions per exercise. Analysis showed that athletes training at ≥80 percent of 1RM had significantly greater strength improvements than athletes training at <80 percent of their 1RM. Strength gains in this study were not related to competitive performance, but others have reported improvements in sled pushing and blocking as a result of improved strength.

Putting together a training program for an individual or a team can be a daunting task for any coach. We recommend a systematic approach to designing both individualized and group programs. Here is a four-step plan you could use to develop a successful training plan for your athletes.

1. Set goals, both team, and individual.
2. Perform a needs analysis.
   a. Analyze muscular and energy fitness needs of the sport.
   b. Perform team and individual assessment.
3. Periodize the training plan.
   a. Define the season.
   b. Plan weekly cycles.
   c. Write daily plans.
4. Monitor the progress and health of your athletes.

We have discussed steps 1, 2, and 4 elsewhere in this book and will review them only briefly before showing you how to periodize the training plan.

Goal Setting and Needs Analysis

Goals provide a destination: They give direction, drive, and motivation. Your first step in planning a season is to work with your athletes to define individual and team goals. A training plan may need to include specific goals for muscular and energy fitness, technique, tactics, and psychological skills. Consult a book on psychological skills that includes good sections on goal setting (see appendix H, "Suggested Readings").

Goal setting and assessment are an ongoing process involving the athletes and coaches.

TABLE 10.1

Muscular and Energy Fitness Groups for Different Sports

Muscular fitness	Energy fitness	Sport or event
Low	Low	Archery, boccie, bowling, equestrian, golf, shooting sports
Medium	Low	Baseball, bobsled, cricket, fencing, football (kicker/punter), ice hockey goalkeeper, javelin, lacrosse goalkeeper, luge, sailing, ski jumping, soccer goalkeeper, softball
High	Low	Discus, diving, football (linemen), high jump, shot, weightlifting
Medium	Medium	Aikido, badminton, basketball, football (offensive and defensive backs, receivers), field hockey, Frisbee, ice hockey, judo, karate, lacrosse, long jump, motocross, racquetball, roller skating, rugby, skiing (Alpine), soccer, speed skating, sprints, squash, surfing, swimming (sprints), synchronized swimming, table tennis, tae kwon do, team handball, tennis, triple jump, volleyball
High	Medium	Boxing, decathlon, figure skating, gymnastics, hurdles (100 m/400 m), modern pentathlon, pole vault, wrestling
Medium	High	Biathlon, canoeing, cycling, distance running, kayaking, mountaineering, orienteering, rowing, skiing (cross-country), swimming (above 200 m), triathlon, water polo

Reprinted, by permission, from B.J. Sharkey, 1986, *Coaches guide to sport physiology* (Champaign, IL: Human Kinetics), 155.

As for needs assessment, think of it this way: When planning a trip, you need to know where you are, where you want to go, and how you can get there. When you are planning training, athlete assessment defines where you are. Goal setting determines where you want to go, and an analysis of the muscular and energy fitness needs will guide your route to your destination. See table 10.1 for the muscular and energy fitness training requirements for different sports. Once the assessment and goals are completed, it is time to develop the training plan.

Periodize the Training Plan

Periodization of training is the process of incorporating systematic variation into the training plan. This variation is programmed at several levels: daily, weekly, seasonal, and career (chapter 12).

Define the Training Season

The training season includes the calendar period during which you expect your athletes to train. The periods of training include recovery, basic, precompetition, early competition, and peak performance training.

The season may include an "official season," when you work with the athletes, and an unofficial training season, when the athletes are expected to train on their own. In many instances, school or college organizations define the duration of the **official training season.** For academic programs, the official season generally corresponds with the precompetition, early competition, and peak performance seasons. Most coaches understand that it takes time to develop strength and endurance, so they encourage their athletes to participate in running and strength training during the basic training period prior to their official season. Having athletes complete their basic training prior to the start of the official season allows appropriate time to systematically develop the physiological systems necessary to achieve individual and team goals. This additional training requires self-motivation on the part of the athlete and an effort by coaches to encourage participation.

We will teach you how to use the periods of the training season to systematically periodize training programs. You will use that knowledge to develop reasonable training programs for your athletes. Table 10.2 summarizes the training and duration of these periods. In chapter 11 we present sample programs showing the relative volumes of muscular fitness training (strength, power, power endurance, and speed) and energy fitness training (EZ, PZ, and MZ).

As you read this section, consult a calendar and decide on appropriate dates to start and end each period. Assigning dates to the training periods will help you begin planning your season. Use a pencil and make adjustments as you go.

Recovery Period

The **recovery training period** is generally a time of recovering from the previous competitive season and includes nonspecific, nonstructured activities done at a low intensity. Many athletes do physical activities that they enjoy but that may be unrelated to their competitive sport(s). Vedar Ulvang, one of the great Norwegian cross-country skiers, was well known for his sea kayaking and mountaineering exploits during the recovery period. This postcompetitive recovery period is generally a few weeks to a few months in length. Sport-specific conditioning declines during this period. Research has shown that without this period of recovery, athletes do not perform as well in the following year.

Basic Training

The **basic training period** marks the beginning of planned training. Due to the short official season of many leagues, this training may need to be completed before the official season begins. The focus of this period is to build a strength

TABLE 10.2

Periods of the Training Season

Period	Duration (weeks)		General muscular fitness guidelines	General energy fitness guidelines
	Annual program	Minimum duration		
Recovery	4-8	1-2	Non-sport-specific, low-intensity activities. Focus is on both physical and mental recovery.	
Basic strength and endurance	20-24	10-12	Light preparatory resistance training Strength training (main focus) Maintenance of power and PE Begin transition to power training (basic skill development)	Build aerobic endurance base Focus on EZ training (90-95%*) Gradual increase in PZ intervals (5-10%*)
Precompetition	12-14	6-8	Power training (main focus) Maintenance of strength and PE Begin transition to PE (Skill and tactics development)	Reduction in EZ training volume (80-85%*) Increase in PZ intervals (10-15%*) Small MZ speed intervals (<1%)
Early competition	8-10	4-5	Power endurance (main focus) Speed training and plyometrics Maintenance of strength and power	Reduction in total energy fitness training Further reduction in EZ training (70-80%) Focus on PZ intervals (18-28%) MZ speed intervals (2%)
High performance	3-8	2-4	Gradually reduced training load Increased recovery time Competitions (main focus) Maintenance of power, PE, and speed	Gradual reduction in training volume to 50% of previous period Maintenance of PZ and MZ intervals Focus on competitions and recovery

*Percentage of energy fitness training duration.

and energy fitness foundation to support the higher-intensity work that will be done as the competition season approaches. High-intensity training is minimal during this period.

Strength is the focus of muscular fitness training during this period. After a preparatory adaptation, the training is done using heavy weights, slow movement speeds, and long rest periods between sets. Most sports require some measure of muscular strength and power. This is true across the continuum from endurance to sprint and explosive events. Obviously the 100-meter sprinter needs more power than does the endurance athlete, who requires greater endurance; but power underlies the ability of all athletes to move quickly. Since nearly all sports require skilled and coordinated movements, improved power is often a training goal. Power is the rate of doing work, and work requires force. Thus, to improve power it is often necessary to improve strength and then learn how to use the strength rapidly (power).

Most aerobic fitness training will be done as EZ (easy distance or recovery) training with maintenance amounts of higher-intensity PZ (performance zone) intervals. Speed and power, which are critical for all sports, can be maintained with short-duration work once or twice a week. Most sports require various degrees of aerobic energy. This is true across the continuum from long-distance endurance sports to sprint and explosive events. It is clear that the endurance athlete requires more aerobic fitness than does the sprint or power athlete. Improved aerobic function will enhance recovery and allow the sprinter and power athlete to do more repetitions and improve their performance. Changes in aerobic fitness often take extended periods of training; thus training the aerobic energy system is begun in this cycle.

Skill development using imagery, drills, and easy practice may also be a priority. Sport psychologists have shown that changing technique requires patience, feedback, and visualization. The off-season is a great time to work on technique, allowing athletes time to master new skills before competitions begin.

The basic training period lasts a minimum of eight weeks in order to optimize muscle recruitment for strength and to lay an adequate foundation for aerobic energy. For sports with limited training periods, such as high school and college programs, the training during this period is generally done individually and not as a part of the structured season. For academic programs, this period needs to be at least 8 to 12 weeks. For year-round sports, the basic training period may last four to five months. During this period there will be a gradual increase in total training hours and intensity of training as the precompetition period approaches.

Precompetition Period

The **precompetition training period** is the transition from basic training to competition. It lasts 6 to 8 weeks for school sports and 12 to 14 weeks for year-round programs, extending up to the date of the first competition. The muscular fitness training, which focused mostly on strength during basic training, transitions to power and power endurance training. Energy fitness training includes more PZ intervals. Moderate competition simulations begin, with an emphasis on speed, skill, and tactics.

This period is the start of the training season for many school-based programs. If no systematic training has been done in advance of the official season, individuals will begin without adequate strength and endurance. Coaches need to help athletes understand the need to prepare before the season. Without basic training, it is more likely that athletes will become injured and that their performances will suffer. Athletes who have not adequately prepared should not do high-intensity training until they are ready.

Athletes who compete in multiple sports may finish one competitive cycle in the fall and head into a new sport during the winter. Some even compete in a spring sport as well. In these cases, athletes need one to two weeks for rest and recovery before beginning the new season.

Early Competition Period

This period of early-season competitions generally lasts 4 to 5 weeks for school sports and as long as 8 to 10 weeks for yearlong programs. It may include scrimmages and early competitions that are used to prepare for important league and championship contests. Generally, overall training volume decreases while sport-specific speed and power endurance increase. Competitive success can be high, but the emphasis remains on power endurance, speed, and PZ interval training. Speed training is generally increased while resistance training is reduced to maintenance status. More time is spent on technique and tactics. Athletes need to recover more between training sessions in order to be able to do high-quality, competition-specific training. Toward the end of this period, training volume will begin to decline (taper) in preparation for major competitions. The **early competition training period** ends about two to four weeks before peak performance.

Peak Performance Period

In the **peak performance period,** depending on the type of sport, training continues with high-intensity work; but the total volume of training is reduced so that athletes are rested for important competitions. The main training emphasis is on speed, skill, and tactics while athletes maintain power, power endurance, and aerobic fitness.

In the one to two weeks leading to the most important event, athletes may taper their training in an effort to peak for an important event. Effective tapers gradually reduce training volume and stress by 40 to 60 percent. This allows athletes time for complete recovery between high-quality workouts and competitions. In addition to the taper, coaches may wish to focus on slight adaptations to the PZ intervals. For aerobic, middle-distance, and intermittent events, we recommend PZ intervals at about 4 to 5 percent above current race speed, along with a shortening of the interval length and the recovery period. If athletes were doing 400-meter (437-yard) repeats in 72 seconds with 2 minutes of rest, consider 200 meters (219 yards) in 35 seconds with 1 minute of rest, decreasing to 30 seconds of rest the final week before the championship event. Reduce the number of intervals and stop before speed deteriorates. This slight increase in intensity and reduction in rest will help the athlete "peak." We do not recommend increases in MZ (maximal zone) training, as many athletes respond negatively to increased anaerobic work. For sprint athletes, combining a taper with assisted MZ speed intervals, while maintaining long rest periods, will help

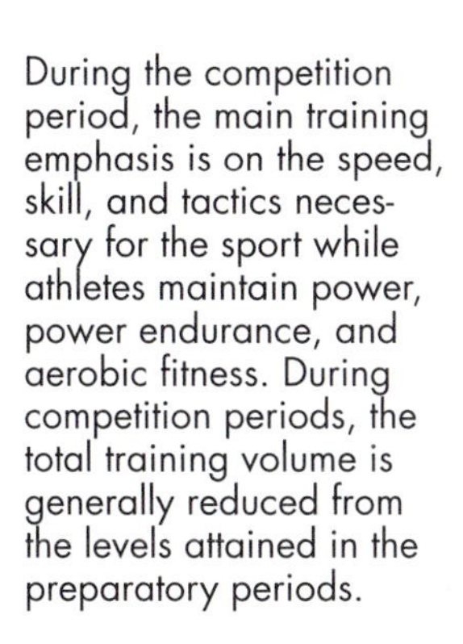

During the competition period, the main training emphasis is on the speed, skill, and tactics necessary for the sport while athletes maintain power, power endurance, and aerobic fitness. During competition periods, the total training volume is generally reduced from the levels attained in the preparatory periods.

to improve performance. Peaking can improve performance by 2 to 4 percent.

The timing of the training periods will depend on your sport, your season, and the needs of your athletes. At this point, you should have established start and end dates for each period and have recorded them on your calendar.

Weekly Periodization

Within the four structured training periods, systematic variation in physical training stress needs to be planned. We recommend using a three-week cycle of medium, hard, and recovery weeks. Within a period, the training in each cycle becomes progressively a little harder. Figure 10.1 shows a typical planned progression using a three-week cycle.

One difficulty for coaches is determining the stress of the training program. We can offer two methods: one to help you estimate stress, and one to help you monitor athlete fatigue. The first method involves the concept of **training impulses (TRIMPS).** The TRIMPS concept is described in the sidebar on page 185 and will help you control the weekly training stress. The second tool is the fatigue index, described in chapter 12 (p. 237). The fatigue index is a simple morning evaluation during which the sum of resting, exercise, and recovery heart rates is used to monitor fatigue.

An easy way to plan your training is to lay out an average week, appropriate for the training period. This average week can then be adjusted for the hard or easy weeks. An average week should tire your athletes on hard days, but they should not feel overly tired at the end of the week. Generally, the second week of a three-week cycle will be a hard week during which you should increase the overall stress so that your athletes are very tired by the end of the week (this may require individual adjustments). The third week should be physically easy and allow the athletes to recover. During easy weeks, spend more time on technique and tactics and less on high-intensity or long-duration training.

Remember, more is not always better. Adding easy weeks may seem unwise, especially with short official seasons, but their value has been shown repeatedly. Coaches who fail to allow

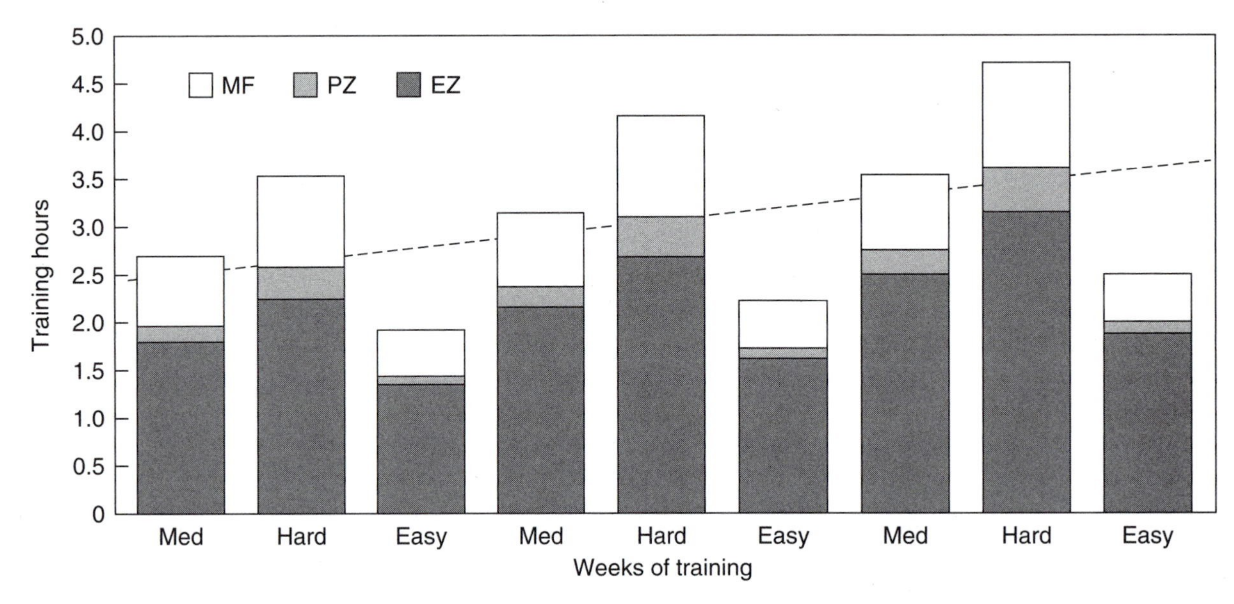

FIGURE 10.1 Weekly periodization. The graph represents a theoretical model of weekly progression during the "basic endurance and strength" period using a three-week periodization. The dotted line shows the average progression of training stress. During moderate weeks, performance zone (PZ) training is about 10 percent of energy fitness training (easy zone [EZ] and PZ). Muscular fitness training, three days per week, repeats three sets of strength training (3- to 6-repetition maximum). During hard weeks, PZ training is increased to 15 percent, and muscular fitness takes place on three days, comprising two sets for strength and one set for power training. For easy weeks, PZ training is reduced to 7 percent, and muscular fitness is reduced to two days, with two sets for strength and one set for muscular endurance with higher repetitions.

TRIMPS

In 1980, Bannister and Calvert proposed a model to combine intensity and duration of training into a single unit. Since then, others have proposed similar models. All the models show a nonlinear increase in training stress or impulses (TRIMPS) as intensity of exercise increases. We have modified the concept to help coaches use TRIMPS without all of the mathematics. Using the graph in figure 10.2, you can estimate hourly values for training in each of the energy fitness intensity zones and then make estimates for the stress of EZ, PZ, and MZ training. We have summarized the relationship of TRIMPS, including muscular fitness, to training intensity in table 10.3.

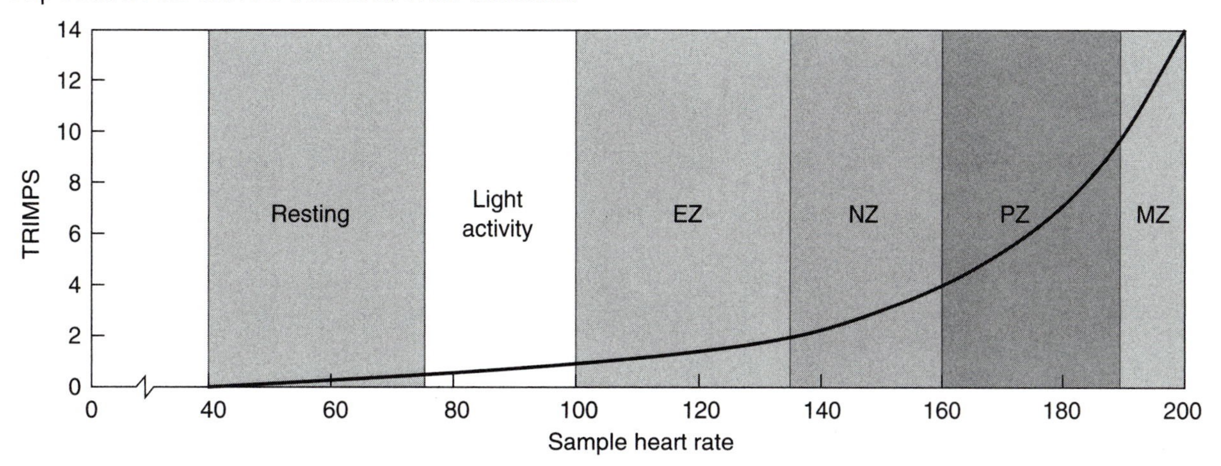

FIGURE 10.2 Relationship of TRIMPS to training intensity. Training impulses (TRIMPS) are shown per hour of training. The heart rate values associated with the training zones are average values only and should not be used for individual athletes. Each athlete will determine appropriate training intensity for each zone based on performance speed, heart rate, or rating of perceived exertion. The performance training zone (PZ) depicted covers the possible range from short- to long-duration events. The individual performance threshold will depend on the athlete and the event. Short events are performed at higher speed, heart rate, and rating of perceived exertion values; aerobic events are performed at lower speed, heart rate, and rating of perceived exertion values. EZ = easy training zone; NZ = no-training zone; MZ = maximal training zone.

TABLE 10.3

Modified TRIMP Values for Each Training Intensity Zone

Intensity zone	TRIMPS per hour including rest*
ENERGY FITNESS	
EZ	1.5
PZ	6
MZ	9
MUSCULAR FITNESS	
Strength	5
Power	7
Power endurance	6
Plyometrics	8

These values are averages and can be used for planning training to ensure daily and weekly periodization.

*TRIMPS per hour include rest periods for PZ and MZ intervals and for muscular fitness training.

The TRIMP values are for total time spent in an intensity zone, including the rest between PZ and MZ intervals. Thus, 1 hour of PZ intervals might include 30 minutes of 3-minute intervals plus 30 minutes of recovery. The total workout would be 5 to 7 TRIMPS. An EZ distance workout at 2.5 TRIMPS per hour would need to last about 2-1/2 hours to provide an equal training stress. The TRIMP values in table 10.3 are general estimates. You will learn to adjust the values depending on the actual intensity and the activity. The values that we present are basic guidelines to help you to balance the weekly stress. When periodizing training, a 25 to 50 percent TRIMP reduction is adequate to allow recovery, while increases of 25 to 50 percent will be very fatiguing.

The use of TRIMPS was developed for energy fitness training. Although no similar scale has been scientifically developed for muscular fitness training, we have given TRIMP estimates in table 10.3. In our experience, each hour of resistance training is equal to 5 to 7 TRIMPS.

If you wish to use TRIMPS in the development of your training program, see further information in appendix E.

time for recovery pay for it with poor performance, illness, and injuries.

Weekly periodization should be programmed into all of the training periods. On your calendar, divide each period into weeks. The first week of each period can be designated as medium, with the following week designated as hard, then easy (recovery). Continue this cycle through each training period. The peak performance period requires a more structured periodization to ensure that major competitions follow easier weeks of training. If you know the dates of competitions in both the early competition and peak performance periods, mark them on your calendar. Classify the weeks before important competitions as easy weeks and allow two weeks for peaking before the most important competition.

Note that not all competitions can be considered important, especially during the early competition period. Some less important competitions may follow medium- or even high-stress training weeks. Figure 10.3 demonstrates the weekly periodization of sample precompetition and peak performance periods. The normal cycle of medium–hard–easy is not followed consistently during the peak performance period so that athletes are able to be rested for the more important competitions.

Finally, make a list of the training goals for each period. What specific training do you need for muscular and energy fitness? (Hint: See table 10.3.) Sample programs are shown in the next chapter. The figures on pages 188 and 189 have a form and instructions to guide you through your season plan. Use of the seasonal planning forms (figures 10.4 and 10.5) will help you prepare to write your daily training plan.

The Weekly Plan

By now, you have defined your training season and have set dates for each period. Within each period, you will have assigned weeks as medium, high, and low stress. You are now about ready to write a daily schedule for a few weeks. Don't do this too far in advance, because you will end up changing your plan when unforeseen circumstances arise. When you write plans for each week, keep in mind the following ideas:

- Vary the stress of the training day.
- High intensity requires more rest.
- Move from general to specific.

We'll discuss each of these in more detail in the following sections.

Vary the Stress of Training Days

Within each training week, be sure to vary the physical stress of the training days. This weekly periodization will keep your athletes healthy and improve training on high-intensity days. The training stress needs to be appropriate for your athletes' maturity. The best gauge will be how the athletes respond to your plan. At the end of hard weeks they should be tired but should recover easily during the following easy week. Vary the overall stress of the training days so that you seldom have more than two high-stress

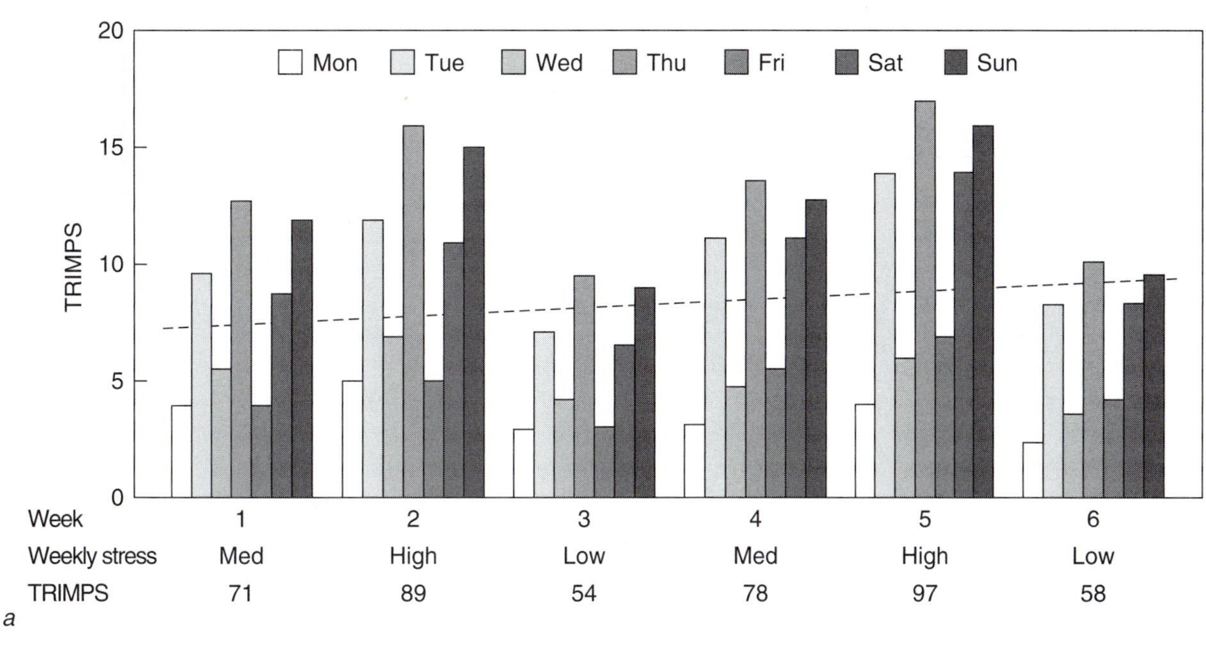

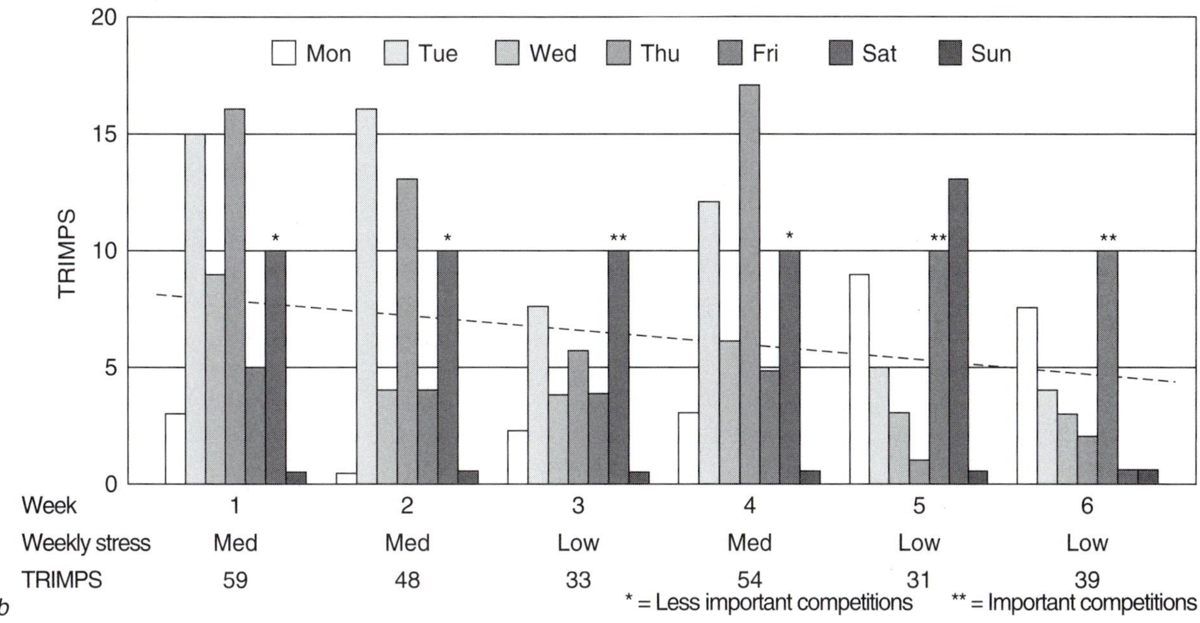

FIGURE 10.3 Daily periodization. Theoretical sample weeks are shown for *(a)* precompetition and *(b)* high-performance periods. Each day of the week is shown for two three-week cycles of medium-, high-, and low-stress training. The training impulse (TRIMP) values are a measure of total stress (see sidebar on p. 185). The dotted line shows the average progression of training stress over the six-week period. Note variation in the daily stress. During the high-performance period, each high-stress day is followed by an easy day. Less important competitions are noted by a single asterisk (*) and important competitions by double asterisks (**).

days in a row. Figure 10.6 shows sample weeks for a high school soccer season during three of the training periods. During the precompetition and competition periods, it is better to follow every hard day with an easier day. Remember; Overtraining is much more dangerous and damaging than undertraining.

Stress comes in many forms for athletes. When they are in the middle of finals, the mental stress and lack of sleep may require a reduction in the training load. Coaches also forget that endurance days, even if the pace is slow, may be stressful if the duration is long. Pay attention to your athletes. If athletes are

1. Athletes' ages:_________ Major age group focus: _________________ ___________

2. Team goals:

3. Official season dates: Start training:_____________ First competition: _______________

   Start of major competitions:_____________ Final competition: ______________

4. Define start and end dates for each training period (see p. 181). Mark these dates on a calendar. Next, fill in the dates on the season plan form (figure 10.5) at the top of each training period. In the fourth row from the bottom of the form, fill in the date of each Monday and complete the next row designating the training periods.

5. On the season plan form in the second row from the bottom, designate the weekly stress. Start with the first week and then alternate each week as medium, hard, or easy.

6. Set athlete assessment dates: ______________ to ______________ (mark on calendar). List tests:

7. Using copies of figure 10.7 (Weekly Planning Form), develop weekly plans for the first moderate week of each training period. Sample plans for five sports are shown in the next chapter. The following are steps for using the weekly planning sheets:

   a. From the season plan form, look up the weekly stress planned for the week (easy, medium, or hard) and enter that in the top left-hand box of the form.

   b. Designate each day as easy, medium, or hard. Vary the days and allow for recovery.

   c. Write the daily plan to meet the goals for the training period you defined on the season planning form by filling in the appropriate boxes.

   d. Adjust the weekly program as needed to meet your planned weekly stress level from the season planning sheet. The sample programs in chapter 11 should serve as a template.

At the beginning of each training period, you should complete the weekly plans for the remainder of the period. Near the end of each period, begin to gradually adjust the training to transition to the next period. If you wish to use TRIMP guidelines to help you in this process, age-based guides are in tables E.1 and E.2 in appendix E.

FIGURE 10.4 Season planning steps

Coach_______________ Year______ Sport_______________

Recovery Period: Dates_______________ – _______________ # weeks_____ (Minimum 2 weeks)	
Major Muscular-Fitness Goals of Period	Major Energy-Fitness Goals of Period
Basic Training Period: Dates_______________ – _______________ # weeks_____ (Minimum 9 weeks)	
Major Muscular-Fitness Goals of Period	Major Energy-Fitness Goals of Period
Precompetition Period: Dates_______________ – _______________ # weeks_____ (Minimum 6 weeks)	
Major Muscular-Fitness Goals of Period	Major Energy-Fitness Goals of Period
Early Competition Period: Dates_______________ – _______________ # weeks_____ (Minimum 3 weeks)	
Major Muscular-Fitness Goals of Period	Major Energy-Fitness Goals of Period
Peak Performance Period: Dates_______________ – _______________ # weeks_____ (3-6 weeks)	
Major Muscular-Fitness Goals of Period	Major Energy-Fitness Goals of Period

Week Starting With Basic Strength and Endurance Period	Week 1	Week 2	Week 3	Week 4	Week 5	Week 6	Week 7	Week 8	Week 9	Week 10	Week 11	Week 12	Week 13	Week 14	Week 15	Week 16	Week 17	Week 18	Week 19	Week 20	Week 21	Week 22	Week 23	Week 24	Week 25
Start Date of Week (Monday)																									
Training Season* R, B, P, E, PK																									
Weekly Stress**: E, M, H																									
Guideline TRIMPS (From Table F-1)																									

*R = recovery period, B = basic training period, P = precompetition period, E = early competition period, PK = peak performance period

**E = easy week (recovery), M = medium-stress week, H = high-stress week

FIGURE 10.5 Season planning form

From *Sport Physiology for Coaches* by Brian J. Sharkey and Steven E. Gaskill, 2006, Champaign, IL: Human Kinetics.

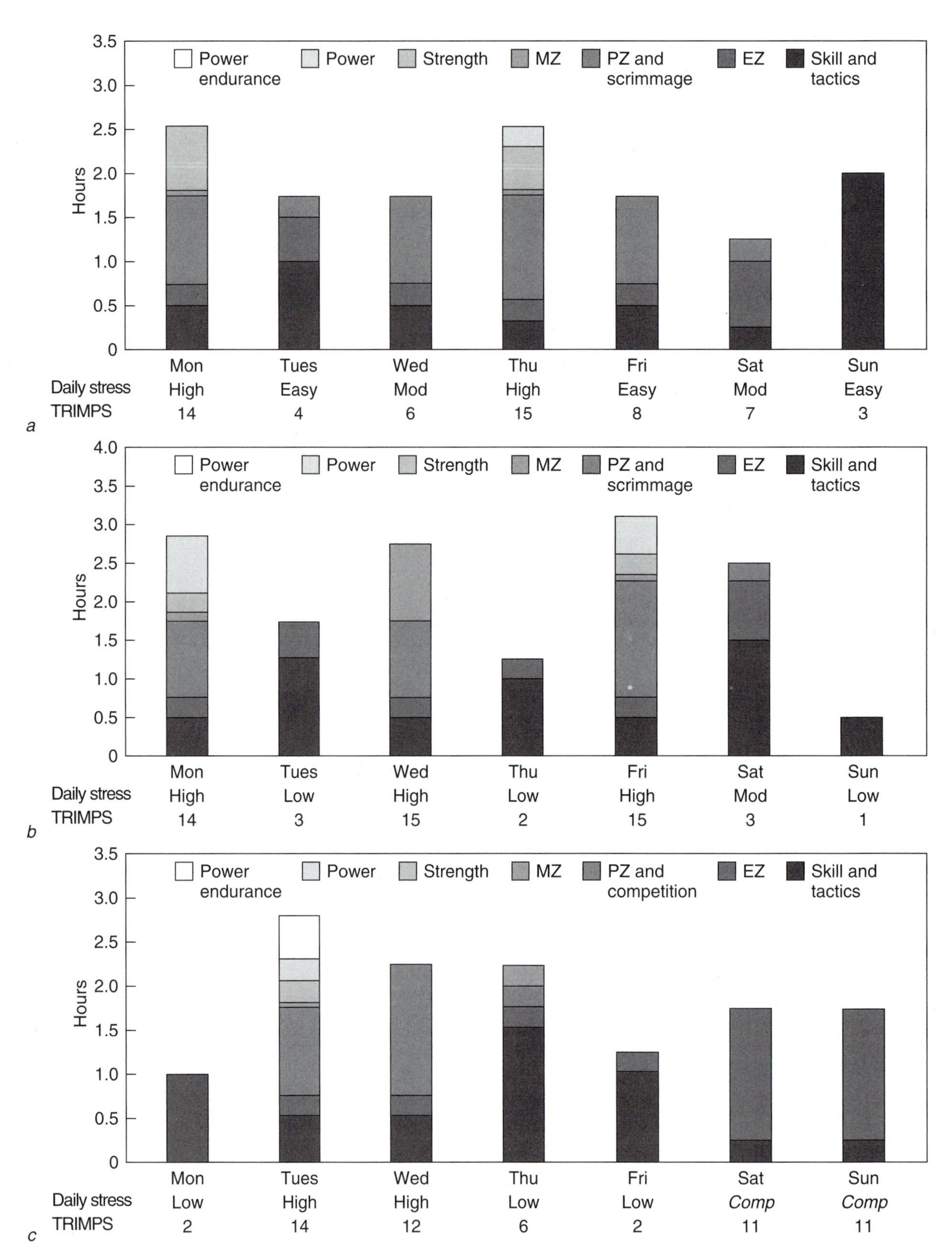

FIGURE 10.6 Sample training weeks. Theoretical hours of training for categories of muscular and energy fitness are shown for an average week from three periods of the training season for a high school soccer team: *(a)* basic training period, *(b)* precompetition period, *(c)* early competition period. The relative daily stress is shown as training impulse values (TRIMPS).

more tired than you anticipated, adjust your practice for the day.

Easy days do not mean that you don't do a full practice or that you need to spend less time; they simply mean that you need to use activities that will allow your athletes to recover. These are great days to do technique and skill work, watch game films, practice psychological skills, work on tactics, or even do game simulations with plenty of recovery time.

High Intensity Requires More Rest

The closer you are to competition, the greater the need for high-intensity training. This added stress requires added rest. Don't try to get your athletes into shape coming into the major competitions. That work needs to be done much earlier. In the few weeks leading up to competitions, focus on sport-specific tasks and fine-tuning. The high-intensity training that you do should build on the foundations laid in the months earlier. Athletes who have not laid an appropriate foundation are not ready for higher-intensity training and may become overtrained or injured. Lower-fit athletes should be encouraged to continue basic fitness training through the competitions to lay the foundation for future years. The fatigue index (p. 237) is a great tool to use to ensure that your athletes are getting adequate rest.

Move From General to Specific

Apply the basic concept of starting with general training and moving toward sport-specific training. During periods of transition from one period to the next, include multiple types of training within each week. When moving from strength to power training, it is appropriate to include both, just as it is common for aerobic fitness to require days of long-duration training mixed with days of speed work.

At the beginning of each training period, you should plan sample weeks. If a period is nine weeks, you will have three cycles of medium, hard, and easy weeks. Try to plan each of the three medium weeks before the period starts. Based on your needs analysis, team and individual goals, and the work that you have done as described in the previous sections, you should have the training needs well defined. Each three-week cycle should flow logically from the precious cycle. At the start of a period, transition the training gradually. During the training period, increase stress slowly. As the end of a period approaches, begin to incorporate transitional training, getting ready for the next period.

After planning the medium week for each three-week cycle, it is easy to increase the workload over the medium week for the second week (25-50 percent) and decrease it for the third week (25-50 percent). The training plan that you develop should be designed for your team in general. You will need to work with each athlete to help each one individualize his or her training within the overall plan. Figure 10.7 is a weekly planning form (version B includes room to use TRIMPS). Sample weekly planning sheets for each training period are provided for five sports in chapter 11.

Monitor Your Athletes

The most important coaching concern is to maintain the health of your athletes. Maintenance of health and optimal performance can best be accomplished with a properly periodized program incorporating variation in stress, intensity, duration, and activities. Pay attention to athletes who appear tired. Chapter 12 contains more information on monitoring athletes. Encourage your athletes to use the fatigue index test daily (chapter 12, p. 237).

Following the outline at the start of this chapter will help your athletes thrive, maintain their health, and improve performance. Once again:

1. Set individual and team goals.
2. Analyze needs.
3. Periodize the training program.
4. Monitor health and progress.

In chapter 11 we provide sample programs to help you get started.

Week rating (circle) Easy Med Hard		Monday	Tuesday	Wednesday	Thursday	Friday	Saturday	Sunday
		Day rating (circle) Easy Med Hard	Day rating (circle) Easy Med Hard	Day rating (circle) Easy Med Hard	Day rating (circle) Easy Med Hard	Day rating (circle) Easy Med Hard	Day rating (circle) Easy Med Hard	Day rating (circle) Easy Med Hard
MUSCULAR FITNESS	Strength							
	Power							
	Power endurance							
	Plyometrics and other muscular fitness							
ENERGY FITNESS	EZ							
	PZ							
	MZ							
Skill, tactics, and other training								

FIGURE 10.7a Weekly planning form

From *Sport Physiology for Coaches* by Brian J. Sharkey and Steven E. Gaskill, 2006, Champaign, IL: Human Kinetics.

Season of training:_____________ Season dates:______-______ Week:_____-______

	Suggested weekly TRIMPS	Weekly rating Easy Med Hard	Monday		Tuesday		Wednesday		Thursday		Friday		Saturday		Sunday	
			Suggested TRIMPS	Daily rating Easy Med Hard	Suggested TRIMPS	Daily rating Easy Med Hard	Suggested TRIMPS	Daily rating Easy Med Hard	Suggested TRIMPS	Daily rating Easy Med Hard	Suggested TRIMPS	Daily rating Easy Med Hard	Suggested TRIMPS	Daily rating Easy Med Hard	Suggested TRIMPS	Daily rating Easy Med Hard
MUSCULAR FITNESS	Strength 4-6 TRIMPS/hr		Hours: : TRIMPS:		Hours: : TRIMPS:		Hours: : TRIMPS:		Hours: : TRIMPS:		Hours: : TRIMPS:		Hours: : TRIMPS:		Hours: : TRIMPS:	
	Power 6-8 TRIMPS/hr		Hours: : TRIMPS:		Hours: : TRIMPS:		Hours: : TRIMPS:		Hours: : TRIMPS:		Hours: : TRIMPS:		Hours: : TRIMPS:		Hours: : TRIMPS:	
	Power endurance 5-7 TRIMPS/hr		Hours: : TRIMPS:		Hours: : TRIMPS:		Hours: : TRIMPS:		Hours: : TRIMPS:		Hours: : TRIMPS:		Hours: : TRIMPS:		Hours: : TRIMPS:	
	Plyometrics and other 4-7 TRIMPS/hr		Hours: : TRIMPS:		Hours: : TRIMPS:		Hours: : TRIMPS:		Hours: : TRIMPS:		Hours: : TRIMPS:		Hours: : TRIMPS:		Hours: : TRIMPS:	
ENERGY FITNESS	EZ 1.5-3 TRIMPS/hr		Hours: : TRIMPS:		Hours: : TRIMPS:		Hours: : TRIMPS:		Hours: : TRIMPS:		Hours: : TRIMPS:		Hours: : TRIMPS:		Hours: : TRIMPS:	
	PZ 5-7 TRIMPS/hr		Hours: : TRIMPS:		Hours: : TRIMPS:		Hours: : TRIMPS:		Hours: : TRIMPS:		Hours: : TRIMPS:		Hours: : TRIMPS:		Hours: : TRIMPS:	
	MZ 7-9 TRIMPS/hr		Hours: : TRIMPS:		Hours: : TRIMPS:		Hours: : TRIMPS:		Hours: : TRIMPS:		Hours: : TRIMPS:		Hours: : TRIMPS:		Hours: : TRIMPS:	
Skills, psychological, tactics, and other training			Hours: : TRIMPS:		Hours: : TRIMPS:		Hours: : TRIMPS:		Hours: : TRIMPS:		Hours: : TRIMPS:		Hours: : TRIMPS:		Hours: : TRIMPS:	
Daily hours																
Daily TRIMPS																

FIGURE 10.7b Weekly planning form

From *Sport Physiology for Coaches* by Brian J. Sharkey and Steven E. Gaskill, 2006, Champaign, IL: Human Kinetics.

SUMMARY

This chapter presented a four-step plan for developing successful training programs. The plan includes setting goals, completing a needs analysis, periodizing the plan, and monitoring the progress and health of the athletes. The periodizing of a training plan requires that coaches define the periods of training including recovery, basic training, precompetition, early competition, and peak performance.

- The recovery period is generally a time of recovery from the previous competitive season and includes nonspecific, nonstructured activities done at a low intensity.
- The focus of the basic training period is to build a strength and energy fitness foundation to support the higher-intensity work that will be done as the competition season approaches. High-intensity training is minimal during the basic training period.
- The precompetition period is the transition from basic training to competition. It lasts 6 to 8 weeks for school sports and 12 to 14 weeks for year-round programs, extending up to the date of the first competition.
- The period of early-season competitions generally lasts 4 to 5 weeks for school sports and as long as 8 to 10 weeks for yearlong programs. It may include scrimmages and early competitions that are used to prepare for important league and championship contests. Generally, overall training volume decreases while sport-specific speed and power endurance increase.
- The peak performance period includes the important competitions of the season, consisting of league play, championship series, or season-culminating events. Generally, the peak performance period lasts three to six weeks including one to two weeks of taper (reduced training volume) prior to the most important competitions. Sleep, recovery, and reduction of outside stress all help to improve performance.

Within each period of training, the weeks are periodized such that an athlete may be given progressively more stress for three weeks, then be given an easier week to begin a new cycle ending at a higher fitness level. Several methods, including training impulses (TRIMPS), are proposed as methods to monitor the training load. Within each week, the training load and emphasis of each day should be varied to allow for specific overload, then recovery. Overall, a training plan can be said to move from general training to training that is more specific. Finally, a set of forms is provided to help the coach plan training for his or her team or for individuals.

KEY TERMS

basic training period (p. 181)
early competition period (p. 183)
official training season (p. 180)
peak performance period (p. 183)
precompetition period (p. 182)
recovery training period (p. 181)
training impulses (TRIMPS) (p. 184)

REVIEW QUESTIONS

Fill in the blank(s) with the appropriate word or words to complete the following statements.

1. ________ , ________ , ________ , and ________ are the four steps a coach zshould use to develop a successful training plan.
2. The periods of training include ________ , ________ , ________ , ________ , and ________.
3. The concept of ________, which helps to control weekly training stress, and the ________, which is a simple morning evaluation whereby the sum of resting, exercise, and recovery heart rate is used to monitor fatigue, are two methods used to determine the stress of a training program.

Determine if the following statements are true or false and then circle the correct answer.

4. T/F Maintenance of health and optimal performance can best be accomplished with a properly periodized program incorporating variation in stress, intensity, duration, and activities.
5. T/F Strength is the focus of muscular fitness training during the high-performance period of training.
6. T/F The TRIMPS method was developed for energy fitness training.
7. T/F Overtraining is more dangerous and damaging than undertraining.

Match the periods of training in the left column with the correct general muscular and energy fitness guidelines in the right column.

8. Recovery
9. Basic training
10. Precompetition
11. Early competition
12. High performance

a. Main focus is on power endurance and maintenance of strength and power. There is a reduction in total energy fitness training with a further reduction in EZ training and a focus on PZ intervals.
b. Training load is gradually reduced and recovery time increased as the competitions become the main focus. There is a maintenance of power, power endurance, and speed along with PZ and MZ intervals.
c. Non-sport-specific, low-intensity activities. Focus is on physical and mental recovery.
d. Beginning of planned training with a focus on strength training and skill development; building an aerobic base with a focus on EZ training.
e. Main focus is power training with a transition into power endurance. Starts with a reduction in EZ training and an increase in PZ intervals with small MZ intervals. Skill and tactics are developed.

PRACTICAL ACTIVITIES

Use the guidelines and forms presented in this chapter to design a training program for your sport or for an individual in your sport. (No answers are given in the appendix; instead, compare your training plan with a plan for a similar sport provided in chapter 11.)

11 Sample Training Programs

This chapter presents sample programs for

- a skill sport (high school golf),
- a power sport (high school football—interior lineman),
- a power endurance sport (junior wrestling),
- an intermittent sport (high school soccer), and
- an aerobic endurance sport (high school cross country running).

Lynn was one of those all-around athletes who used competitions as an excuse to train. What she really loved was the long runs, pushing her body during intervals, getting up early for a bike ride with the triathlon team. Her friends were constantly amazed at her training load and her ability to push hard every day. Not surprisingly, Lynn was pretty good at a lot of endurance sports. She competed in running, cycling, swimming, rowing, flat water canoeing, and cross-country skiing. She was nearly always the first finisher during practice but was consistently in the middle of the pack during competitions.

During one particular awards ceremony she watched as her friends received their awards and asked herself why she never won when it counted. Following the awards, a former great athlete talked about his development and the day that he finally learned to train to perform rather than training randomly. He recited the quote "Don't play sports to get in shape, get in shape to play sports." The message sank in for Lynn. She began to read about effective training, attended seminars, and began to ask for advice. Working with a knowledgeable coach, she developed a set of goals and then designed a training plan to help her achieve those goals. It wasn't long before Lynn started to put together winning performances. As with Lynn, learning the importance of planning for training and training to perform is the first step in athletic success.

This chapter provides sample training programs for representative sports. We have chosen to provide examples in each of five sport categories: skill, power, power endurance, intermittent, and aerobic. Each sample program includes an overall plan followed by specific examples for each training period (basic, precompetitive, early competition, and peak performance). Find the category that best fits your sport, then use the sample program to help build a plan specific to your sport. **Skill sports** include golf, archery, and bowling. **Power sports** include football (linemen), weightlifting, and weight events in track and field. **Power endurance sports** include wrestling, downhill (Alpine) skiing, and middle-distance running. **Intermittent (start-and-go) sports** include soccer, basketball, field hockey, and lacrosse. **Aerobic (endurance) sports** include distance running, swimming, cycling, and cross-country skiing. Of course, some sports like tennis don't fit easily into one category. Tennis requires skill, power endurance, and aerobic endurance. Some positions have requirements that are different from those of other players on the team (e.g., goalkeepers). In these cases you may have to draw ideas from two or more programs.

The sample programs include recommended stress based on training impulse (TRIMP) guidelines, discussed in chapter 10. More information about TRIMPS is included in appendix E. You may choose to design your training plans without using TRIMPS. Copies of the weekly planning forms with or without the TRIMP information can be found in chapter 10 (pages 192 and 193) and may be copied as needed.

Skill Sport: Golf

While all sports require some skill, a few can be classified as requiring high levels of skill, with only modest demands for muscular or energy fitness. However, we now recognize that athletes in skill sports improve dramatically with proper physical training. Sports in this category include golf, table tennis, shooting sports (archery and rifle), and bowling. Though a considerable amount of time must be devoted to skill development, each sport has specific muscular and energy requirements. High school, collegiate, and even professional golfers have learned that attention to training translates into improvements in performance. Training allows more effective practice time and improves distance and consistency while reducing the risk of overuse injuries common to the sport. See pages 202 through 206 for sample golf training programs.

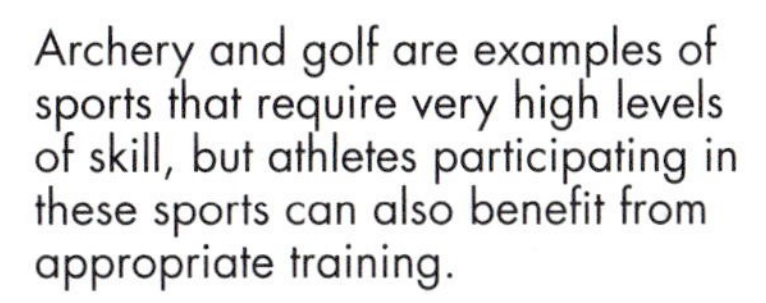

Archery and golf are examples of sports that require very high levels of skill, but athletes participating in these sports can also benefit from appropriate training.

Power Sport: Football Lineman

Football linemen exemplify the athletic needs of a power sport and require a high level of muscular strength and power, especially in the muscle groups involved in blocking and tackling. These total body movements require a strong core, as well as neck, arms, shoulders, and legs. Blocking relies on extension strength from the triceps muscle group in the upper arms, while tackling demands flexion strength from the biceps muscle group. To apply the principle of specificity, training should be individualized for offensive and defensive players.

The most important change in football training has been the adoption of power training, the application of force at high speeds. This sport-specific training contributes to the ability to fire off the ball and drive the opponent backward. Power training has been enhanced by the development of power training apparatus.

Surprisingly, the increase in the size of linemen has increased the need for energy fitness training, aerobic and anaerobic. Football consists of a series of anaerobic bouts, each lasting about 4 to 6 seconds, followed by recovery periods lasting about 25 seconds. Maximal anaerobic power can quickly lead to exhaustion if aerobic power is limited. Improving aerobic power enhances the ability to recover between plays, as well as improving the ability to perform late in the game. See pages 207 through 211 for sample lineman training programs.

Power Endurance Sport: Wrestling

Power endurance is the ability to do repetitions of forceful contractions as quickly as possible or to sustain a contraction. It defines the application of endurance to the world of competitive sport, where force and speed occur together. Some sports involve repetitions with a fairly heavy load and require short-term endurance. Wrestlers, for example, have to contend with an opponent's body weight while attempting to move as quickly as possible. Because the competition lasts only a few minutes, wrestlers need short- to medium-term power endurance. Other

Resistance exercises such as leg press done at high speed can be used to develop power.

endurance sports such as cycling and running involve numerous repetitions with a light load; they require long-term muscular endurance and well-developed aerobic capabilities.

Wrestlers must be prepared to sustain a total effort for several minutes; therefore muscular fitness, especially power endurance, is critical when one is training wrestlers. Since the sport involves the entire body, a systematic program must be designed to exercise upper and lower body and trunk (core) muscles used for controlling, lifting, and throwing an opponent. A fair amount of sport-specific power endurance training takes place on the mat, as the athlete repeats moves while working against the resistance provided by a teammate.

The relatively short duration of wrestling bouts places major demands on anaerobic energy pathways. However, wrestlers need adequate aerobic ability to assist recovery between periods and to sustain energy throughout day-long tournaments. See pages 212 through 216 for sample training programs for wrestling. Use the basic training and precompetition periods to build a strong aerobic base.

Intermittent Sport: Soccer

Intermittent or start-and-go sports, like soccer, basketball, field hockey, and lacrosse, consist of bursts of energy interspersed with periods of less demanding effort. Soccer athletes can range up and down the field, sprint to the ball, then slow down or even stop for moments. In basketball, the pace is frenzied, but the lulls come more often, with defense, foul shots, and ever more frequent time-outs. The games have become so competitive that muscular and energy fitness training are essential to the preparation of athletes for peak performance.

Muscular fitness is required to maximize power and speed and to withstand contact with opponents. Intermittent sport athletes need a strong aerobic foundation to enable them to maintain continuous motion and to recover quickly from more intense bursts of activity. Anaerobic energy is essential to perform a series of sprints. Energy fitness helps sustain the quickness needed to beat an opponent to the ball and to maintain skills late in the contest. See pages 217 through 221 for sample soccer training programs.

Skills must be practiced at game speed if they are to hold up in competition. A knowledgeable soccer coach used high-speed drills to help his players increase the pace of play. Eventually, after they became comfortable performing at a faster pace, they went on to win the National Collegiate Athletic Association soccer title.

Aerobic (Endurance) Sport: Cross Country Running

Cross country running, a sport requiring exceptional aerobic power, involves many hours of energy training to achieve success and is a good example of an aerobic (endurance) sport. Events lasting more than 15 minutes derive over 80 percent of the energy from aerobic energy pathways, so training should emphasize race-pace training in the performance zone. Some anaerobic

Wrestling requires both power and endurance.

or maximal zone work should be done to build leg speed and the ability to kick at the end of a race. And a considerable amount of time should be spent in the easy training zone to develop endurance and metabolic efficiency, as well as to allow recovery from more intense training.

Runners also need sport-specific muscular fitness training in order to improve leg power and to minimize the risk of overuse injuries. Many coaches are convinced that upper body and core training are important to the maintenance of form later in a race. The goal here is not to build strength that adds weight, but to improve power endurance in sport-specific drills. Cross country runners use plyometrics to gain an edge in hill running and during the kick to the finish line.

Since running is a skill that can be improved with coaching and practice, emphasize good form, even during runs in the easy training zone. Studies show that efficient runners can outperform less efficient opponents even when the opponent has an edge in aerobic power. Help athletes minimize extraneous movements, such as excessive arm swing, except in the final portion of the race. Help runners refine their natural stride, the one that feels most efficient. Use the sample training programs in pages 222 through 226.

Sample plan for high school golf.

Recovery Period: Dates Dec. 15 – Dec. 28 # weeks 2 (Minimum 2 weeks)

Major Muscular-Fitness Goals of Period	Major Energy-Fitness Goals of Period
Flexibility Possibly begin a preparatory weight program with light resistance Athletes on own unofficial training	General activities, especially those with components of aerobic energy fitness Light jogging

Basic Training Period: Dates Dec. 29 – Feb. 28 # weeks 9 (Minimum 9 weeks)

Major Muscular-Fitness Goals of Period	Major Energy-Fitness Goals of Period
Major emphasis: Upper body strength (4-8RM) Core strength Maintenance of power (8-12RM, fast movements) Maintenance of power endurance, 20-30RM	EZ-intensity runs, 45 min, 2 or 3 times per week

Precompetition Period: Dates Mar. 1 – Apr. 18 # weeks 6 (Minimum 6 weeks)

Major Muscular-Fitness Goals of Period	Major Energy-Fitness Goals of Period
Major emphasis: Upper body power (8 to 12RM) Maintenance of core strength Maintenance of strength (6RM) Maintenance of power endurance (20-30RM) Flexibility	EZ-intensity runs, 45 min, 1 or 2 times per week PZ intervals (relaxed, RPE 15), 1 time per week Psychological skills training

Early Competition Period: Dates May 10 – May 29 # weeks 3 (Minimum 3 weeks)

Major Muscular-Fitness Goals of Period	Major Energy-Fitness Goals of Period
Major emphasis: Upper body power endurance (20-30RM) Improve core strength Maintenance of power (8-12RM) Flexibility	Maintain aerobic fitness with EZ run 1 or 2 times per week PZ intervals (relaxed, RPE 15), 1 or 2 times per week Psychological skills training

Peak Performance Period: Dates May 10 – May 29 # weeks 3 (3-6 weeks)

Major Muscular-Fitness Goals of Period	Major Energy-Fitness Goals of Period
Maintenance of power and power endurance with 1 set of each per week Maintenance of core strength with 1 session per week Flexibility Focus on rest and recovery	Maintain aerobic fitness with walking during golf play PZ intervals (relaxed, RPE 14), 1 time per week (note reduced intensity) Psychological skills training Focus on rest and recovery

Week starting with basic strength and endurance period	Week 1	Week 2	Week 3	Week 4	Week 5	Week 6	Week 7	Week 8	Week 9	Week 10	Week 11	Week 12	Week 13	Week 14	Week 15	Week 16	Week 17	Week 18	Week 19	Week 20	Week 21	Week 22	Week 23	Week 24	Week 25
Start date of week (Monday)	12-29	1-5	1-12	1-19	1-26	2-2	2-9	2-16	2-23	3-1	3-8	3-15	3-22	3-29	4-5	4-12	4-19	4-26	5-3	5-10	-517	5-24			
Training season* R, B, P, E, PK	B	→	→	→	→	→	→	→	→	P	→	→	→	→	→	E	→	→	→	PK	→	→			
U = unofficial season O = official season	U	→	→	→	→	→	→	→	→	O	→	→	→	→	→	→	→	→	→	→	→	→	→		
Weekly stress**: E, M, H	M	H	E	M	H	E	M	H	E	M	H	E	M	H	E	M	H	E	H	E	M	E			
Guideline TRIMPS (from table 10.3)	47	63	31	49	66	32	52	69	34	70	94	46	63	85	42	56	37	37	76	34	53	31			

*R = recovery period, B = basic training period, P = precompetition period, E = early competition period, PK = peak performance period

**E = easy week (recovery), M = medium-stress week, H = high-stress week

From *Sport Physiology for Coaches* by Brian J. Sharkey and Steven E. Gaskill, 2006, Champaign, IL: Human Kinetics.

Sample moderate week for basic training period in golf.

Suggested weekly TRIMPS 47	Weekly rating Easy Med Hard	Monday Suggested TRIMPS 10	Monday Daily rating Easy Med Hard	Tuesday Suggested TRIMPS 4	Tuesday Daily rating Easy Med Hard	Wednesday Suggested TRIMPS 10	Wednesday Daily rating Easy Med Hard	Thursday Suggested TRIMPS 4	Thursday Daily rating Easy Med Hard	Friday Suggested TRIMPS 10	Friday Daily rating Easy Med Hard	Saturday Suggested TRIMPS 4	Saturday Daily rating Easy Med Hard	Sunday Suggested TRIMPS 7	Sunday Daily rating Easy Med Hard
MUSCULAR FITNESS	Strength 4-6 TRIMPS/hr	Hours: 1:00 TRIMPS: 5.0	1 × 4 1 × 6 1 × 8 Mainly arms	Hours: : TRIMPS:		Hours: 1:00 TRIMPS: 5.0	1 × 4 1 × 6 1 × 8 Mainly arms	Hours: : TRIMPS:		Hours: :40 TRIMPS: 3.5	1 × 4 1 × 6 Mainly arms	Hours: : TRIMPS:		Hours: : TRIMPS:	
	Power 6-8 TRIMPS/hr	Hours: :20 TRIMPS: 2.3	1 × 10 Fast Mainly arms	Hours: : TRIMPS:		Hours: :20 TRIMPS: 2.3	1 × 10 Fast Mainly arms	Hours: : TRIMPS:		Hours: :40 TRIMPS: 5.0	1 × 8 1 × 12 Fast Mainly arms	Hours: : TRIMPS:		Hours: : TRIMPS:	
	Power endurance 5-7 TRIMPS/hr	Hours: : TRIMPS:		Hours: : TRIMPS:		Hours: : TRIMPS:		Hours: : TRIMPS:		Hours: :20 TRIMPS: 2.0	1 × 25 Arms only	Hours: : TRIMPS:		Hours: : TRIMPS:	
	Plyometrics and other 4-7 TRIMPS/hr	Hours: :15 TRIMPS: 1.5	Golf-specific bungee exercise	Hours: :30 TRIMPS: 3.5	Core strength training	Hours: :15 TRIMPS: 1.5	Golf-specific bungee exercise	Hours: :30 TRIMPS: 3.5	Core strength training	Hours: : TRIMPS:		Hours: : TRIMPS:		Hours: :30 TRIMPS: 3.5	Core strength training
ENERGY FITNESS	EZ 1.5-3 TRIMPS/hr	Hours: : TRIMPS:		Hours: :45 TRIMPS: 2.0	Jogging RPE 10-11	Hours: : TRIMPS:		Hours: :45 TRIMPS: 2.0	Jogging RPE 10-11	Hours: : TRIMPS:		Hours: 2:00 TRIMPS: 3.0	Golf	Hours: 2:45 TRIMPS: 4.0	Golf 2 hr Jog easy 45 min
	PZ 5-7 TRIMPS/hr	Hours: : TRIMPS:		Hours: : TRIMPS:		Hours: : TRIMPS:		Hours: : TRIMPS:		Hours: : TRIMPS:		Hours: : TRIMPS:		Hours: : TRIMPS:	
	MZ 7-9 TRIMPS/hr	Hours: : TRIMPS:		Hours: : TRIMPS:		Hours: : TRIMPS:		Hours: : TRIMPS:		Hours: : TRIMPS:		Hours: : TRIMPS:		Hours: : TRIMPS:	
Skills, psychological, tactics, and other training		Hours: : TRIMPS:		Hours: : TRIMPS:		Hours: : TRIMPS:		Hours: : TRIMPS:		Hours: : TRIMPS:		Hours: : TRIMPS:		Hours: : TRIMPS:	
Daily hours		1:35		2:15		1:35		1:15		1:40		2:00		3:15	
Daily TRIMPS		8.8		5.0		8.8		5.0		10.5		3.0		7.5	

Note: Resistance training should be done separately from energy fitness.

Total hours for week 12:35 Total TRIMPS for week 48.6

Sample moderate week for precompetition period in high school golf.

Suggested weekly TRIMPS 70	Weekly rating Easy Med Hard	Monday Suggested TRIMPS 15	Monday Daily rating Easy Med Hard	Tuesday Suggested TRIMPS 5	Tuesday Daily rating Easy Med Hard	Wednesday Suggested TRIMPS 10	Wednesday Daily rating Easy Med Hard	Thursday Suggested TRIMPS 5	Thursday Daily rating Easy Med Hard	Friday Suggested TRIMPS 15	Friday Daily rating Easy Med Hard	Saturday Suggested TRIMPS 5	Saturday Daily rating Easy Med Hard	Sunday Suggested TRIMPS 5	Sunday Daily rating Easy Med Hard
MUSCULAR FITNESS	Strength 4-6 TRIMPS/hr	Hours: :20 TRIMPS: 1.7	1 × 6 Mainly arms	Hours: : TRIMPS:		Hours: 2:00 TRIMPS: 1.7	1 × 5 Mainly arms	Hours: : TRIMPS:		Hours: : TRIMPS:		Hours: : TRIMPS:		Hours: : TRIMPS:	
MUSCULAR FITNESS	Power 6-8 TRIMPS/hr	Hours: :40 TRIMPS: 5.0	1 × 10 1 × 12 Fast Mainly arms	Hours: : TRIMPS:		Hours: :40 TRIMPS: 5.1	1 × 8 1 × 8 Fast Mainly arms	Hours: : TRIMPS:		Hours: 1:00 TRIMPS: 7.0	1 × 8 1 × 10 1 × 12 Fast	Hours: : TRIMPS:		Hours: : TRIMPS:	
MUSCULAR FITNESS	Power endurance 5-7 TRIMPS/hr	Hours: : TRIMPS:		Hours: : TRIMPS:		Hours: : TRIMPS:		Hours: : TRIMPS:		Hours: :20 TRIMPS: 2.0	1 × 25 Arms only	Hours: : TRIMPS:		Hours: : TRIMPS:	
MUSCULAR FITNESS	Plyometrics and other 4-7 TRIMPS/hr	Hours: : TRIMPS:		Hours: :30 TRIMPS: 2.5	Core strength training	Hours: : TRIMPS:		Hours: :30 TRIMPS: 2.5	Core strength training	Hours: : TRIMPS:		Hours: : TRIMPS:		Hours: : TRIMPS:	
ENERGY FITNESS	EZ 1.5-3 TRIMPS/hr	Hours: 1:30 TRIMPS: 3.0	Driving range	Hours: 2:00 TRIMPS: 4.0	Golf Walk briskly	Hours: 2:00 TRIMPS: 4.00	Golf Driving range	Hours: 1:30 TRIMPS: 3.0	Golf Walk briskly	Hours: 1:30 TRIMPS: 3.00	Golf Walk briskly	Hours: 3:00 TRIMPS: 6.0	Golf and driving range	Hours: 2:0 TRIMPS: 4.0	Golf -1:15 Jog easy 45 min
ENERGY FITNESS	PZ 5-7 TRIMPS/hr	Hours: :30 TRIMPS: 3.00	PZ intervals at RPE 15 (relaxed) 4 min long	Hours: : TRIMPS:		Hours: : TRIMPS:		Hours: : TRIMPS:		Hours: : TRIMPS:		Hours: : TRIMPS:		Hours: : TRIMPS:	
ENERGY FITNESS	MZ 7-9 TRIMPS/hr	Hours: : TRIMPS:		Hours: : TRIMPS:		Hours: : TRIMPS:		Hours: : TRIMPS:		Hours: : TRIMPS:		Hours: : TRIMPS:		Hours: : TRIMPS:	
Skills, psychological, tactics, and other training		Hours: :15 TRIMPS:		Hours: :15 TRIMPS:		Hours: :15 TRIMPS:		Hours: :30 TRIMPS:		Hours: :15 TRIMPS:		Hours: : TRIMPS:		Hours: : TRIMPS:	
Daily hours		3:15		2:30		3:00		2:00		3:20		3:00		2:00	
Daily TRIMPS		12.7		6.5		10.7		5.5		15.0		6.0		4.0	

Note: Resistance training should be done on own in the morning.

Total hours for week 19:05 Total TRIMPS for week 60.4

From *Sport Physiology for Coaches* by Brian J. Sharkey and Steven E. Gaskill, 2006, Champaign, IL: Human Kinetics

Sample moderate week for early competition period in high school golf.

Suggested weekly TRIMPS 56		Weekly rating Easy Med Hard	Monday		Tuesday		Wednesday		Thursday		Friday		Saturday		Sunday	
			Suggested TRIMPS 12	Daily rating Easy Med Hard	Suggested TRIMPS 4	Daily rating Easy Med Hard	Suggested TRIMPS 8	Daily rating Easy Med Hard	Suggested TRIMPS 8	Daily rating Easy Med Hard	Suggested TRIMPS 12	Daily rating Easy Med Hard	Suggested TRIMPS 4	Daily rating Easy Med Hard	Suggested TRIMPS 4	Daily rating Easy Med Hard
MUSCULAR FITNESS	Strength 4-6 TRIMPS/hr		Hours: : TRIMPS:5.		Hours: : TRIMPS:		Hours: : TRIMPS:		Hours: : TRIMPS:		Hours: : TRIMPS:		Hours: : TRIMPS:		Hours: : TRIMPS:	
	Power 6-8 TRIMPS/hr		Hours: :20 TRIMPS: 2.5	1 × 12 Fast Mainly arms	Hours: : TRIMPS:		Hours: : TRIMPS:		Hours: : TRIMPS:		Hours: :20 TRIMPS: 2.5	1 × 10 Fast	Hours: : TRIMPS:		Hours: : TRIMPS:	
	Power endurance 5-7 TRIMPS/hr		Hours: :40 TRIMPS: 3.0	1 × 25 1 × 30 Arms only	Hours: : TRIMPS:		Hours: :40 TRIMPS: 3.0	1 × 25 1 × 30 Arms only	Hours: : TRIMPS:		Hours: :40 TRIMPS: 3.0	1 × 25 1 × 30 Arms only	Hours: :40 TRIMPS: 3.0		Hours: : TRIMPS:	
	Plyometrics and other 4-7 TRIMPS/hr		Hours: :20 TRIMPS: 2.5	Core strength training	Hours: : TRIMPS:		Hours: :20 TRIMPS: 2.5	Golf-specific bungee exercises	Hours: :20 TRIMPS: 2.5	Core strength training	Hours: : TRIMPS:		Hours: : TRIMPS:		Hours: : TRIMPS:	
ENERGY FITNESS	EZ 1.5-3 TRIMPS/hr		Hours: 1:30 TRIMPS: 2.5	Golf driving range	Hours: 1:30 TRIMPS: 2.5	Golf driving range	Hours: 1:30 TRIMPS: 2.5	Golf and driving range	Hours: 1:30 TRIMPS: 2.5	Golf and driving range	Hours: 1:30 TRIMPS: 2.5	Golf and driving range	Hours: 2:00 TRIMPS: 2.0	Golf	Hours: 1:00 TRIMPS: 1	Jog easy 45 min plus warm-up
	PZ 5-7 TRIMPS/hr		Hours: :15 TRIMPS: 1.5	PZ intervals at RPE 14 (relaxed) 3-4 min long	Hours: : TRIMPS:		Hours: : TRIMPS:		Hours: : TRIMPS:		Hours: :30 TRIMPS: 3.0	PZ intervals at RPE 14 (relaxed) 3-4 min long	Hours: : TRIMPS:		Hours: : TRIMPS:	
	MZ 7-9 TRIMPS/hr		Hours: : TRIMPS:		Hours: : TRIMPS:		Hours: : TRIMPS:		Hours: : TRIMPS:		Hours: : TRIMPS:		Hours: : TRIMPS:		Hours: : TRIMPS:	
Skills, psychological, tactics, and other training			Hours: : TRIMPS:		Hours: :30 TRIMPS:	Psycho-logical skills and flexibility	Hours: :30 TRIMPS:	Psycho-logical skills and flexibility	Hours: :30 TRIMPS:	Psycho-logical skills and flexibility	Hours: : TRIMPS:		Hours: : TRIMPS:		Hours: : TRIMPS:	
Daily hours			3:25		2:00		3:00		3:00		3:00		2:00		1:00	
Daily TRIMPS			12.0		2.5		8.0		5.0		11.0		3.0		1.5	

Note: Resistance training should be done on own in the morning.

Total hours for week 17:25 Total TRIMPS for week 43

Sample moderate week for peak performance period in high school golf.

Suggested weekly TRIMPS 43		Weekly rating Easy Med Hard	Monday		Tuesday		Wednesday		Thursday		Friday		Saturday		Sunday	
			Suggested TRIMPS 10	Daily rating Easy Med Hard	Suggested TRIMPS 2	Daily rating Easy Med Hard	Suggested TRIMPS 6	Daily rating Easy Med Hard	Suggested TRIMPS 2	Daily rating Easy Med Hard	Suggested TRIMPS 6	Daily rating Easy Med Hard	Suggested TRIMPS 2	Daily rating Easy Med Hard	Suggested TRIMPS 6	Daily rating Easy Med Hard
MUSCULAR FITNESS	Strength 4-6 TRIMPS/hr		Hours: : TRIMPS:		Hours: : TRIMPS:		Hours: : TRIMPS:		Hours: : TRIMPS:		Hours: : TRIMPS:		Hours: : TRIMPS:		Hours: : TRIMPS:	
	Power 6-8 TRIMPS/hr		Hours: :30 TRIMPS: 3.5	2 × 12 Fast Mainly arms	Hours: : TRIMPS:		Hours: : TRIMPS:		Hours: : TRIMPS:		Hours: : TRIMPS:		Hours: : TRIMPS:		Hours: : TRIMPS:	
	Power endurance 5-7 TRIMPS/hr		Hours: :30 TRIMPS: 3.0	1 × 25 1 × 30 Arms only	Hours: : TRIMPS:		Hours: : TRIMPS:		Hours: : TRIMPS:		Hours: : TRIMPS:		Hours: : TRIMPS:		Hours: : TRIMPS:	
	Plyometrics and other 4-7 TRIMPS/hr		Hours: : TRIMPS:		Hours: : TRIMPS:		Hours: : TRIMPS:		Hours: : TRIMPS:		Hours: : TRIMPS:		Hours: : TRIMPS:		Hours: : TRIMPS:	
ENERGY FITNESS	EZ 1.5-3 TRIMPS/hr		Hours: :30 TRIMPS: 1.0	Easy jog	Hours: 1:30 TRIMPS: 2.5	Golf	Hours: :30 TRIMPS: 2.0	Golf 30 min Jog 30 min	Hours: 1:00 TRIMPS: 1.5	Golf and driving range	Hours: 2:30 TRIMPS: 4.0	Golf match	Hours: :20 TRIMPS: 1.0	Jog on own	Hours: 3:00 TRIMPS: 4.5	Golf or other light activity
	PZ 5-7 TRIMPS/hr		Hours: :30 TRIMPS: 3.0	PZ intervals at RPE 14-15 (relaxed) 5 min long	Hours: : TRIMPS:		Hours: : TRIMPS:		Hours: : TRIMPS:		Hours: : TRIMPS:		Hours: : TRIMPS:		Hours: : TRIMPS:	
	MZ 7-9 TRIMPS/hr		Hours: : TRIMPS:		Hours: : TRIMPS:		Hours: : TRIMPS:		Hours: : TRIMPS:		Hours: : TRIMPS:		Hours: : TRIMPS:		Hours: : TRIMPS:	
Skills, psychological, tactics, and other training			Hours: 1:00 TRIMPS: 0	Technique, psycho-logical skills, and flexibility	Hours: : TRIMPS:		Hours: :30 TRIMPS:	Psycho-logical skills and flexibility	Hours: 1:00 TRIMPS: 0	Technique, psycho-logical skills, and flexibility	Hours: : TRIMPS:		Hours: : TRIMPS:		Hours: : TRIMPS:	
Daily hours			3:00		2:00		2:00		2:00		2:00		0:30		3:00	
Daily TRIMPS			10.5		2.5		4.0		1.5		3.0		1.0		4.5	

Note: Resistance training should be done on own in the morning. Matches on Fridays.

Total hours for week 14:30 Total TRIMPS for week 27

From *Sport Physiology for Coaches* by Brian J. Sharkey and Steven E. Gaskill, 2006, Champaign, IL: Human Kinetics

Sample plan for high school football lineman.

Recovery Period: Dates May 24 – June 13 # weeks 3 (Minimum 2 weeks)	
Major Muscular-Fitness Goals of Period	Major Energy-Fitness Goals of Period
Flexibility Athletes on own: unofficial training	No specific goals Light jogging

Basic Training Period: Dates June 14 – Aug. 15 # weeks 9 (Minimum 9 weeks)	
Major Muscular-Fitness Goals of Period	Major Energy-Fitness Goals of Period
Major emphasis: Upper and lower body strength (4-8RM) Core strength Maintenance of power (8-12RM, fast movements) Flexibility Note: Start first few weeks with light weights and build to heavy, 4-8 RM Resistance training 6 days per week, alternating upper and lower body Ensure adequate rest between sets (2-4 min)	EZ-intensity runs, 20-45 min, 3 times per week

Precompetition Period: Dates Aug. 16 – Sept. 26 # weeks 6 (Minimum 6 weeks)	
Major Muscular-Fitness Goals of Period	Major Energy-Fitness Goals of Period
Major emphasis: Body power (8-12RM) Plyometrics 2 times per week Core strength Maintenance of strength (6RM) Flexibility Resistance training 3 days per week with upper and lower body Ensure adequate rest between sets (3-4 min)	EZ-intensity runs, 15-25 min, 3 times per week PZ intervals (relaxed, RPE 15), 1 time per week (2-3 min x 5) Psychological skills training

Early Competition Period: Dates Sept. 27 – Oct. 24 # weeks 4 (Minimum 3 weeks)	
Major Muscular-Fitness Goals of Period	Major Energy-Fitness Goals of Period
Major emphasis: Maintain power with limited resistance training and plyometrics Maintain core strength Speed training Flexibility	Maintain aerobic fitness with EZ jog, 15 min, 2 times per week PZ intervals (relaxed, RPE 15), 1 per week (1-2 min x 7) MZ intervals, 1 per week (5 sec maximal effort, 2 min rest x 15) Psychological skills training

Peak Performance Period: Dates Oct. 25 – Nov 13 # weeks 3 (3-6 weeks)	
Major Muscular-Fitness Goals of Period	Major Energy-Fitness Goals of Period
Maintenance of power with plyometrics Maintenance of core strength with 1 session per week Flexibility Maintenance of speed with 1 session per week Focus on rest and recovery	MZ intervals, 1 per week (5 sec maximal effort, 1 min rest x 20) Psychological skills training Focus on rest and recovery

Week starting with basic strength and endurance period	Week 1	Week 2	Week 3	Week 4	Week 5	Week 6	Week 7	Week 8	Week 9	Week 10	Week 11	Week 12	Week 13	Week 14	Week 15	Week 16	Week 17	Week 18	Week 19	Week 20	Week 21	Week 22	Week 23	Week 24	Week 25
Start date of week (Monday)	5-24	5-31	6-7	6-14	6-21	6-28	7-5	7-12	7-19	7-26	8-2	8-9	8-16	8-23	8-30	9-6	9-13	9-20	9-27	10-4	10-11	10-18	10-25	11-1	11-8
U = unofficial season O = official season	U	→	→	→	→	→	→	→	→	→	→	→	O	→	→	→	→	→	→	→	→	→	→	→	→
Training Season* R, B, P, E, PK	R	→	→	B	→	→	→	→	→	→	→	→	P	→	→	→	→	→	E	→	→	→	PK	→	→
Guideline TRIMPS (from Table F-1)	47	63	31	49	66	32	52	69	34	70	94	46	63	85	42	56	37	37	76	34	53	31			

*R = recovery period, B = basic training period, P = precompetition period, E = early competition period, PK = peak performance period

**E = easy week (recovery), M = medium-stress week, H = high-stress week

Sample moderate week for basic training period for high school football lineman.

Suggested weekly TRIMPS 47		Weekly rating Easy Med Hard	Monday		Tuesday		Wednesday		Thursday		Friday		Saturday		Sunday	
			Suggested TRIMPS 10	Daily rating Easy Med Hard	Suggested TRIMPS 4	Daily rating Easy Med Hard	Suggested TRIMPS 10	Daily rating Easy Med Hard	Suggested TRIMPS 4	Daily rating Easy Med Hard	Suggested TRIMPS 10	Daily rating Easy Med Hard	Suggested TRIMPS 7	Daily rating Easy Med Hard	Suggested TRIMPS	Daily rating Easy Med Hard
MUSCULAR FITNESS	Strength 4-6 TRIMPS/hr		Hours: 1:30 TRIMPS: 7.5	1 × 3 1 × 5 1 × 8 Legs and arms	Hours: : TRIMPS:		Hours: 1:30 TRIMPS: 7.5	1 × 3 1 × 5 1 × 8 Legs and arms	Hours: : TRIMPS:		Hours: 1 :15 TRIMPS: 5.5	1 × 3 1 × 5 1 × 8 Legs	Hours: 1:00 TRIMPS: 4.5	1 × 5 1 × 7 1 × 9 Arms	Hours: : TRIMPS:	
	Power 6-8 TRIMPS/hr		Hours: :30 TRIMPS: 4.0	1 × 10 Fast Full body	Hours: : TRIMPS:		Hours: :30 TRIMPS: 4.0	1 × 10 Fast Full body	Hours: : TRIMPS:		Hours: :30 TRIMPS: 4.0	1 × 10 Fast Full body	Hours: : TRIMPS:		Hours: : TRIMPS:	
	Power endurance 5-7 TRIMPS/hr		Hours: : TRIMPS:		Hours: : TRIMPS:		Hours: :30 TRIMPS: 3.0	1 × 25 Legs	Hours: : TRIMPS:		Hours: : TRIMPS:		Hours: : TRIMPS:		Hours: : TRIMPS:	
	Plyometrics and other 4-7 TRIMPS/hr		Hours: : TRIMPS:		Hours: : TRIMPS:		Hours: : TRIMPS:		Hours: :30 TRIMPS: 3.0	Core strength training	Hours: : TRIMPS:		Hours: :30 TRIMPS: 3.0	Core strength training	Hours: : TRIMPS:	
ENERGY FITNESS	EZ 1.5-3 TRIMPS/hr		Hours: : TRIMPS:		Hours: :45 TRIMPS: 2.0	Jogging RPE 11	Hours: : TRIMPS:		Hours: :45 TRIMPS: 2.0	Jogging RPE 11	Hours: : TRIMPS:		Hours: : TRIMPS:		Hours: :45 TRIMPS: 2.0	Jogging RPE 11
	PZ 5-7 TRIMPS/hr		Hours: : TRIMPS:		Hours: : TRIMPS:		Hours: : TRIMPS:		Hours: : TRIMPS:		Hours: : TRIMPS:		Hours: : TRIMPS:		Hours: : TRIMPS:	
	MZ 7-9 TRIMPS/hr		Hours: : TRIMPS:		Hours: : TRIMPS:		Hours: : TRIMPS:		Hours: : TRIMPS:		Hours: : TRIMPS:		Hours: : TRIMPS:		Hours: : TRIMPS:	
Skills, psychological, tactics, and other training			Hours: : TRIMPS:		Hours: : TRIMPS:		Hours: : TRIMPS:		Hours: : TRIMPS:		Hours: : TRIMPS:		Hours: : TRIMPS:		Hours: : TRIMPS:	
Daily hours			2:00		:45		2:30		1:15		1:45		1:30		:45	
Daily TRIMPS			11.5		2.0		14.5		5.0		9.5		7.5		2.0	

Note: These workouts are during the unofficial period and to be completed by athletes on their own.

Total hours for week 10:30 Total TRIMPS for week 52.0

Sample moderate week for precompetition period for football lineman.

Suggested weekly TRIMPS 70	Weekly rating Easy Med Hard	Monday		Tuesday		Wednesday		Thursday		Friday		Saturday		Sunday	
		Suggested TRIMPS 15	Daily rating Easy Med Hard	Suggested TRIMPS 10	Daily rating Easy Med Hard	Suggested TRIMPS 15	Daily rating Easy Med Hard	Suggested TRIMPS 5	Daily rating Easy Med Hard	Suggested TRIMPS 15	Daily rating Easy Med Hard	Suggested TRIMPS 5	Daily rating Easy Med Hard	Suggested TRIMPS 0	Daily rating Easy Med Hard
MUSCULAR FITNESS	Strength 4-6 TRIMPS/hr	Hours: :15 TRIMPS: 1.5	1 × 6 Legs	Hours: :15 TRIMPS: 1.5	1 × 6 Arms	Hours: : TRIMPS:		Hours: :15 TRIMPS: 1.5		Hours: :15 TRIMPS: 1.5	1 × 3 Legs	Hours: : TRIMPS:		Hours: : TRIMPS:	
	Power 6-8 TRIMPS/hr	Hours: :45 TRIMPS: 5.0	1 × 8 1 × 10 1 × 12 Legs	Hours: :30 TRIMPS: 3.5	1 × 10 1 × 12 Arms	Hours: : TRIMPS:		Hours: : TRIMPS:		Hours: 1:00 TRIMPS: 7.0	1 × 8 1 × 10 Arms and legs	Hours: : TRIMPS:		Hours: : TRIMPS:	
	Power endurance 5-7 TRIMPS/hr	Hours: : TRIMPS:		Hours: : TRIMPS:		Hours: : TRIMPS:		Hours: : TRIMPS:		Hours: : TRIMPS:		Hours: : TRIMPS:		Hours: : TRIMPS:	
	Plyometrics and other 4-7 TRIMPS/hr	Hours: :30 TRIMPS: 3.5	Leg plyometrics	Hours: :30 TRIMPS: 2.5	Core strength training	Hours: :30 TRIMPS: 3.5	Leg plyometrics	Hours: :30 TRIMPS: 2.5	Core strength training	Hours: :30 TRIMPS: 3.5	Leg plyometrics	Hours: : TRIMPS:		Hours: : TRIMPS:	
ENERGY FITNESS	EZ 1.5-3 TRIMPS/hr	Hours: :15 TRIMPS: .5	Relaxed jog	Hours: : TRIMPS:		Hours: :15 TRIMPS: .5	Relaxed jog	Hours: : TRIMPS:		Hours: :15 TRIMPS: .5	Relaxed jog	Hours: :30 TRIMPS: 1.0	Relaxed jog	Hours: : TRIMPS:	
	PZ 5-7 TRIMPS/hr	Hours: :30 TRIMPS: 3.0	Intervals: 15 sec at 90%, RPE 17, 3 min rest	Hours: : TRIMPS:		Hours: : TRIMPS:		Hours: :30 TRIMPS: 3	Intervals: 2 min with 2 min rest, RPE 16	Hours: : TRIMPS:		Hours: : TRIMPS:		Hours: : TRIMPS:	
	MZ 7-9 TRIMPS/hr	Hours: : TRIMPS:		Hours: : TRIMPS:		Hours: :15 TRIMPS: 2.0	SLEDS: 5 sec on, 1 min rest	Hours: : TRIMPS:		Hours: :30 TRIMPS: 4.0	SLEDS: 7 sec on, 1.5 min rest	Hours: : TRIMPS:		Hours: : TRIMPS:	
Skills, psychological, tactics, and other training		Hours: 1:15 TRIMPS: 2.0	Technical and tactical drills, psychological skills	Hours: 2:00 TRIMPS: 3.0	Light technical and tactical drills, psychological skills	Hours: 1:30 TRIMPS: 4.0	Scrimmage and psychological skills	Hours: 1:00 TRIMPS: 2.0	Light technical and tactical drills, psychological skills	Hours: 1:00 TRIMPS: 2.0	Scrimmage and technique	Hours: : TRIMPS:		Hours: : TRIMPS:	
Daily hours		3:30		3:15		3:30		2:00		3:30		:30			
Daily TRIMPS		15.5		10.5		16.5		7.5		18.5		1.0			

Note: Resistance training should be done on own in the morning.

Total hours for week 16:15 Total TRIMPS for week 69.5

From *Sport Physiology for Coaches* by Brian J. Sharkey and Steven E. Gaskill, 2006, Champaign, IL: Human Kinetics

Sample moderate week for early competition period for high school football lineman.

Suggested weekly TRIMPS 56	Weekly rating Easy Med Hard	Monday Suggested TRIMPS 12	Monday Daily rating Easy Med Hard	Tuesday Suggested TRIMPS 8	Tuesday Daily rating Easy Med Hard	Wednesday Suggested TRIMPS 8	Wednesday Daily rating Easy Med Hard	Thursday Suggested TRIMPS 4	Thursday Daily rating Easy Med Hard	Friday Suggested TRIMPS 8	Friday Daily rating Easy Med Hard	Saturday Suggested TRIMPS 8	Saturday Daily rating Easy Med Hard	Sunday Suggested TRIMPS 0	Sunday Daily rating Easy Med Hard
MUSCULAR FITNESS	Strength 4-6 TRIMPS/hr	Hours: :30 TRIMPS: 3.0	1 × 6 Arms and legs	Hours: : TRIMPS:		Hours: : TRIMPS:		Hours: : TRIMPS:		Hours: : TRIMPS:		Hours: : TRIMPS:		Hours: : TRIMPS:	Day off
MUSCULAR FITNESS	Power 6-8 TRIMPS/hr	Hours: : TRIMPS:	1 × 10 1 × 12 Arms and legs	Hours: : TRIMPS:		Hours: :30 TRIMPS: 3.5	1 × 12 Arms and legs	Hours: : TRIMPS:		Hours: : TRIMPS:		Hours: 1:00 TRIMPS: 7.0	1 × 10 1 × 12 Arms and legs	Hours: : TRIMPS:	
MUSCULAR FITNESS	Power endurance 5-7 TRIMPS/hr	Hours: : TRIMPS:		Hours: : TRIMPS:		Hours: : TRIMPS:		Hours: : TRIMPS:		Hours: : TRIMPS:		Hours: : TRIMPS:		Hours: : TRIMPS:	
MUSCULAR FITNESS	Plyometrics and other 4-7 TRIMPS/hr	Hours: : TRIMPS:		Hours: :30 TRIMPS: 3.5	Leg plyometrics and speed training	Hours: : TRIMPS:		Hours: : TRIMPS:		Hours: : TRIMPS:		Hours: : TRIMPS:		Hours: : TRIMPS:	
ENERGY FITNESS	EZ 1.5-3 TRIMPS/hr	Hours: :15 TRIMPS: .5	Jog	Hours: :15 TRIMPS: .5	Jog	Hours: : TRIMPS:		Hours: : TRIMPS:		Hours: : TRIMPS:		Hours: :20 TRIMPS: 1.0	Jog or run, RPE 11	Hours: : TRIMPS:	Day off
ENERGY FITNESS	PZ 5-7 TRIMPS/hr	Hours: :15 TRIMPS: 2.0	3 sec on with 55 sec rest (15 repeats)	Hours: : TRIMPS:		Hours: :15 TRIMPS: 2.0	Speed and sled explo- sive power	Hours: : TRIMPS:		Hours: : TRIMPS:		Hours: : TRIMPS:		Hours: : TRIMPS:	
ENERGY FITNESS	MZ 7-9 TRIMPS/hr	Hours: : TRIMPS:		Hours: :15 TRIMPS: 2.0	5 sec max intervals with 55 sec rest	Hours: : TRIMPS:		Hours: : TRIMPS:		Hours: : TRIMPS:		Hours: : TRIMPS:		Hours: : TRIMPS:	
Skills, psychological, tactics, and other training		Hours: 1:30 TRIMPS: 2.0	Film, light technique, and psy- chological skills	Hours: 1:15 TRIMPS: 3.0	Scrimmage, psychologi- cal skills	Hours: 1:15 TRIMPS: 3.0	Scrimmage, psychologi- cal skills	Hours: 2:00 TRIMPS: 2.0	Easy drills and scrim- mage, psychologi- cal skills	Hours: 2:00 TRIMPS: 6.0	**Game**	Hours: : TRIMPS:		Hours: : TRIMPS:	
Daily hours		3:00		2;15		2:00		2:00		2:00		1:30		0	
Daily TRIMPS		11.5		9.0		8.5		2.0		6.0		8.0		0	

Note: Resistance training should be done on own in the morning.

Total hours for week 12.45 Total TRIMPS for week 45

Sample moderate week for peak performance period for high school football lineman.

Suggested weekly TRIMPS 43	Weekly rating Easy Med Hard	Monday		Tuesday		Wednesday		Thursday		Friday		Saturday		Sunday	
		Suggested TRIMPS 10	Daily rating Easy Med Hard	Suggested TRIMPS 6	Daily rating Easy Med Hard	Suggested TRIMPS 3	Daily rating Easy Med Hard	Suggested TRIMPS 3	Daily rating Easy Med Hard	Suggested TRIMPS 6	Daily rating Easy Med Hard	Suggested TRIMPS 6	Daily rating Easy Med Hard	Suggested TRIMPS 0	Daily rating Easy Med Hard
MUSCULAR FITNESS	Strength 4-6 TRIMPS/hr	Hours: : TRIMPS:		Hours: : TRIMPS:		Hours: : TRIMPS:		Hours: : TRIMPS:		Hours: : TRIMPS:		Hours: : TRIMPS:		Hours: : TRIMPS:	
	Power 6-8 TRIMPS/hr	Hours: 1:00 TRIMPS: 7.0	1 × 10 1 × 12 Arms and legs	Hours: : TRIMPS:		Hours: : TRIMPS:		Hours: : TRIMPS:		Hours: : TRIMPS:		Hours: : TRIMPS:		Hours: : TRIMPS:	
	Power endurance 5-7 TRIMPS/hr	Hours: :30 TRIMPS: 3.0	2 × 25 Legs	Hours: : TRIMPS:		Hours: : TRIMPS:		Hours: : TRIMPS:		Hours: : TRIMPS:		Hours: : TRIMPS:		Hours: : TRIMPS:	
	Plyometrics and other 4-7 TRIMPS/hr	Hours: : TRIMPS:		Hours: :30 TRIMPS: 3.5	Leg plyometrics	Hours: : TRIMPS:		Hours: : TRIMPS:		Hours: : TRIMPS:		Hours: :30 TRIMPS: 3.5	Speed and plyometrics	Hours: : TRIMPS:	
ENERGY FITNESS	EZ 1.5-3 TRIMPS/hr	Hours: :15 TRIMPS: .5	Jog	Hours: : TRIMPS:		Hours: :15 TRIMPS: .5	Jog	Hours: : TRIMPS:		Hours: : TRIMPS:		Hours: :15 TRIMPS: .5	Jog	Hours: : TRIMPS:	
	PZ 5-7 TRIMPS/hr	Hours: : TRIMPS:		Hours: : TRIMPS:		Hours: : TRIMPS:		Hours: : TRIMPS:		Hours: : TRIMPS:		Hours: : TRIMPS:		Hours: : TRIMPS:	
	MZ 7-9 TRIMPS/hr	Hours: :30 TRIMPS: 3.5	Speed work with some plyometrics	Hours: : TRIMPS:		Hours: :15 TRIMPS: 2.0	Sled 5 sec on with 55 sec rest	Hours: : TRIMPS:		Hours: : TRIMPS:		Hours: : TRIMPS:		Hours: : TRIMPS:	
Skills, psychological, tactics, and other training		Hours: 1:15 TRIMPS: 2.0	Film, light technique, and psychological skills	Hours: 1:30 TRIMPS: 3.0	Scrimmage, psychological skills	Hours: 1:30 TRIMPS: 2.0	Light technique, psychological skills	Hours: 1:30 TRIMPS: 2.0	Light technique, psychological skills	Hours: 2:00 TRIMPS: 6.0	**Game**	Hours: : TRIMPS:		Hours: : TRIMPS:	
Daily hours		3:30		2:00		2:00		1:30		2:00		:45		0	
Daily TRIMPS		16.0		6.5		4.5		2.0		6.0		4.0		0	

Note: Resistance training should be done on own in morning.

Total hours for week 11:45 Total TRIMPS for week 39

From *Sport Physiology for Coaches* by Brian J. Sharkey and Steven E. Gaskill, 2006, Champaign, IL: Human Kinetics

Sample plan for junior wrestling.

Recovery Period: Dates Jan. 25 – Feb. 8 # weeks 2 (Minimum 2 weeks)

Major Muscular-Fitness Goals of Period	Major Energy-Fitness Goals of Period
No training is defined for this period. Young athletes should be encouraged to remain active in a variety of sports and activities.	

Basic Training Period: Dates Feb. 9 – Mar. 14 # weeks 5 (Minimum 9 weeks)

Major Muscular-Fitness Goals of Period	Major Energy-Fitness Goals of Period
The major focus of this age group is tecnique. Strength, power, and power endurance are developed through body weight and partner resistance and activities. Some resistance training is shown but will be limited Major emphasis: Strength (9-12RM, no overhead lifts) Core strength (body weight and partner resistance) Maintenance of power using easy plyometrics Maintenance of power (endurance, circuits with body resistance)	Aerobic fitness through a variety of activities that are fun and moderately sustained and increase heart rate. Examples are ultimate Frisbee, jogging, soccer, cycling, and related activities. These should be encouraged both at practice and on one's own.

Precompetition Period: Dates Mar. 15 – Apr. 18 # weeks 4 (Minimum 6 weeks)

Major Muscular-Fitness Goals of Period	Major Energy-Fitness Goals of Period
Major emphasis: Power: plyometrics and wrestling Core strength (body weight and partner resistance) Maintenance of power endurance: circuits with body resistance Flexibility	EZ-intensity runs, 45 min, 1 or 2 times per week PZ intervals (relaxed, RPE 15), relay races, uphill jogging (1-3 min long) Psychological skills training

Early Competition Period: Dates Apr. 19 – May 23 # weeks 6 (Minimum 3 weeks)

Major Muscular-Fitness Goals of Period	Major Energy-Fitness Goals of Period
For this age group, the training will remain similar to that in the precompetition period with a slight reduction in training load and a shift from power to power endurance. Power endurance is accomplished with body and partner resistance exercises.	Maintain aerobic fitness with EZ-intensity activities daily PZ intervals (relaxed, RPE 15), relay races, uphill jogging (1-3 min) Anaerobic ability with high intensity, short relays, and sprints—keep it fun Psychological skills training

Peak Performance Period: Dates May 24 – May 30 # weeks 1 (3-6 weeks)

Major Muscular-Fitness Goals of Period	Major Energy-Fitness Goals of Period
The focus of this age group is not on peak performance but on skills, technique, and foundation training. The peak performance period is shown as only one week and will consist of reduced training, fun activities, and technique work with a focus on rest and recovery for the final competition.	Psychological skills training Maintain light, fun activities Focus on rest and recovery

Week starting with basic strength and endurance period	Week 1	Week 2	Week 3	Week 4	Week 5	Week 6	Week 7	Week 8	Week 9	Week 10	Week 11	Week 12	Week 13	Week 14	Week 15	Week 16	Week 17	Week 18	Week 19	Week 20	Week 21	Week 22	Week 23	Week 24	Week 25
Start date of week (Monday)	2-9	2-16	2-23	3-1	3-8	3-15	3-22	3-29	4-5	4-12	4-19	4-26	5-3	5-10	5-17	5-24									
U = unofficial season O = official season	O →	→	→	→	→	→	→	→	→	→	→	→	→	→	→	→									
Training season* R, B, P, E, PK	B →	→	→	→	→	P →	→	→	→	E →	→	→	→	→	→	PK									
Weekly stress**: E, M, H	M	H	E	M	H	E	M	H	E	M	H	E	M	H	E	M									
Guideline TRIMPS (from table 10.4	21	28	14	22	30	21	28	38	19	25	34	17	25	34	17	15									

*R = recovery period, B = basic training period, P = precompetition period, E = early competition period, PK = peak performance period

**E = easy week (recovery), M = medium-stress week, H = high-stress week

Sample moderate week for basic training period in youth wrestling.

Suggested weekly TRIMPS 26	Weekly rating Easy Med Hard	Monday		Tuesday		Wednesday		Thursday		Friday		Saturday		Sunday	
		Suggested TRIMPS 6	Daily rating Easy Med Hard	Suggested TRIMPS 4	Daily rating Easy Med Hard	Suggested TRIMPS 6	Daily rating Easy Med Hard	Suggested TRIMPS 4	Daily rating Easy Med Hard	Suggested TRIMPS 6	Daily rating Easy Med Hard	Suggested TRIMPS	Daily rating Easy Med Hard	Suggested TRIMPS	Daily rating Easy Med Hard
MUSCULAR FITNESS	Strength 4-6 TRIMPS/hr	Hours: : TRIMPS:		Hours: : TRIMPS:		Hours: : TRIMPS:		Hours: : TRIMPS:		Hours: : TRIMPS:		Hours: : TRIMPS:		Hours: : TRIMPS:	
	Power 6-8 TRIMPS/hr	Hours: : TRIMPS:		Hours: : TRIMPS:		Hours: : TRIMPS:		Hours: : TRIMPS:		Hours: : TRIMPS:		Hours: : TRIMPS:		Hours: : TRIMPS:	
	Power endurance 5-7 TRIMPS/hr	Hours: :30 TRIMPS: 2.5	Body and partner resistance circuits	Hours: : TRIMPS:		Hours: :30 TRIMPS: 2.5	Body and partner resistance circuits	Hours: : TRIMPS:		Hours: :30 TRIMPS: 2.5	Body and partner resistance circuits	Hours: : TRIMPS:		Hours: : TRIMPS:	
	Plyometrics and other 4-7 TRIMPS/hr	Hours: : TRIMPS:		Hours: : TRIMPS:		Hours: : TRIMPS:		Hours: : TRIMPS:		Hours: : TRIMPS:		Hours: : TRIMPS:		Hours: : TRIMPS:	
ENERGY FITNESS	EZ 1.5-3 TRIMPS/hr	Hours: :30 TRIMPS:		Hours: :45 TRIMPS: 2.0	Aerobic activities on own	Hours: : TRIMPS:		Hours: :45 TRIMPS: 2.0	Aerobic activities on own	Hours: : TRIMPS:		Hours: :45 TRIMPS: 2.0	Aerobic activities on own	Hours: :45 TRIMPS: 2.0	Aerobic activities on own
	PZ 5-7 TRIMPS/hr	Hours: : TRIMPS:		Hours: : TRIMPS:		Hours: : TRIMPS:		Hours: : TRIMPS:		Hours: : TRIMPS:		Hours: : TRIMPS:		Hours: : TRIMPS:	
	MZ 7-9 TRIMPS/hr	Hours: : TRIMPS:		Hours: : TRIMPS:		Hours: : TRIMPS:		Hours: : TRIMPS:		Hours: : TRIMPS:		Hours: : TRIMPS:		Hours: : TRIMPS:	
Skills, psychological, tactics, and other training		Hours: 1:30 TRIMPS: 4.0	Technical skills, flexi-biliy, and wrestling	Hours: : TRIMPS:		Hours: 1:30 TRIMPS: 4.0	Technical skills, flexi-biliy, and wrestling	Hours: : TRIMPS:	T	Hours: 1:30 TRIMPS: 4.0	Technical skills, flexi-biliy, and wrestling	Hours: : TRIMPS:		Hours: : TRIMPS:	
Daily hours		2:00		:45		2:00		:45		2:00		:45		:45	
Daily TRIMPS		6.5		2.0		6.5		2.0		6.5		2.0		2.0	

Note: Practices on Monday, Wednesday, Friday.

Total hours for week 9:00 Total TRIMPS for week 27.5

Sample moderate week for precompetition period in youth wrestling.

Suggested weekly TRIMPS 40	Weekly rating Easy Med Hard	Monday		Tuesday		Wednesday		Thursday		Friday		Saturday		Sunday	
		Suggested TRIMPS 8	Daily rating Easy Med Hard	Suggested TRIMPS 4	Daily rating Easy Med Hard	Suggested TRIMPS 8	Daily rating Easy Med Hard	Suggested TRIMPS 4	Daily rating Easy Med Hard	Suggested TRIMPS 8	Daily rating Easy Med Hard	Suggested TRIMPS 6	Daily rating Easy Med Hard	Suggested TRIMPS 2	Daily rating Easy Med Hard
MUSCULAR FITNESS	Strength 4-6 TRIMPS/hr	Hours: :45 TRIMPS: 4.0	3 × 8 Arms and legs	Hours: : TRIMPS:		Hours: :45 TRIMPS: 4.0	3 × 8 Arms and legs	Hours: : TRIMPS:		Hours: :45 TRIMPS: 4.0	3 × 8 Arms and legs	Hours: : TRIMPS:		Hours: : TRIMPS:	
	Power 6-8 TRIMPS/hr	Hours: : TRIMPS:		Hours: : TRIMPS:		Hours: : TRIMPS:		Hours: : TRIMPS:		Hours: : TRIMPS:		Hours: : TRIMPS:		Hours: : TRIMPS:	
	Power endurance 5-7 TRIMPS/hr	Hours: : TRIMPS:		Hours: :45 TRIMPS: 3.0	Arm and leg circuits	Hours: : TRIMPS:		Hours: :45 TRIMPS: 3.0	Arm and leg circuits	Hours: : TRIMPS:		Hours: :45 TRIMPS: 3.0	Arm and leg circuits	Hours: : TRIMPS:	
	Plyometrics and other 4-7 TRIMPS/hr	Hours: :15 TRIMPS: 1.5	Core strength	Hours: : TRIMPS:		Hours: :15 TRIMPS: 1.5	Core strength	Hours: : TRIMPS:		Hours: :15 TRIMPS: 1.5	Core strength	Hours: : TRIMPS:		Hours: : TRIMPS:	
ENERGY FITNESS	EZ 1.5-3 TRIMPS/hr	Hours: : TRIMPS:		Hours: :30 TRIMPS: 1.0	Easy aerobic activities on own	Hours: : TRIMPS:		Hours: :30 TRIMPS: 1.0	Easy aerobic activities on own	Hours: : TRIMPS:		Hours: 1:00 TRIMPS: 1.5	Easy aerobic activities on own	Hours: : TRIMPS:	
	PZ 5-7 TRIMPS/hr	Hours: : TRIMPS:		Hours: : TRIMPS:		Hours: : TRIMPS:		Hours: : TRIMPS:		Hours: : TRIMPS:		Hours: : TRIMPS:		Hours: : TRIMPS:	
	MZ 7-9 TRIMPS/hr	Hours: : TRIMPS:		Hours: : TRIMPS:		Hours: : TRIMPS:		Hours: : TRIMPS:		Hours: : TRIMPS:		Hours: : TRIMPS:		Hours: : TRIMPS:	
Skills, psychological, tactics, and other training		Hours: 1:30 TRIMPS: 4.0	Skills, technique, flexibility, and wrestling	Hours: : TRIMPS:		Hours: 1:30 TRIMPS: 4.0	Skills, technique, flexibility, and wrestling	Hours: : TRIMPS:		1:30 TRIMPS: 4.0	Skills, technique, flexibility, and wrestling	Hours: : TRIMPS:		Hours: : TRIMPS:	
Daily hours		2:30		1:15		2:30		1:15		2:30		1:45			
Daily TRIMPS		9.5		4.0		9.5		4.0		9.5		4.5			

Note: Resistance training should be done on own in the morning.

Total hours for week 11:45 Total TRIMPS for week 41.0

Sample moderate week for early competition period in youth wrestling.

Suggested weekly TRIMPS 33	Weekly rating Easy Med Hard	Monday		Tuesday		Wednesday		Thursday		Friday		Saturday		Sunday	
		Suggested TRIMPS 8	Daily rating Easy Med Hard	Suggested TRIMPS 3	Daily rating Easy Med Hard	Suggested TRIMPS 8	Daily rating Easy Med Hard	Suggested TRIMPS 5	Daily rating Easy Med Hard	Suggested TRIMPS 5	Daily rating Easy Med Hard	Suggested TRIMPS 5	Daily rating Easy Med Hard	Suggested TRIMPS 0	Daily rating Easy Med Hard
MUSCULAR FITNESS	Strength 4-6 TRIMPS/hr	Hours: : TRIMPS:		Hours: : TRIMPS:		Hours: : TRIMPS:		Hours: : TRIMPS:		Hours: : TRIMPS:		Hours: : TRIMPS:		Hours: : TRIMPS:	
	Power 6-8 TRIMPS/hr	Hours: :45 TRIMPS: 3.5	2 × 12 Arms and legs	Hours: : TRIMPS:		Hours: :45 TRIMPS: 3.5	2 × 12 Arms and legs	Hours: : TRIMPS:		Hours: :20 TRIMPS: 2.0	1 × 12 Arms and legs	Hours: : TRIMPS:		Hours: : TRIMPS:	
	Power endurance 5-7 TRIMPS/hr	Hours: : TRIMPS:		Hours: :45 TRIMPS: 3.0	Arm and leg circuits	Hours: : TRIMPS:		Hours: :45 TRIMPS: 3.0	Arm and leg circuits	Hours: : TRIMPS:		Hours: : TRIMPS:		Hours: : TRIMPS:	
	Plyometrics and other 4-7 TRIMPS/hr	Hours: : TRIMPS:		Hours: : TRIMPS:		Hours: :15 TRIMPS: 1.5	Core strength	Hours: : TRIMPS:		Hours: : TRIMPS:		Hours: : TRIMPS:		Hours: : TRIMPS:	
ENERGY FITNESS	EZ 1.5-3 TRIMPS/hr	Hours: :30 TRIMPS:		Hours: :30 TRIMPS: 1.0	Easy aerobic activities on own	Hours: : TRIMPS:		Hours: :30 TRIMPS: 1.0	Easy aerobic activities on own	Hours: : TRIMPS:		Hours: : TRIMPS:		Hours: : TRIMPS:	
	PZ 5-7 TRIMPS/hr	Hours: : TRIMPS:		Hours: : TRIMPS:		Hours: : TRIMPS:		Hours: : TRIMPS:		Hours: : TRIMPS:		Hours: : TRIMPS:		Hours: : TRIMPS:	
	MZ 7-9 TRIMPS/hr	Hours: : TRIMPS:		Hours: : TRIMPS:		Hours: : TRIMPS:		Hours: : TRIMPS:		Hours: : TRIMPS:		Hours: : TRIMPS:		Hours: : TRIMPS:	
Skills, psychological, tactics, and other training		Hours: 1:30 TRIMPS: 4.0	Skills technique, flexibility, and wrestling	Hours: : TRIMPS:		Hours: 1:30 TRIMPS: 4.0	Skills technique, flexibility, and wrestling	Hours: : TRIMPS:		Hours: 1:00 TRIMPS: 3.0:	Skills technique, flexibility, and wrestling	Hours: :40 TRIMPS: 5.0	MATCH	Hours: : TRIMPS:	
Daily hours		2:15		1:15		2:30		1:15		1:20		:40			
Daily TRIMPS		7.5		4.0		9.5		4.0		5.0		5.0			

Note: Resistance training should be done on own in the morning.

Total hours for week 9:15 Total TRIMPS for week 35.0

Sample moderate week for peak performance period in youth wrestling.

Suggested weekly TRIMPS 25		Weekly rating Easy Med Hard	Monday		Tuesday		Wednesday		Thursday		Friday		Saturday		Sunday	
			Suggested TRIMPS 6	Daily rating Easy Med Hard	Suggested TRIMPS 2	Daily rating Easy Med Hard	Suggested TRIMPS 4	Daily rating Easy Med Hard	Suggested TRIMPS 2	Daily rating Easy Med Hard	Suggested TRIMPS 6	Daily rating Easy Med Hard	Suggested TRIMPS 6	Daily rating Easy Med Hard	Suggested TRIMPS 0	Daily rating Easy Med Hard
MUSCULAR FITNESS	Strength 4-6 TRIMPS/hr		Hours: : TRIMPS:		Hours: : TRIMPS:		Hours: : TRIMPS:		Hours: : TRIMPS:	Day off	Hours: : TRIMPS:		Hours: : TRIMPS:		Hours: : TRIMPS:	
	Power 6-8 TRIMPS/hr		Hours: :30 TRIMPS: 3.5	1 × 2 Arms and legs	Hours: : TRIMPS:		Hours: : TRIMPS:		Hours: : TRIMPS:		Hours: : TRIMPS:		Hours: : TRIMPS:		Hours: : TRIMPS:	
	Power endurance 5-7 TRIMPS/hr		Hours: :30 TRIMPS: 2.0	1 × 25 Arms and legs	Hours: : TRIMPS:		Hours: :30 TRIMPS: 2.0	1 × 25 Arms and legs	Hours: : TRIMPS:		Hours: : TRIMPS:		Hours: :30 TRIMPS: 2.0	1 × 25 Arms and legs	Hours: : TRIMPS:	
	Plyometrics and other 4-7 TRIMPS/hr		Hours: : TRIMPS:		Hours: : TRIMPS:		Hours: :15 TRIMPS: 1.5	Core strength	Hours: : TRIMPS:		Hours: : TRIMPS:		Hours: :15 TRIMPS: 1.5	Core strength	Hours: : TRIMPS:	
ENERGY FITNESS	EZ 1.5-3 TRIMPS/hr		Hours: :30 TRIMPS:		Hours: :30 TRIMPS: 1.0	Easy aerobic activities on own	Hours: : TRIMPS:		Hours: : TRIMPS:		Hours: : TRIMPS:		Hours: : TRIMPS:		Hours: :30 TRIMPS: 1.0	Easy aerobic activities on own
	PZ 5-7 TRIMPS/hr		Hours: : TRIMPS:		Hours: : TRIMPS:		Hours: : TRIMPS:		Hours: : TRIMPS:		Hours: : TRIMPS:		Hours: : TRIMPS:		Hours: : TRIMPS:	
	MZ 7-9 TRIMPS/hr		Hours: :15 TRIMPS: 1.0	5 sec sprints with 55 sec rest, 10-15 repeats	Hours: : TRIMPS:		Hours: : TRIMPS:		Hours: : TRIMPS:		Hours: : TRIMPS:		Hours: : TRIMPS:		Hours: : TRIMPS:	
Skills, psychological, tactics, and other training			Hours: 1:15 TRIMPS: 3.0	Skills, tech- nique, flex- ibility., and wrestling	Hours: : TRIMPS:		Hours: 1;15 TRIMPS: 3.0	Skills, tech- nique, flex- ibility., and wrestling	Hours: : TRIMPS:		Hours: :40 TRIMPS: 5.0	MATCH	Hours: : TRIMPS:		Hours: : TRIMPS:	
Daily hours			2:30		:30		2:00		1:15		:45		:45			
Daily TRIMPS			9.5		1.0		6.5		0		5.0		3.5			

Note: Meets on Friday.

Total hours for week 7:45 Total TRIMPS for week 25.5

Sample plan for high school soccer (midfield).

Recovery Period: Dates May 24 – June 13 # weeks 3 (Minimum 2 weeks)

Major Muscular-Fitness Goals of Period	Major Energy-Fitness Goals of Period
Flexibility Begin a weight program with light resistance Athletes on own—unofficial training	General activities, especially those with aerobic energy fitness components Light jogging

Basic Training Period: Dates June 14 – Aug. 15 # weeks 9 (Minimum 9 weeks)

Major Muscular-Fitness Goals of Period	Major Energy-Fitness Goals of Period
Major emphases: Strength: 4-8RM (focus on legs and specific throwing for arms) Core strength Maintenance of power, 8-12RM (fast movements, arms and legs) Maintenance of power endurance, 20-30RM (legs only) Flexibility	EZ-intensity continuous activities, 30-60 min, 3 times per week Running is best, but occasional cycling, stair stepping, or rowing machines can provide variety

Precompetition Period: Dates Aug. 15 – Sept. 5 # weeks 3 (Minimum 6 weeks)

Major Muscular-Fitness Goals of Period	Major Energy-Fitness Goals of Period
Competitions begin after 3 weeks of official season. Thus, maintain this period into the early competitions for 2-3 weeks if possible. Major emphases: Power, 8-12RM (arms and legs), mainly plyometrics Core strength Maintenance of power endurance, 20-30RM (legs only) Flexibility	EZ-intensity runs, 30-45 min daily PZ intervals (relaxed, RPE 15) 2 times per week, 2-3 min with 3 min rest Games and scrimmages will provide adequate MZ intervals at this time Psychological skills training

Early Competition Period: Dates Sept. 6 – Oct. 10 # weeks 5 (Minimum 3 weeks)

Major Muscular-Fitness Goals of Period	Major Energy-Fitness Goals of Period
Major emphases: Power endurance, 20-30RM (legs only) Continued plyometrics, but tapering during this period Maintain core strength Speed training Flexibility	Maintain aerobic fitness with EZ run 1 to 2 times per week EZ intervals (relaxed, RPE 15), 1 time per week, 1-3 min with 2 min rest MZ intervals (3-10 sec, 1.5 min rest, 20 repeats) Psychological skills training

Peak Performance Period: Dates Oct. 11 – Oct. 31 # weeks 3 (3-6 weeks)

Major Muscular-Fitness Goals of Period	Major Energy-Fitness Goals of Period
Maintenance of power (plyometrics) Maintenance of power endurance (body and partner resistance) Maintenance of core strength with 1 session per week Flexibility Focus on rest and recovery	Maintain aerobic fitness with EZ jog, 30 min, 1-2 times per week EZ intervals (relaxed, RPE 14), 1 time per week (note reduced intensity) MZ intervals (3-10 sec, 30 sec rest, 20 repeats) Psychological skills training Focus on rest and recovery

Week starting with basic strength and endurance period	Week 1	Week 2	Week 3	Week 4	Week 5	Week 6	Week 7	Week 8	Week 9	Week 10	Week 11	Week 12	Week 13	Week 14	Week 15	Week 16	Week 17	Week 18	Week 19	Week 20	Week 21	Week 22	Week 23	Week 24	Week 25
Start date of week (Monday)	5-24	5-31	6-7	6-14	6-21	6-28	7-5	7-12	7-19	7-26	8-2	8-9	8-16	8-23	8-30	9-6	9-13	9-20	9-27	10-4	10-11	10-18	10-25		
U = unofficial season O = official season	U →	→	→	→	→	→	→	→	→	→	→	→	O →	→	→	→	→	→	→	→	→	→	→		
Training season* R, B, P, E, PK	R →	→	→	B →	→	→	→	→	→	→	→	→	P →	→	→	E →	→	→	→	→	PK →	→	→		
Weekly stress**: E, M, H	M	H	E	M	H	E	M	H	E	M	H	E	M	H	E	M	H	E	M	H	M	E	E		
Guideline TRIMPS (from table 10.3	47	63	31	49	66	32	52	69	34	53	71	35	70	94	46	56	76	37	50	66	48	34	30		

*R = recovery period, B = basic training period, P = precompetition period, E = early competition period, PK = peak performance period

**E = easy week (recovery), M = medium-stress week, H = high-stress week

From *Sport Physiology for Coaches* by Brian J. Sharkey and Steven E. Gaskill, 2006, Champaign, IL: Human Kinetics

Sample moderate week for basic training period in high school soccer.

Suggested weekly TRIMPS 47		Weekly rating Easy Med Hard	Monday		Tuesday		Wednesday		Thursday		Friday		Saturday		Sunday	
			Suggested TRIMPS 10	Daily rating Easy Med Hard	Suggested TRIMPS 7	Daily rating Easy Med Hard	Suggested TRIMPS 7	Daily rating Easy Med Hard	Suggested TRIMPS 4	Daily rating Easy Med Hard	Suggested TRIMPS 10	Daily rating Easy Med Hard	Suggested TRIMPS 7	Daily rating Easy Med Hard	Suggested TRIMPS	Daily rating Easy Med Hard
MUSCULAR FITNESS	Strength 4-6 TRIMPS/hr		Hours: 1:00 TRIMPS: 5.0	1 × 5 1 × 8 Legs and arms	Hours: : TRIMPS:		Hours: 1:00 TRIMPS: 5.0	1 × 3 1 × 5 Legs and arms	Hours: : TRIMPS:		Hours: 1 :30 TRIMPS: 8.0	1 × 3 1 × 5 1 × 8 Legs and arms	Hours: : TRIMPS:		Hours: : TRIMPS:	Day off
MUSCULAR FITNESS	Power 6-8 TRIMPS/hr		Hours: :30 TRIMPS: 3.0	1 × 10 Fast Mainly arms	Hours: : TRIMPS:		Hours: : TRIMPS:		Hours: : TRIMPS:		Hours: : TRIMPS:		Hours: : TRIMPS:		Hours: : TRIMPS:	
MUSCULAR FITNESS	Power endurance 5-7 TRIMPS/hr		Hours: : TRIMPS:		Hours: : TRIMPS:		Hours: :30 TRIMPS: 3.0	1 × 25	Hours: : TRIMPS:		Hours: : TRIMPS:		Hours: : TRIMPS:		Hours: : TRIMPS:	
MUSCULAR FITNESS	Plyometrics and other 4-7 TRIMPS/hr		Hours: : TRIMPS:		Hours: :30 TRIMPS: 3.5	Leg plyometrics	Hours: : TRIMPS:		Hours: :30 TRIMPS: 3.0	Core strength training	Hours: : TRIMPS:		Hours: :30 TRIMPS: 3.0	Core strength training	Hours: : TRIMPS:	
ENERGY FITNESS	EZ 1.5-3 TRIMPS/hr		Hours: :45 TRIMPS: 2.0	Jogging RPE 11	Hours: : TRIMPS:		Hours: :30 TRIMPS: 1.0	Any aero- bic activity: bike, jog, or swim	Hours: : TRIMPS:		Hours: 1:00 TRIMPS: 2.5	Jogging RPE 10-11	Hours: : TRIMPS:		Hours: : TRIMPS:	
ENERGY FITNESS	PZ 5-7 TRIMPS/hr		Hours: : TRIMPS:		Hours: :30 TRIMPS: 3.0	2-min intervals RPE 16 with 2 min rest	Hours: : TRIMPS:		Hours: : TRIMPS:		Hours: : TRIMPS:		Hours: : TRIMPS:		Hours: : TRIMPS:	
ENERGY FITNESS	MZ 7-9 TRIMPS/hr		Hours: : TRIMPS:		Hours: : TRIMPS:		Hours: : TRIMPS:		Hours: : TRIMPS:		Hours: : TRIMPS:		Hours: : TRIMPS:		Hours: : TRIMPS:	
Skills, psychological, tactics, and other training			Hours: 1:00 TRIMPS: 4.0	Soccer	Hours: : TRIMPS:		Hours: 1:00 TRIMPS: 4.0	Soccer	Hours: :45 TRIMPS: 3.0	Soccer	Hours: : TRIMPS:		Hours: 2:00 TRIMPS: 4.0	Soccer	Hours: : TRIMPS:	
Daily hours			2:45		2:00		3:00		1:15		2:30		2:30		0	
Daily TRIMPS			11.0		6.5		13.0		6.0		10.5		7.0		0	

Note: Resistance training should be done separately from energy fitness.

Total hours for week 14:00 Total TRIMPS for week 54.0

Sample moderate week for precompetition period in high school soccer.

Suggested weekly TRIMPS 70	Weekly rating (Easy Med Hard)	Monday		Tuesday		Wednesday		Thursday		Friday		Saturday		Sunday	
		Suggested TRIMPS 15	Daily rating (Easy Med Hard)	Suggested TRIMPS 5	Daily rating (Easy Med Hard)	Suggested TRIMPS 15	Daily rating (Easy Med Hard)	Suggested TRIMPS 5	Daily rating (Easy Med Hard)	Suggested TRIMPS 15	Daily rating (Easy Med Hard)	Suggested TRIMPS 5	Daily rating (Easy Med Hard)	Suggested TRIMPS 0	Daily rating (Easy Med Hard)
MUSCULAR FITNESS	Strength 4-6 TRIMPS/hr	Hours: : TRIMPS:		Hours: : TRIMPS:		Hours: : TRIMPS:		Hours: : TRIMPS:		Hours: : TRIMPS:		Hours: : TRIMPS:		Hours: : TRIMPS:	
MUSCULAR FITNESS	Power 6-8 TRIMPS/hr	Hours: 1:00 TRIMPS: 7.0	1 × 10 1 × 12 Fast Legs and arms	Hours: : TRIMPS:		Hours: 1:00 TRIMPS: 7.0	1 × 10 1 × 12 Fast Legs and arms	Hours: : TRIMPS:		Hours: 1:00 TRIMPS: 7.0	1 × 10 1 × 12 Fast Legs and arms	Hours: : TRIMPS:		Hours: : TRIMPS:	
MUSCULAR FITNESS	Power endurance 5-7 TRIMPS/hr	Hours: :30 TRIMPS: 3.0	1 × 25 Legs and arms	Hours: : TRIMPS:		Hours: : TRIMPS:		Hours: : TRIMPS:		Hours: :30 TRIMPS: 3.0	1 × 25 Legs and arms	Hours: : TRIMPS:		Hours: : TRIMPS:	
MUSCULAR FITNESS	Plyometrics and other 4-7 TRIMPS/hr	Hours: : TRIMPS:		Hours: :30 TRIMPS: 2.5	Core strength training	Hours: :30 TRIMPS: 3.5	Leg plyometrics	Hours: : TRIMPS:		Hours: :30 TRIMPS: 2.5	Core strength training	Hours: :30 TRIMPS: 3.5	Leg plyometrics	Hours: : TRIMPS:	
ENERGY FITNESS	EZ 1.5-3 TRIMPS/hr	Hours: :30 TRIMPS:		Hours: :30 TRIMPS: 1.0	Relaxed jog	Hours: : TRIMPS:		Hours: : TRIMPS:		Hours: : TRIMPS:		Hours: :30 TRIMPS: 1.0	Relaxed jog	Hours: :45 TRIMPS: a.5	Relaxed jog
ENERGY FITNESS	PZ 5-7 TRIMPS/hr	Hours: :30 TRIMPS: 3.0	Intervals, 1.5 min with 2 min rest, RPE 17	Hours: : TRIMPS:		Hours: : TRIMPS:		Hours: :30 TRIMPS: 3.0	Intervals, 2 min with 2 min rest, RPE 16	Hours: : TRIMPS:		Hours: : TRIMPS:		Hours: : TRIMPS:	
ENERGY FITNESS	MZ 7-9 TRIMPS/hr	Hours: : TRIMPS:		Hours: :10 TRIMPS: 1	5 sec max sprints with 55 sec rest	Hours: : TRIMPS:		Hours: : TRIMPS:		Hours: : TRIMPS:		Hours: : TRIMPS:		Hours: : TRIMPS:	
Skills, psychological, tactics, and other training		Hours: 1:30 TRIMPS: 2.0	Soccer drills and scrimmage, psychological skills	Hours: 1:30 TRIMPS: 4.0	Soccer drills and scrimmage, psychological skills	Hours: 2:00 TRIMPS: 6.0	Soccer drills and scrimmage, psychological skills	Hours: 1:00 TRIMPS: 2.0	Easy drills and small-sided games	Hours: 1:30 TRIMPS: 5.0	4 × 20 min scrimmages	Hours: : TRIMPS:		Hours: : TRIMPS:	
Daily hours		3:30		2:40		3:30		1:30		3:30		1:00		0:45	
Daily TRIMPS		15		8.5		16.5		5.0		17.5		4.5		0	

Note: Resistance training should be done on own in the morning.

Total hours for week 16.25 Total TRIMPS for week 67.0

Sample moderate week for early competition period in high school soccer.

Suggested weekly TRIMPS 56	Weekly rating Easy Med Hard	Monday Suggested TRIMPS 12	Monday Daily rating Easy Med Hard	Tuesday Suggested TRIMPS 4	Tuesday Daily rating Easy Med Hard	Wednesday Suggested TRIMPS 8	Wednesday Daily rating Easy Med Hard	Thursday Suggested TRIMPS 8	Thursday Daily rating Easy Med Hard	Friday Suggested TRIMPS 12	Friday Daily rating Easy Med Hard	Saturday Suggested TRIMPS 8	Saturday Daily rating Easy Med Hard	Sunday Suggested TRIMPS 0	Sunday Daily rating Easy Med Hard
MUSCULAR FITNESS Strength 4-6 TRIMPS/hr		Hours: : TRIMPS:		Hours: : TRIMPS:		Hours: : TRIMPS:		Hours: : TRIMPS:		Hours: : TRIMPS:		Hours: : TRIMPS:		Hours: : TRIMPS:	Day off
MUSCULAR FITNESS Power 6-8 TRIMPS/hr		Hours: :20 TRIMPS: 2.0	2 × 10 Fast Arms	Hours: : TRIMPS:		Hours: : TRIMPS:		Hours: : TRIMPS:		Hours: : TRIMPS:		Hours: :20 TRIMPS: 2.0	2 × 10 Fast Arms	Hours: : TRIMPS:	
MUSCULAR FITNESS Power endurance 5-7 TRIMPS/hr		Hours: :40 TRIMPS: 4.0	1 × 25 1 × 30 Legs	Hours: : TRIMPS:		Hours: : TRIMPS:		Hours: :40 TRIMPS: 4.0	1 × 25 1 × 30 Legs	Hours: : TRIMPS:		Hours: :40 TRIMPS: 4.0	1 × 25 1 × 30 Legs	Hours: : TRIMPS:	
MUSCULAR FITNESS Plyometrics and other 4-7 TRIMPS/hr		Hours: :30 TRIMPS: 3.5	Leg plyometrics	Hours: : TRIMPS:		Hours: : TRIMPS:		Hours: :30 TRIMPS: 3.5	Leg plyometrics	Hours: : TRIMPS:		Hours: :30 TRIMPS: 3.5	Leg plyometrics	Hours: : TRIMPS:	
ENERGY FITNESS EZ 1.5-3 TRIMPS/hr		Hours: :30 TRIMPS:		Hours: : TRIMPS:		Hours: : TRIMPS:		Hours: : TRIMPS:		Hours: : TRIMPS:		Hours: 1:00 TRIMPS: 2.0	Jog or run, RPE 11	Hours: : TRIMPS:	Day off
ENERGY FITNESS PZ 5-7 TRIMPS/hr		Hours: :15 TRIMPS: 2.0	30 sec intervals, RPE 18 (90%) 1:30 rest	Hours: : TRIMPS:	PZ occurs naturally during soccer	Hours: : TRIMPS:	PZ occurs naturally during soccer	Hours: : TRIMPS:	PZ occurs naturally during soccer	Hours: : TRIMPS:	PZ occurs naturally during soccer	Hours: : TRIMPS:		Hours: : TRIMPS:	
ENERGY FITNESS MZ 7-9 TRIMPS/hr		Hours: : TRIMPS:		Hours: :15 TRIMPS: 2.0	5 sec max intervals with 55 sec rest	Hours: : TRIMPS:		Hours: : TRIMPS:		Hours: : TRIMPS:		Hours: : TRIMPS:		Hours: : TRIMPS:	
Skills, psychological, tactics, and other training		Hours: 1:15 TRIMPS: 2.0	Soccer drills and scrimmage, psychologi- cal skills	Hours: 1:15 TRIMPS: 4.0	3 × 15 min scrimmages, psychologi- cal skills	Hours: 1:30 TRIMPS: 6.0	Game	Hours: 1:30 TRIMPS: 4.0	Easy drills and small-sided games (3/4 speed)	Hours: 1:30 TRIMPS: 6.0	Game	Hours: : TRIMPS:		Hours: : TRIMPS:	
Daily hours		3:00		1:30		1:30		2:40		1:30		2:30		0	
Daily TRIMPS		13.5		6.0		6.0		11.5		6.0		11.5		0	

Note: Resistance training should be done on own in the morning. Reduce Thursday intensity if Friday games are more important.

Total hours for week 11:40 Total TRIMPS for week 54.5

Sample moderate week for peak performance period in high school soccer.

	Suggested weekly TRIMPS 43	Weekly rating Easy Med Hard	Monday Suggested TRIMPS 10	Monday Daily rating Easy Med Hard	Tuesday Suggested TRIMPS 6	Tuesday Daily rating Easy Med Hard	Wednesday Suggested TRIMPS 3	Wednesday Daily rating Easy Med Hard	Thursday Suggested TRIMPS 6	Thursday Daily rating Easy Med Hard	Friday Suggested TRIMPS 3	Friday Daily rating Easy Med Hard	Saturday Suggested TRIMPS 6	Saturday Daily rating Easy Med Hard	Sunday Suggested TRIMPS 3	Sunday Daily rating Easy Med Hard
MUSCULAR FITNESS	Strength 4-6 TRIMPS/hr		Hours: : TRIMPS:		Hours: : TRIMPS:		Hours: : TRIMPS:		Hours: : TRIMPS:		Hours: : TRIMPS:		Hours: : TRIMPS:		Hours: : TRIMPS:	
MUSCULAR FITNESS	Power 6-8 TRIMPS/hr		Hours: :15 TRIMPS: 2.0	1 × 10 Fast Arms	Hours: : TRIMPS:		Hours: : TRIMPS:		Hours: : TRIMPS:		Hours: : TRIMPS:		Hours: : TRIMPS:		Hours: : TRIMPS:	
MUSCULAR FITNESS	Power endurance 5-7 TRIMPS/hr		Hours: :30 TRIMPS: 2.0	1 × 25 Legs	Hours: : TRIMPS:		Hours: : TRIMPS:		Hours: : TRIMPS:		Hours: : TRIMPS:		Hours: : TRIMPS:		Hours: : TRIMPS:	
MUSCULAR FITNESS	Plyometrics and other 4-7 TRIMPS/hr		Hours: :30 TRIMPS: 3.5	Leg plyometrics	Hours: : TRIMPS:		Hours: : TRIMPS:		Hours: : TRIMPS:		Hours: : TRIMPS:		Hours: : TRIMPS:		Hours: : TRIMPS:	
ENERGY FITNESS	EZ 1.5-3 TRIMPS/hr		Hours: : TRIMPS:		Hours: : TRIMPS:		Hours: : TRIMPS:		Hours: : TRIMPS:		Hours: : TRIMPS:		Hours: 1:00 TRIMPS: 2.0	Jog or run, RPE 11	Hours: 1:00 TRIMPS: 3.0	Jog or run, RPE 11
ENERGY FITNESS	PZ 5-7 TRIMPS/hr		Hours: : TRIMPS:	PZ occurs naturally during soccer	Hours: : TRIMPS:	PZ occurs naturally during soccer	Hours: : TRIMPS:	PZ occurs naturally during soccer	Hours: : TRIMPS:	PZ occurs naturally during soccer	Hours: : TRIMPS:	PZ occurs naturally during soccer	Hours: : TRIMPS:		Hours: : TRIMPS:	
ENERGY FITNESS	MZ 7-9 TRIMPS/hr		Hours: :15 TRIMPS: 2.0	5 sec max intervals with 55 sec rest	Hours: : TRIMPS:		Hours: : TRIMPS:		Hours: : TRIMPS:		Hours: : TRIMPS:		Hours: : TRIMPS:		Hours: : TRIMPS:	
Skills, psychological, tactics, and other training			Hours: 1:00 TRIMPS: 1.0	Tactics, very easy, psychological skills	Hours: 1:30 TRIMPS: 6.0	4 × 15 min hard scrimmages, psychological skills	Hours: 1:30 TRIMPS: 4.0	4 × 15 min hard games with 5 min rest, 30 min 3/4-speed game	Hours: 1:30 TRIMPS: 6.0	Game	Hours: 1:30 TRIMPS: 3.0	Psychological skills, video, tactics, and 20 min easy scrimmage	Hours: 1:30 TRIMPS: 6.0	Game	Hours: : TRIMPS:	
Daily hours			2:30		1:30		1:30		1:30		1:30		2:30		0	
Daily TRIMPS			11.5		6.0		4.0		6.0		3.0		6.0		0	

Note: Resistance training should be done on own, Monday mornings.

Total hours for week 11:00 Total TRIMPS for week 36.5

Sample plan for high school cross country running.

Recovery Period: Dates May 24 – June 13 # weeks 3 (Minimum 2 weeks)

Major Muscular-Fitness Goals of Period	Major Energy-Fitness Goals of Period
Athletes on own—unofficial training	General activities, especially nonweight bearing, such as cycling or swimming

Basic Training Period: Dates June 14 – Aug. 15 # weeks 9 (Minimum 9 weeks)

Major Muscular-Fitness Goals of Period	Major Energy-Fitness Goals of Period
Begin resistance training with a light preparatory program Major emphases: Strength, 6-8RM (focus on legs, 2 to 3 times per week) Core strength Maintenance of power, 8-12RM (fast movements, legs) Maintenance of power endurance, 20-30RM (legs) Flexibility with a focus on hamstrings and lower back	Large volume of EZ-intensity training; add miles of slow running PZ intervals once per week, 2-3% above target performance speed, 4-6 min in duration with full recovery (3-5 min) Run in a variety of terrain similar to race courses; occasionally add longer hills, but keep intensity low (EZ)

Precompetition Period: Dates Aug. 15 – Sept. 12 # weeks 4 (Minimum 6 weeks)

Major Muscular-Fitness Goals of Period	Major Energy-Fitness Goals of Period
Some early races are included in this period Major emphases: Power, 8-12RM (legs), mainly plyometrics Core strength Power endurance, 20-30RM (2 times per week, legs only) Flexibility	EZ-intensity runs; see daily schedule (80% of volume) PZ intervals twice per week, 3-4% above target performance speed, 3-4 min in duration with full recovery (3-5 min) MZ intervals added as short bursts of speed (10-15 sec) occasionally during one or two days of EZ training; limit these to 4-5 per day with long periods (5-10 min) between intervals

Early Competition Period: Dates Sept. 13 – Oct. 10 # weeks 4 (Minimum 3 weeks)

Major Muscular-Fitness Goals of Period	Major Energy-Fitness Goals of Period
Major emphases: Power endurance, 20-30RM (legs only, body and partner resistance) Continued plyometrics, but tapering during this period Maintain core strength Speed training Flexibility	Maintain aerobic fitness with EZ run 1 to 2 times per week PZ intervals twice per week, 4-5% above target performance speed, 3-8 min in duration with shorter recovery (1-2 min) MZ intervals twice per week (5-30 sec, 1.5 min rest, 8-10 repeats) Psychological skills training

Peak Performance Period: Dates Oct. 11 – Oct. 31 # weeks 3 (3-6 weeks)

Major Muscular-Fitness Goals of Period	Major Energy-Fitness Goals of Period
Maintenance of power (plyometrics, light and short sets) Maintenance of power endurance (body and partner resistance) Flexibility Focus on rest and recovery	Maintain aerobic fitness with EZ run 1 to 2 times per week PZ intervals once per week, 2-3% above target performance speed, 5-10 min in duration with short recovery (1-2 min) MZ intervals once per week (10-20 sec, 1 min rest, 8-10 repeats) Psychological skills training Focus on rest and recovery

Week starting with basic strength and endurance period	Week 1	Week 2	Week 3	Week 4	Week 5	Week 6	Week 7	Week 8	Week 9	Week 10	Week 11	Week 12	Week 13	Week 14	Week 15	Week 16	Week 17	Week 18	Week 19	Week 20	Week 21	Week 22	Week 23	Week 24	Week 25
Start date of week (Monday)	5-24	5-31	6-7	6-14	6-21	6-28	7-5	7-12	7-19	7-26	8-2	8-9	8-16	8-23	8-30	9-6	9-13	9-20	9-27	10-4	10-11	10-18	10-25		
U = unofficial season O = official season	U	→	→	→	→	→	→	→	→	→	→	→	O	→	→	→	→	→	→	→	→	→	→	→	
Training season*: R, B, P, E, PK	R	→	→	B	→	→	→	→	→	→	→	→	P	→	→	→	E	→	→	→	PK	→	→		
Weekly stress**: E, M, H	M	H	E	M	H	E	M	H	E	M	H	E	M	H	E	M	H	E	H	E	M	E	E		
Guideline TRIMPS (from table 10.4	47	63	31	49	66	32	52	69	34	53	71	35	70	94	46	56	76	37	76	40	48	34	30		

*R = recovery period, B = basic training period, P = precompetition period, E = early competition period, PK = peak performance period

**E = easy week (recovery), M = medium-stress week, H = high-stress week

Sample moderate week for basic training period in high school cross country running.

Suggested weekly TRIMPS 47		Weekly rating (Easy Med Hard)	Monday		Tuesday		Wednesday		Thursday		Friday		Saturday		Sunday	
			Suggested TRIMPS 10	Daily rating (Easy Med Hard)	Suggested TRIMPS 7	Daily rating (Easy Med Hard)	Suggested TRIMPS 10	Daily rating (Easy Med Hard)	Suggested TRIMPS 4	Daily rating (Easy Med Hard)	Suggested TRIMPS 10	Daily rating (Easy Med Hard)	Suggested TRIMPS 4	Daily rating (Easy Med Hard)	Suggested TRIMPS 7	Daily rating (Easy Med Hard)
MUSCULAR FITNESS	Strength 4-6 TRIMPS/hr		Hours: 1:00 TRIMPS: 5.0	1 × 6 1 × 8 1 × 10 Legs	Hours: : TRIMPS:		Hours: :40 TRIMPS: 4.0	1 × 8 1 × 10 Legs	Hours: : TRIMPS:		Hours: :40 TRIMPS: 4.0	1 × 4 1 × 6 Legs	Hours: : TRIMPS:		Hours: : TRIMPS:	
	Power 6-8 TRIMPS/hr		Hours: :20 TRIMPS: 2.0	1 × 12 Fast Legs	Hours: : TRIMPS:		Hours: :20 TRIMPS: 2.0	1 × 12 Fast Legs	Hours: : TRIMPS:		Hours: : TRIMPS:		Hours: : TRIMPS:		Hours: : TRIMPS:	
	Power endurance 5-7 TRIMPS/hr		Hours: : TRIMPS:		Hours: : TRIMPS:		Hours: :30 TRIMPS: 3.0	1 × 25 Legs	Hours: : TRIMPS:		Hours: :20 TRIMPS: 2.0	1 × 25 Legs	Hours: : TRIMPS:		Hours: : TRIMPS:	
	Plyometrics and other 4-7 TRIMPS/hr		Hours: : TRIMPS:		Hours: :30 TRIMPS: 3.0	Core strength training	Hours: : TRIMPS:		Hours: :15 TRIMPS: 1.5	Core strength training	Hours: : TRIMPS:		Hours: :15 TRIMPS: 1.5	Core strength training	Hours: : TRIMPS:	
ENERGY FITNESS	EZ 1.5-3 TRIMPS/hr		Hours: 2:00 TRIMPS: 4.0	Run RPE 11	Hours: 1:00 TRIMPS: 4.0	Nonspecific (bike, etc.) aerobic exercise	Hours: :20 TRIMPS: .5	Run RPE 10	Hours: 1:30 TRIMPS: 3.0	Run RPE 11	Hours: 1:30 TRIMPS: 5.0	Run in hills, vary RPE (mostly easy)	Hours: :30 TRIMPS: .1	Run RPE 11	Hours: :20 TRIMPS: .5	Run RPE 10
	PZ 5-7 TRIMPS/hr		Hours: : TRIMPS:		Hours: : TRIMPS:		Hours: :30 TRIMPS: 3.0	5 min @ race + 3% with 3 min rest	Hours: : TRIMPS:		Hours: : TRIMPS:		Hours: : TRIMPS:		Hours: 1:00 TRIMPS: 6.0	4 min @ race + 2% with 3 min rest
	MZ 7-9 TRIMPS/hr		Hours: : TRIMPS:		Hours: : TRIMPS:		Hours: : TRIMPS:		Hours: :05 TRIMPS: 1.0	5 × 10-20 sec max-effort sprints during run	Hours: : TRIMPS:		Hours: : TRIMPS:		Hours: : TRIMPS:	
Skills, psychological, tactics, and other training			Hours: : TRIMPS:		Hours: : TRIMPS:		Hours: : TRIMPS:		Hours: : TRIMPS:		Hours: : TRIMPS:		Hours: : TRIMPS:		Hours: : TRIMPS:	
Daily hours			3:20		2:00		2:20		1:50		2:30		:45		1:20	
Daily TRIMPS			11.0		7.0		12.5		5.5		11.0		2.5		6.5	

Note: Resistance training should be done separately from energy fitness.

Total hours for week 14.05 Total TRIMPS for week 56.0

Sample moderate week for precompetition period in high school cross country running.

	Suggested weekly TRIMPS 70	Weekly rating (Easy Med Hard)	Monday Suggested TRIMPS 15	Monday Daily rating (Easy Med Hard)	Tuesday Suggested TRIMPS 5	Tuesday Daily rating (Easy Med Hard)	Wednesday Suggested TRIMPS 15	Wednesday Daily rating (Easy Med Hard)	Thursday Suggested TRIMPS 5	Thursday Daily rating (Easy Med Hard)	Friday Suggested TRIMPS 10	Friday Daily rating (Easy Med Hard)	Saturday Suggested TRIMPS 5	Saturday Daily rating (Easy Med Hard)	Sunday Suggested TRIMPS 10	Sunday Daily rating (Easy Med Hard)
MUSCULAR FITNESS	Strength 4-6 TRIMPS/hr		Hours: : TRIMPS:		Hours: : TRIMPS:		Hours: : TRIMPS:		Hours: : TRIMPS:		Hours: : TRIMPS:		Hours: : TRIMPS:		Hours: : TRIMPS:	
MUSCULAR FITNESS	Power 6-8 TRIMPS/hr		Hours: :40 TRIMPS: 7.0	1 × 10 1 × 12 Fast Legs	Hours: : TRIMPS:		Hours: : TRIMPS:		Hours: : TRIMPS:		Hours: : TRIMPS:		Hours: : TRIMPS:		Hours: : TRIMPS:	
MUSCULAR FITNESS	Power endurance 5-7 TRIMPS/hr		Hours: :20 TRIMPS: 2.0	1 × 25 Legs	Hours: : TRIMPS:		Hours: :40 TRIMPS: 4.0	2 × 25 Legs	Hours: : TRIMPS:		Hours: :30 TRIMPS: 3.0	1 × 25 Legs and arms	Hours: : TRIMPS:		Hours: : TRIMPS:	
MUSCULAR FITNESS	Plyometrics and other 4-7 TRIMPS/hr		Hours: :30 TRIMPS: 3.5	Leg plyometrics	Hours: :30 TRIMPS: 2.5	Core strength training	Hours: :30 TRIMPS: 3.5	Leg plyometrics	Hours: :30 TRIMPS: 2.5	Core strength training	Hours: : TRIMPS:		Hours: : TRIMPS:		Hours: : TRIMPS:	
ENERGY FITNESS	EZ 1.5-3 TRIMPS/hr		Hours: 2:30 TRIMPS: 5.0	Distance run, RPE 11	Hours: 1:00 TRIMPS: 2.0	Nonspecific (bike or swim)	Hours: :20 TRIMPS: 1.0	Warm-up	Hours: :45 TRIMPS: 1.0	Easy run	Hours: :20 TRIMPS: 1.0	Warm-up	Hours: 1:55 TRIMPS: 5	Distance run RPE 11	Hours: : TRIMPS:	
ENERGY FITNESS	PZ 5-7 TRIMPS/hr		Hours: : TRIMPS:		Hours: : TRIMPS:		Hours: 1:30 TRIMPS: 9.0	3-5 min @ race + 4% with 3+ min rest	Hours: : TRIMPS:		Hours: 1:00 TRIMPS: 6.0	5 min @ race + 3% with 3+ min rest (part in hills)	Hours: : TRIMPS:		Hours: 1:30 TRIMPS: 9.0	3 × 10 min @ race + 2% with long rest
ENERGY FITNESS	MZ 7-9 TRIMPS/hr		Hours: : TRIMPS:		Hours: : TRIMPS:		Hours: : TRIMPS:		Hours: : TRIMPS:		Hours: : TRIMPS:		Hours: :05 TRIMPS: 1.0	1 × 10-20 sec max effort sprints during run	Hours: : TRIMPS:	
Skills, psychological, tactics, and other training			Hours: : TRIMPS:		Hours: :30 TRIMPS:	Psycho-logical skills	Hours: : TRIMPS:		Hours: :30 TRIMPS:	Psycho-logical skills	Hours: :40 TRIMPS:	Psycho-logical skills	Hours: : TRIMPS:		Hours: : TRIMPS:	
Daily hours			4:00		2:00		3:00		1:45		2:30		2:00		2:00	
Daily TRIMPS			17.5		3.5		17.5		4.5		10.0		6.0		10	

Note: Resistance training should be done on own in the morning.

Total hours for week 17:15 Total TRIMPS for week 69.0

Sample moderate week for early competition period in high school cross country running.

Suggested weekly TRIMPS 56		Weekly rating (Easy Med Hard)	Monday		Tuesday		Wednesday		Thursday		Friday		Saturday		Sunday	
			Suggested TRIMPS 4	Daily rating (Easy Med Hard)	Suggested TRIMPS 8	Daily rating (Easy Med Hard)	Suggested TRIMPS 12	Daily rating (Easy Med Hard)	Suggested TRIMPS 4	Daily rating (Easy Med Hard)	Suggested TRIMPS 8	Daily rating (Easy Med Hard)	Suggested TRIMPS 8	Daily rating (Easy Med Hard)	Suggested TRIMPS 4	Daily rating (Easy Med Hard)
MUSCULAR FITNESS	Strength 4-6 TRIMPS/hr		Hours: : TRIMPS:		Hours: : TRIMPS:		Hours: : TRIMPS:		Hours: : TRIMPS:		Hours: : TRIMPS:		Hours: : TRIMPS:		Hours: : TRIMPS:	
	Power 6-8 TRIMPS/hr		Hours: : TRIMPS:		Hours: : TRIMPS:		Hours: : TRIMPS:		Hours: : TRIMPS:		Hours: : TRIMPS:		Hours: : TRIMPS:		Hours: :20 TRIMPS: 3.0	1 × 12 Legs Fast
	Power endurance 5-7 TRIMPS/hr		Hours: : TRIMPS:		Hours: : TRIMPS:		Hours: : TRIMPS:		Hours: : TRIMPS:		Hours: : TRIMPS:		Hours: : TRIMPS:		Hours: :40 TRIMPS: 5.0	1 × 25 1 × 30 Legs
	Plyometrics and other 4-7 TRIMPS/hr		Hours: :30 TRIMPS: 3.5	Leg plyometrics	Hours: : TRIMPS:		Hours: :20 TRIMPS: 3.0	Leg plyometrics	Hours: : TRIMPS:		Hours: : TRIMPS:		Hours: : TRIMPS:		Hours: : TRIMPS:	
ENERGY FITNESS	EZ 1.5-3 TRIMPS/hr		Hours: 1:45 TRIMPS: 6.0	Distance run RPE 10-11 in mod. hills	Hours: :20 TRIMPS: .5	Warm-up run	Hours: :20 TRIMPS: .5	Warm-up run	Hours: 1:15 TRIMPS: 4.0	Easy run	Hours: :20 TRIMPS: .5	Warm-up	Hours: 2:00 TRIMPS: 5.0	Distance run RPE 11	Hours: : TRIMPS:	
	PZ 5-7 TRIMPS/hr		Hours: : TRIMPS:		Hours: :30 TRIMPS: 3.0	3 min @ race + 5% with 2 min rest	Hours: 1:00 TRIMPS: 6.0	4 min @ race + 3% with 3 min rest	Hours: : TRIMPS:		Hours: :25 TRIMPS: 5.0	RACE	Hours: : TRIMPS:		Hours: : TRIMPS:	
	MZ 7-9 TRIMPS/hr		Hours: :05 TRIMPS: 1.0	5 × 10-20 sec max effort sprints during run	Hours: : TRIMPS:		Hours: : TRIMPS:		Hours: :05 TRIMPS: 1.0	5 × 10-20 sec max effort sprints during run	Hours: : TRIMPS:		Hours: : TRIMPS:		Hours: : TRIMPS:	
Skills, psychological, tactics, and other training			Hours: : TRIMPS:		Hours: :30 TRIMPS:	Psycho-logical skills	Hours: :40 TRIMPS:	Psycho-logical skills	Hours: :30 TRIMPS:	Psycho-logical skills	Hours: : TRIMPS:		Hours: : TRIMPS:		Hours: : TRIMPS:	
Daily hours			2:20		1:20		2:20		1:50		:45		2:00		1:00	
Daily TRIMPS			10.5		3.5		9.5		5.0		5.5		5.0		8.0	

Note: Races on Fridays.

Total hours for week 11:45 Total TRIMPS for week 47.0

Sample moderate week for peak performance period in high school cross country running.

Suggested weekly TRIMPS 43	Weekly rating (Easy Med Hard)	Monday		Tuesday		Wednesday		Thursday		Friday		Saturday		Sunday	
		Suggested TRIMPS 10	Daily rating (Easy Med Hard)	Suggested TRIMPS 3	Daily rating (Easy Med Hard)	Suggested TRIMPS 6	Daily rating (Easy Med Hard)	Suggested TRIMPS 6	Daily rating (Easy Med Hard)	Suggested TRIMPS 3	Daily rating (Easy Med Hard)	Suggested TRIMPS 6	Daily rating (Easy Med Hard)	Suggested TRIMPS 0	Daily rating (Easy Med Hard)
MUSCULAR FITNESS	Strength 4-6 TRIMPS/hr	Hours: : TRIMPS:		Hours: : TRIMPS:		Hours: : TRIMPS:		Hours: : TRIMPS:		Hours: : TRIMPS:		Hours: : TRIMPS:		Hours: : TRIMPS:	
MUSCULAR FITNESS	Power 6-8 TRIMPS/hr	Hours: : TRIMPS:		Hours: : TRIMPS:		Hours: : TRIMPS:		Hours: : TRIMPS:		Hours: : TRIMPS:		Hours: : TRIMPS:		Hours: : TRIMPS:	
MUSCULAR FITNESS	Power endurance 5-7 TRIMPS/hr	Hours: : TRIMPS:		Hours: : TRIMPS:		Hours: : TRIMPS:		Hours: : TRIMPS:		Hours: : TRIMPS:		Hours: : TRIMPS:		Hours: : TRIMPS:	
MUSCULAR FITNESS	Plyometrics and other 4-7 TRIMPS/hr	Hours: :30 TRIMPS: 3.5	Leg plyometrics	Hours: : TRIMPS:		Hours: : TRIMPS:		Hours: : TRIMPS:		Hours: : TRIMPS:		Hours: : TRIMPS:		Hours: : TRIMPS:	
ENERGY FITNESS	EZ 1.5-3 TRIMPS/hr	Hours: 1:30 TRIMPS: 5.0	Distance run with 1-2 min pickups	Hours: 1:00 TRIMPS: 3.0	Easy run with pick-ups	Hours: :20 TRIMPS: .5	Warm-up run	Hours: 1:20 TRIMPS: 6.0	Distance run with 1-2 min pickups	Hours: :50 TRIMPS: 2.5	Easy run with pickups	Hours: : TRIMPS:		Hours: : TRIMPS:	
ENERGY FITNESS	PZ 5-7 TRIMPS/hr	Hours: : TRIMPS: 1.0	Fartlek (1-2 min) race-pace pickups during run	Hours: : TRIMPS:		Hours: 1:00 TRIMPS: 6.0	6 min @ race + 3% with 2 min rest	Hours: 1:00 TRIMPS: 1.0	Fartlek (1-2 min) race-pace pickups during run	Hours: : TRIMPS:		Hours: :25 TRIMPS: 6.0	RACE	Hours: : TRIMPS:	
ENERGY FITNESS	MZ 7-9 TRIMPS/hr	Hours: : TRIMPS:		Hours: : TRIMPS: 1.0	5 × 10-20 sec max-effort sprints during run	Hours: : TRIMPS:		Hours: : TRIMPS:		Hours: : TRIMPS:		Hours: : TRIMPS:		Hours: : TRIMPS:	
Skills, psychological, tactics, and other training		Hours: : TRIMPS:		Hours: :30 TRIMPS:	Psycho-logical skills	Hours: : TRIMPS:		Hours: :30 TRIMPS:	Psycho-logical skills	Hours: :40 TRIMPS:	Psycho-logical skills	Hours: : TRIMPS:		Hours: : TRIMPS:	
Daily hours		2:00		1:30		1:20		2:50		1:30		:45		0	
Daily TRIMPS		9.5		4.0		6.5		7.0		2.5		6.5		0	

Note: Races on Saturday.

Total hours for week 9:00 Total TRIMPS for week 36.0

From *Sport Physiology for Coaches* by Brian J. Sharkey and Steven E. Gaskill, 2006, Champaign, IL: Human Kinetics

SUMMARY

Designing effective training plans takes time and thought. The sample plans in this chapter should help you in the process but should be taken only as guidelines. Each coach must evaluate the specific needs of his or her athletes and design or modify a program to meet those needs. Successful coaches plan for training and plan training for performance.

KEY TERMS

aerobic (endurance) sport (p. 198)
intermittent (start-and-go) sport (p. 198)
power endurance sport (p. 198)
power sport (p. 198)
skill sport (p. 198)

REVIEW QUESTIONS

Fill in the blank(s) with the appropriate word or words to complete the following statements.

1. Golf, archery, and bowling are examples of _________ sports.
2. Football, weightlifting, and weight events in track and field are examples of _________ sports.
3. Wrestling, cross-country skiing, and middle-distance running are examples of _________ sports.
4. Soccer, basketball, field hockey, and lacrosse are examples of _________ sports.
5. Distance running, swimming, cycling, and cross-country skiing are examples of _________ sports.

Determine if the following statements are true or false and then circle the correct answer.

6. T/F A sport can require high levels of skill with only modest demands for muscular or energy fitness.
7. T/F Improving muscular power enhances the ability to recover between plays and improves the ability to perform late in the game.
8. T/F Energy fitness is required to maximize power and speed and to withstand contact with opponents.
9. T/F Intermittent sport athletes do not need a strong aerobic foundation to enable them to maintain continuous motion or to recover quickly from more intense busts of activity.
10. T/F In aerobic sports, a considerable amount of time should be spent in the easy training zone (EZ) to develop endurance and metabolic efficiency and to allow recovery from more intense training.

PRACTICAL ACTIVITIES

There is nothing better for coaches than to attempt to design a training plan for a sport that they know little about.

1. For one of the following sports, team up with another coach or a fellow student and design a program for that sport. Though the process will take some time, by working together you will both grow in your understanding of how to objectively put together a training program. Some challenging sports might include these:
   a. Cheerleading
   b. Pole-vaulting
   c. Figure skating
2. Work with another coach in your school or from your area, or with a fellow student. Offer to design a training program for your partner if he or she designs one for you. Once you are done, get together and discuss your ideas and thoughts. This process will generate a lot of discussion, and our experience is that you will have great ideas for the other program, just as your partner will have great ideas for your athletes.

12 Performance and Health

This chapter will help you

- understand important health behaviors associated with success in sport;
- identify signs of overwork, overtraining, and illness;
- guide athletes to the position, event, or sport best suited to their capabilities; and
- understand your role in the career development of the athlete.

Ed was a fine young distance runner, the best on his college team, with the potential of becoming an elite national athlete. But he was not responding favorably to the arduous training favored by his coach and tolerated by more mature members of the team. He told the coach that the intense interval training was wearing him down and that the fatigue was affecting his enjoyment of the sport, as well as his performance in the classroom. He requested the coach's permission to substitute other forms of training for some of the intervals. The coach, an old-style authoritarian, disagreed and told him it was "my way or the highway." As his performance continued to deteriorate, Ed chose the highway, and the coach lost an emerging talent.

But Ed didn't drop out of school or stop running. He continued to train and compete in open meets and road races. Months later he was invited to participate in an Olympic Development Camp, an honor restricted to those who demonstrate the potential to compete in national and international competitions. Ed was a unique athlete, one who had learned to listen to his body. He understood the signs of overwork and overtraining and had the conviction to express his feelings to his coach. Unfortunately, the coach was insensitive to the signs of overtraining, even when they were presented by an athlete.

Recent studies indicate that overtraining may result from successive and cumulative alterations in metabolism that become chronic during training (Petibois et al. 2003). Alterations in carbohydrate and lipid metabolism lead to higher use of amino acids, resulting in the catabolism of tissue protein. Thus the metabolism shifts from the main energy sources for exercise (carbohydrates and lipids) to a source not generally used for skeletal muscle contractions, muscle protein. Catabolism of muscle protein undermines training and the ability to perform.

Overtraining is a syndrome in which an athlete trains excessively yet performance declines. It is accompanied by mood and behavior changes and a variety of biochemical and physiological alterations. Smith (2000) argues that high-volume, high-intensity training with insufficient rest will produce muscle and joint trauma; injury-related cytokines and proinflammatory interleukins, leading to systemic inflammation; a whole-body response including central nervous system-related sickness behaviors; alterations in liver function including increased protein breakdown; and a hormone-activated reduction in immune function and resistance to infection. Overtraining can be viewed as a protective mechanism that occurs in response to excessive physiological stress.

Overtraining is not an inevitable consequence of training. Throughout this book we have presented strategies to help you tailor training to individual capabilities. We've emphasized the need to balance work with rest, high-intensity with easy training. Training periods are planned to increase quantity and then quality of training. Progression is tempered with periods of recovery (periodization), and peak performances are preceded by a carefully planned taper. Do these things, or most of them, and overtraining should not be a problem. To put this another way, "Moderation in all things" (Terence, 190-159 B.C.).

A number of behaviors contribute to athletes' physical and mental health and therefore their ability to train and achieve success. In this section we'll discuss those behaviors related to physical health. The most important—good nutrition, hydration, and adequate rest—are outside the direct control of the coach. It is important to encourage these behaviors and to use every opportunity to teach and model good behaviors. The coach can have some control over athletes' level of acclimatization, however, and we discuss strategies to prepare athletes for performing in the heat and at altitude.

Nutrition

Athletes don't always make good food choices, even when they live and eat at home. You can influence those choices and help athletes understand the subtleties of sport nutrition.

In strength training, both resistance exercise and optimal nutrition are major independent stimuli of muscle protein synthesis and muscle growth. Appropriate protein eaten before or after resistance training promotes amino acid uptake, when the synthesis capability of the muscle fiber is highly activated. Consuming protein before exercise is more effective than after, and provision of essential amino acids seems to help diminish muscle breakdown due to hard training (essential amino acids are protein fragments that must be supplied in the diet; they cannot be synthesized in the body).

Similarly, the provision of solid or liquid carbohydrate energy during long-duration practice or competition has been shown to reduce fatigue, increase work output, and improve mood and decision making; and the energy supplement helps to maintain the function of the immune system. We are not advocating the use of expensive supplements; these benefits can be achieved with a sports drink or an energy bar available at most supermarkets. The carbohydrate provides energy to muscles and the nervous system, including the brain. At the same time it decreases the formation of the hormone cortisol, the stress hormone associated with depressed immune function.

One way to monitor overall nutrition is to keep track of an athlete's body weight. One should never lose more than 1 to 2 pounds (0.45 to 0.9 kilograms) per week, and an athlete-in-training shouldn't lose more than 1 pound per week. Rapid weight loss, associated with rigorous training, is a recipe for failure. Encourage adequate intake of complex carbohydrates for energy. Complex carbohydrates include whole-grain bread and pasta, corn, rice, beans, and potatoes. Athletes need carbohydrate energy to sustain vigorous training. Fat intake should be low to moderate, 20 to 25 percent of total calories, with emphasis on polyunsaturated fats. As we said in chapter 7, protein intake should be at least 15 percent of daily caloric intake, or 1.2 to 1.6 grams of protein per kilogram of body weight.

When you travel with the team, discuss good food choices at restaurants, and demonstrate good nutrition in your own selections. Essential vitamins and antioxidants are found in adequate servings of fruits and vegetables. Encourage your athletes to consume four servings (2 cups) of fruits and five servings (2-1/2 cups) of vegetables daily. If you are concerned about their eating behaviors, consider fresh fruit or juice breaks at practice, or ask the team physician about the need for daily multiple vitamin and mineral supplementation. But remember, vitamins in food have proven more effective than those in supplements.

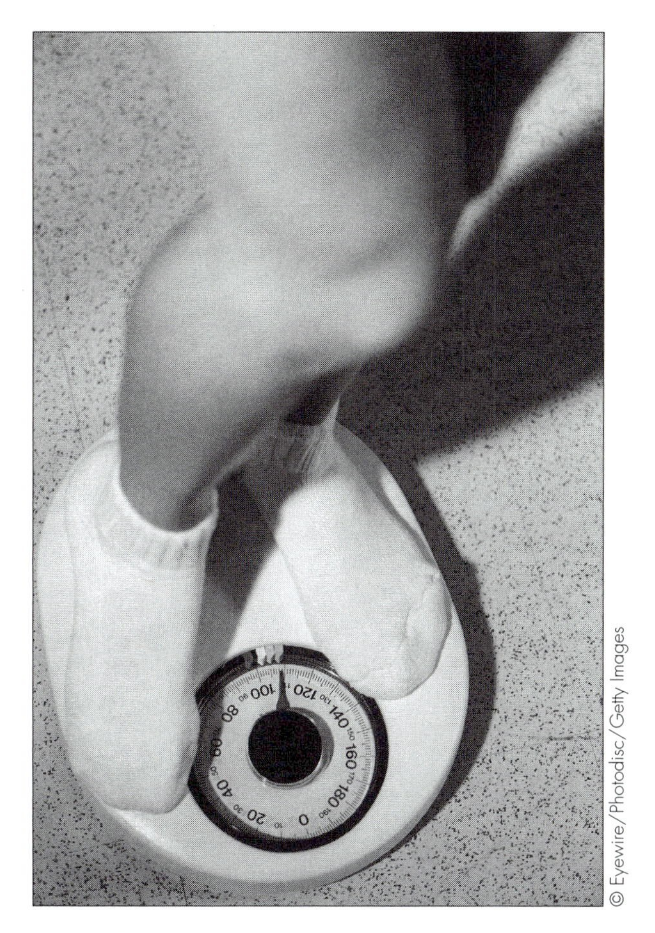

While athletes shouldn't obsess about their weights, monitoring body weight is a way to monitor nutrition.

Hydration

Many athletes live in a state of voluntary dehydration brought on by physical activity and the consumption of caffeinated drinks, including colas, coffee, and so-called energy drinks. Caffeine is a mild diuretic, as is alcohol, another diuretic that contributes to dehydration. While some of these beverages are acceptable when used in moderation, they should not be relied on to meet fluid needs. The body loses fluid in sweat, urine, feces, and respiration. Daily fluid needs vary from 1 liter (1 quart) to over 10 liters (10.6 quarts) according to activity, temperature, and humidity. While many foods provide some water, they cannot provide for the elevated requirements of exertion in a hot environment.

Teach athletes to replace fluids regularly using water or approved sports drinks. Provide fluids at practices and competitions, and urge fluid intake on a regular basis (200-400 milliliters [around 7-14 ounces] every 20 minutes, up to 1 liter [1 quart] per hour). In hot weather, when practice or competition exceeds 1 hour in length, we recommend that half of the fluid requirement be in the form of a sports drink (carbohydrate/electrolyte drink). The flavor encourages fluid intake while providing energy for muscles and the nervous system. The carbohydrate maintains work output, elevates mood and cognitive behavior, and boosts the function of the immune system. The electrolytes replace sodium and potassium losses and help reduce urinary fluid loss. When mixed from powder concentrates, the sports drinks are cheaper than bottled water!

After practice or competition, continue fluid intake to ensure adequate rehydration. Providing some carbohydrate and protein in the hour after exertion helps to replace muscle glycogen and repair minor muscle damage. Recent studies suggest that 100 grams of carbohydrate and 25 grams of protein (a ratio of 4 grams carbohydrate to 1 gram of protein) will aid recovery after vigorous effort. Commercial rehydration products are available, but you can meet athletes' needs with sports drinks and other supplements (energy bar with some protein, beef jerky for protein and salt, fat-free milk for protein and water).

Rest and Sleep

Growing athletes need lots of rest. Many of the adaptations that result from training take place during sleep; so those who miss sleep may miss the benefits they've worked to achieve. Interference with important REM (rapid eye movement) sleep can lead to psychological problems, and persistent lack of sleep can lead to impaired immune function, opening a window of opportunity to upper respiratory tract infections (URTI). Sleep is important for health as well as performance. Adults who get 7 to 8 hours of sleep are healthier than those who get less than 7 or more than 9 hours of sleep. As we said in chapter 3, naps provide a short-term solution to unavoidable sleep loss. Naps less than 20 minutes or more than 90 minutes seem to work best because they avoid having to wake up during a cycle of deep sleep (see figure 12.1).

Schedule time for ample rest in the training program. Observe your athletes closely. If they appear overtired, insist they take time off for rest and recovery.

Acclimatization

We'll take a moment to mention two other important factors that can influence health and

Immune Function

One of the best indicators of good health is a robust immune system. No single measure reflects the overall function of this complex system, but good **immune function** is characterized by a low incidence of upper respiratory and more serious illnesses. Immune function is enhanced with moderate exercise, good nutrition, and adequate rest. Factors that suppress immune function include psychological stress, extreme fatigue, poor nutrition, sleep deprivation, and cigarette smoke, as well as dehydration and air pollution.

Training can cause fatigue, and competition is stressful, so participation in sport has the potential to depress immune function. We have found that one day of hard training causes a suppression of the immune response. Several consecutive days of hard work lead to a depression of immune function that takes days of rest to reverse. After an exhausting competition, such as a marathon run, up to 25 percent of the participants may report an upper respiratory tract infection in the following two weeks. We'll say more about immune function in the section on overtraining.

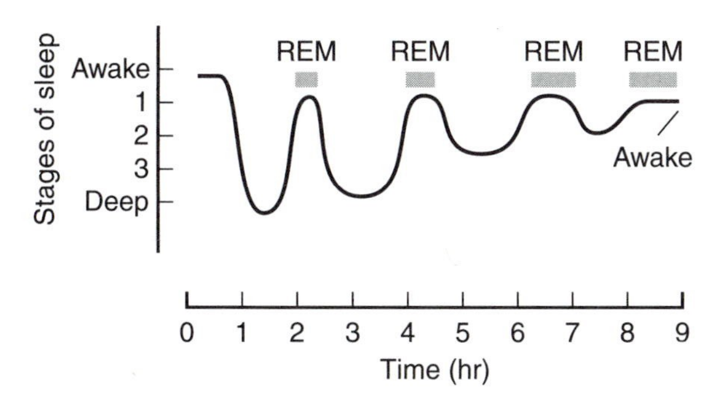

FIGURE 12.1 Stages of Sleep. During the night we alternate between deep (NREM) and light (REM) sleep.

Adapted, by permission, from B.J. Sharkey, 2002, *Fitness & health*, 5th ed. (Champaign, IL: Human Kinetics), 372.

performance and that necessitate **acclimatization:** heat and altitude.

Heat

Failure to acclimatize to the heat can certainly affect performance, and it may threaten life itself. It takes 5 to 10 days to make the physiological adjustments that allow one to perform in a hot environment. Failure to acclimatize can lead to heat cramps, heat exhaustion, or life-threatening heatstroke. The cornerstone of prevention is a high level of aerobic fitness. The fit individual has an increased blood volume and the circulatory capacity to distribute blood to muscles and sweat glands at the same time.

Acclimatization is achieved during periods of exercise in the heat, with attention paid to rest breaks and fluid replacement. After several days, the body begins to adjust to the heat. The adjustments include increased sweat rate, sweating at a lower body temperature, and circulatory adjustments that allow exercise at a lower heart rate. We recommend the acclimatization model recently adopted by the National Collegiate Athletic Association (NCAA) for fall football practice. It includes once-daily practices (without full pads) for the first five days. Thereafter, it limits two-a-day practices to an every-other-day schedule, thereby allowing time for recovery and rehydration. Of course, whenever you exercise in the heat, remember to provide water for athletes to drink before, during, and after practices. Have athletes weigh in to ensure that they have rehydrated before the next practice.

Altitude

It takes longer to adjust or acclimate to altitude. Ideally, you should get five to seven days of exposure for every 1,000 feet (305 meters) above your home elevation. Adjustments include improved air intake, increased red blood cells and hemoglobin, and increased capillaries and muscle myoglobin. These adjustments improve one's ability to take in, transport, and utilize oxygen, reducing but never eliminating the effects of altitude on endurance performance. While living and training at altitude contribute to acclimatization, recent studies indicate that best results are achieved with a live high–train low approach. For example, if you will be performing at 4,500 feet (1,372 meters), live there (sleep, eat, etc.) but train at a lower elevation. Why? Training at the higher elevation is more fatiguing, making it harder to train in the performance zone, and could hurt performance.

Mental Health Behaviors

A number of other behaviors can be loosely categorized as related to mental health and can influence the overall health and performance of athletes. Stress, substance abuse, and disordered eating can reduce immune function and affect the ability to perform in sport. Development of psychological skills can enhance performance. We'll discuss these topics in the following sections.

Stress Management

One's reaction to stress is a learned behavior. Individuals who overreact to stressful situations, such as those encountered while driving in heavy traffic, have been called hot reactors. There is evidence that those who react with hostility and anger exhibit increased blood pressure and stress hormones that may increase their risk of heart disease. Coaches and athletes are by nature aggressive, so it is important that they learn to manage stress and their reaction to it. **Stress management** techniques include simple meditation techniques such as the relaxation response. By taking deep breaths and repeating a soothing word (e.g., "easy"), you can reduce blood pressure and the hormonal response to stressful situations. Meditation has also been shown to improve immune function. Coaches certainly need to learn and practice the relaxation response. Athletes need to learn relaxation techniques so they can play their best game more often. But all that is beyond the scope of this book; consult the suggested readings listed in appendix H for ways to help athletes manage the stress of competition.

Substance Abuse

Everyone knows that excessive use of tobacco products, alcohol, or recreational drugs can interfere with an athlete's ability to practice and perform. For example, most recreational drugs and alcohol interfere with skilled performances and an athlete's ability to perform in the heat, and some have been associated with death from heatstroke. Less well understood are the various compounds lurking in so-called performance supplements and energy drinks. One of these, ephedra, is a stimulant that has been linked with sudden death in high school, college, and professional athletes. The FDA banned ephedra as a dietary supplement in January 2004. The NCAA and the International Olympic Committee have extensive lists of banned substances, and the organizations conduct frequent and random tests of athletes. Banned substances range from stimulants such as ephedra and excess caffeine to the anabolic steroids used to build muscle. Many banned substances carry the risk of serious health problems. Don't risk the health of your athletes by encouraging or allowing the use of unproven and untested supplements.

Disordered Eating

Eating disorders are common in sports such as gymnastics and dance, sports in which excess weight may have a negative effect on appearance or performance. Disordered eating ranges from the "starve and stuff" practices, once common in wrestling, to the serious problems of bulimia and anorexia. **Bulimia** is characterized by binge–purge cycles in which the purge is accomplished with self-induced vomiting. Serious cases combine purging with laxatives and diuretics to avoid weight gains from binge eating. **Anorexia** is a life-threatening problem characterized by a desire to please others, a quest for perfection, and a distorted body image. Victims consider themselves fat and refuse to eat even as starvation erodes essential organs such as the heart. Coaches should know the signs of disordered eating so they can help athletes get the psychological and medical help they need. And coaches must avoid placing undue emphasis on weight loss. Excessive attention to body weight and body fat has the potential to induce disordered eating in vulnerable individuals.

Psychological Skills

Performance depends on more than training, skills, and tactics. Athletes need **psychological skills** so they can play their best game more often. We mention these skills briefly, but important skills include the following:

- Relaxation. You must learn to become relaxed and uncritical so your body can perform the skills it has learned. Practice relaxation at home, in a quiet place. Then transfer to practice and then game situations. Say "easy" on your exhale to focus on your breathing instead of the score. Learn to relax and let the movement flow. Continue to practice the skill in early competitions.
- Concentration. Don't let your mind wander or disassociate during a race or game. Learn to focus on the signals coming from your body during a race: Are you working too hard, not hard enough? Monitor your technique and make adjustments as necessary. Keep track of your competitors; when you lose track of your form or the competition, your pace will suffer. In a sport such as tennis you can focus on the ball to maintain concentration. Say "ball" under your breath to help maintain focus and concentration.
- Imagery. Mentally practice specific skills and movements; visualize yourself performing in good form. See yourself as a skilled and successful performer.
- Activation. When fatigue begins to set in during a grueling event, use emotionally charged words to pep yourself up. Drive, power, charge, or blast up hills; fire off the ball. Focus on the importance of the event and think "go" on every exhale. Avoid negative self-talk.

Who says athletes aren't also thinkers? Relaxation, concentration, imagery, and activation are important mental skills.

Teach athletes how to judge their success in sport. Competitors evaluate their performance in athletic contests strictly on a win–loss basis, giving little regard to the quality of the performance. Performers evaluate the quality of the performance, relative to a standard or ideal, attaching less importance to the final outcome (winning or losing). Athletes can become performers by focusing on the quality of the experience. Analyze but don't judge a performance. Teach athletes to avoid anger; anger inhibits the flow of coordinated movements. Performers seek good competition because they know it is needed in order for them to reach higher levels of performance. Competitors fear good competition because it threatens their win–loss record, their self-esteem. Help athletes set individual goals in terms of performance instead of wins, medals, and trophies. As they achieve good performances, the wins will take care of themselves.

Avoiding Overtraining

When athletes engage in training that involves progressive increases in the intensity and duration of effort, they eventually approach the fine line that separates improved performance from chronic fatigue. Crossing that line leads to overtraining, a syndrome that is characterized by a persistent decline in performance and a host of related symptoms. The symptoms include chronic fatigue, staleness, irritability, decreased interest, and weight loss. Mood changes include a decrease in the sense of vigor and increases in tension, anxiety, dejection, hostility, and confusion. Physiological correlates of overtraining include a rise in the exercise and recovery heart rates, elevated cortisol and decreased testosterone, increased white blood cells, and loss of tissue protein. All of this is associated with decreased strength and endurance, loss of coordination, and a prolonged slump in performance.

The overtraining syndrome develops when insufficient rest accompanies high-volume training. Repetitive microscopic trauma provokes a generalized inflammatory response. Physical and psychological stresses increase secretion of the hormone cortisol, which suppresses immune function. The central nervous system senses the situation and initiates responses that include the breakdown of muscle protein (to provide glucose). The body has switched to the survival mode, and serious training or performance becomes impossible. Factors that contribute to the syndrome are intense training, accumulative fatigue, insufficient sleep, inadequate nutrition, low-grade infections, and excess stress.

Athletes can avoid the overtraining syndrome if they get sufficient rest, eat properly, minimize stressors, and curtail hard training during periods of illness.

Overtraining can be diagnosed with observations of performance, mood, fatigue, and other signs. Some coaches use waking heart rate, body weight, or even body temperature (fever) to try to avoid overtraining. A sensitive but expensive indicator is the cortisol/testosterone ratio. Blood levels of the stress hormone cortisol are compared to levels of testosterone, the anabolic or growth-inducing hormone. When cortisol goes up and testosterone declines, overtraining is likely.

Neck Check

Can you engage in vigorous exercise while you have an upper respiratory infection? Rhinoviruses cause upper respiratory symptoms but do not have much effect on performance. Dr. Randy Eichner suggests the "neck check" to differentiate a harmless virus from a more serious infection (Eichner, personal communication). If the symptoms are above the neck, such as a stuffy nose, sneezing, or a scratchy throat, try a test drive at half speed. If you feel good after 10 minutes you can increase the pace and finish the workout. But if symptoms are below the neck, with aching muscles or coughing, or if you have a fever, nausea, or diarrhea, take the day off. You can return to activity when the fever has been absent for at least 24 hours without the aid of aspirin or other antifever medications.

The fatigue index (pages 237-238) is a tool athletes can use to self-monitor stress and fatigue and avoid overtraining.

Career Development

Athletes go through various stages in their careers, and the training must be appropriate for each stage. Kids (under 10 years) should learn skills and have fun. Competitions should emphasize participation, training should be moderate, and races should be short. Youth sport (11 to 15 years) is the time to refine technique and start serious training. Competitions should emphasize performance over winning, since immature athletes may get discouraged in contests with early developers. Young adults (16 to 19 years) are ready for more serious training and competition. At this stage some athletes will need less training and more rest as they work through a growth spurt. Others will begin to specialize in one sport, and some will undertake year-round training. From age 20 onward, adult athletes are ready for the intensity and volume of training necessary to achieve high levels of performance.

Most coaches work with the youth or young adult stages, then watch as the athletes move on to high school or college programs. It is important that coaches recognize their role and limitations in the development of young athletes. While it occasionally happens in swimming, few young athletes achieve world-class performances early in their development. Coaches need to design training programs suitable to the maturity and development of the athletes.

Multiyear Planning

If any of your athletes are involved in a program that covers multiple years, such as age-group or academic programs, you may wish to define a periodized, multiple-year plan. Set out general performance and training goals for each year. If you are working with young endurance athletes, you will want to increase annual distance slowly and build the aerobic energy base before adding much speed training. Likewise, youth sports tend to focus more on skill and technique before adding strength and power during later adolescence. In general, four principles guide the organization of multiple-year planning:

- Progress from multisport development to specialization as needed.
- From individual skill development, progress to team tactics and strategies.
- Build an aerobic fitness base first, adding no more than a 10 percent annual increase in training volume. Add speed (performance zone [PZ] and maximal zone [MZ] intervals) only after a solid aerobic base has been developed. Figure 12.4 (p. 239) shows periodized multiyear plans for endurance and power athletes
- Advanced skill and coordination may require strength and power. Add strength and power training as needed to meet the skill and coordination demands of the sport. At early ages, strength should be general and should be developed through sport-specific work. As skill and strength demands increase, first light and then heavier weights can be added.

The Fatigue Index

We've used a simple test as a way to detect the early stages of overtraining in athletes. It is an inexpensive but accurate way to judge when an athlete is accumulating fatigue on a collision course with overtraining. The results correlate closely with immune function: When the fatigue index goes up, immune function declines, and vice versa. All you need to conduct the test is a sturdy 8-inch (20-centimeter) bench (or stair step) and a stopwatch. A heart rate monitor helps but is not required. You can teach athletes to self-administer the test at home in the morning. The resting, exercise, and recovery heart rates are summed to provide a fatigue index. After a baseline is established, subsequent tests can be used to gauge the extent of fatigue. Large increases in the fatigue index suggest the need for additional rest or for a break from hard training. Athletes can use figure 12.2 to record their measurements with use of the following procedure.

Athletes can use the following form to record pulse rate in beats per minute for each measurement using the following chart.

1. Resting 10 sec pulse ________ × 6 = ________ bpm
2. Exercise 10 sec pulse ________ × 6 = ________ bpm
3. 30 sec recovery 10 sec pulse ________ × 6 = ________ bpm
4. 60 sec recovery 10 sec pulse ________ × 6 = ________ bpm

Fatigue index (add 1, 2, 3, and 4) together ________ beats

FIGURE 12.2 Fatigue test recording form.

From *Sport Physiology for Coaches* by Brian J. Sharkey and Steven E. Gaskill, 2006, Champaign, IL: Human Kinetics.

TABLE 12.1

The Fatigue Test

Procedure: After rising in the morning but before breakfast or stimulants (tea or coffee):

- Sit quietly for 3 to 5 min until your heart rate is stable. You can read the paper during this time.
- Take the resting heart rate at the wrist for 10 sec; then multiply the number of beats by 6 to get the rate per minute.
- Start the stopwatch and begin stepping (up with one foot then the other, then down with the first foot and then the other; the entire sequence of stepping up and down should take 2 sec and be repeated 30 times in 1 min.
- After 1 min of stepping, stop; and while standing, take the postexercise heart rate, then sit down immediately.
- Sit quietly and relax; at 30 sec after exercise, take your heart rate (10 sec × 6 = bpm).
- At 60 sec after exercise, take the final heart rate (10 sec × 6 = bpm).

To calculate the fitness index, sum all heart rates (resting, postexercise, 30-second recovery, and 60-second recovery). The total is the fitness index. The index is unique for each individual and should be compared to the individual's average index for several days during a rested period. Use table 12.2 to evaluate the index.

TABLE 12.2

Criteria for Evaluating the Fatigue Test

Increase in fitness index	Risk of overtraining
0 to 20 above resting	Not generally a concern unless sustained
20 to 30 above resting	Slightly increased (avoid PZ and MZ training)
30 to 45 above resting	Increased risk (suggest short EZ training only)
More than 45 above resting	High risk (suggest no training)

The higher the index is above baseline values, the greater the likelihood that the athlete has not recovered from prior training. When the fatigue index is more than 20 above baseline, the athlete is at increased risk for depressed immune function and upper respiratory infections.

Figure 12.3 shows data from an athlete during a three-week competition period. The athlete chose to ignore the signs of fatigue appearing at day 8. By day 12 he had developed a serious cold and sore throat and was unable to compete for the next two weeks.

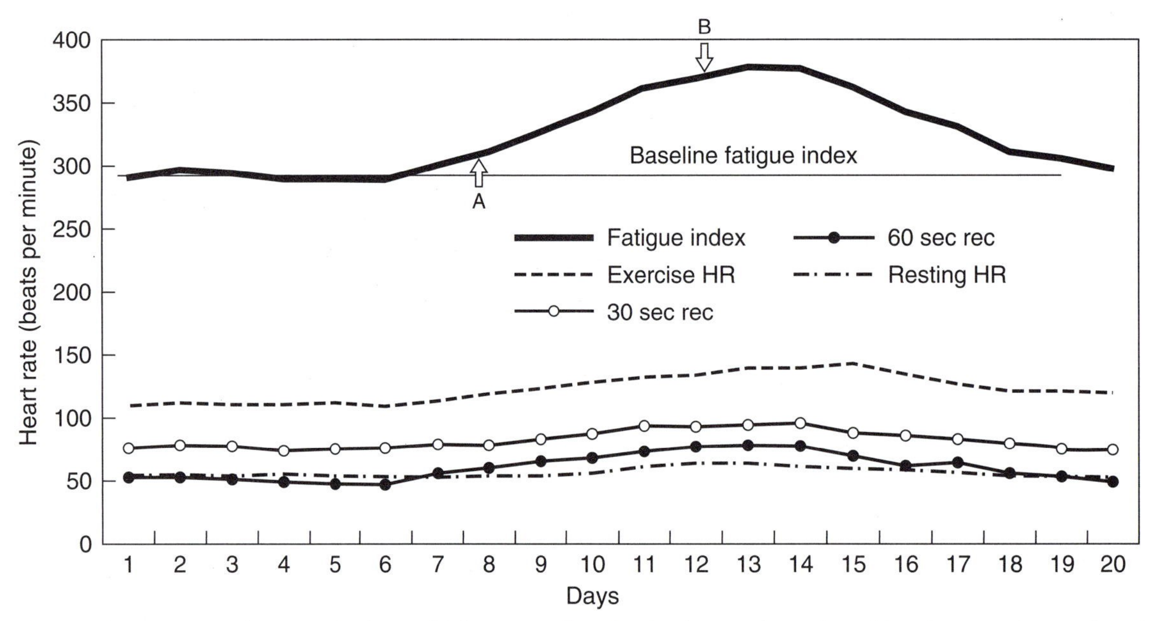

FIGURE 12.3 Fatigue index sample data. The fatigue index clearly shows where the athlete became fatigued and did not recover. The first arrow (A) shows where the index climbed 20 beats above baseline, indicating an increased risk for upper respiratory infection. This athlete did not reduce training and developed a cold and sore throat at the second arrow (B). If the athlete had modified training when the fatigue index exceeded 20 beats above baseline, he might have avoided the upper respiratory infection.

Unfortunately, avoidance of overtraining is as much an art as a science. It requires careful observation of athletes, good communication, and the willingness to prescribe rest—the only effective treatment for overtraining. Mild cases may resolve with reduced training and more rest. More serious cases demand time off or even bed rest. Since the immune system is depressed, the athlete is more likely to succumb to a viral or bacterial infection. Remember, training should be approached as a gentle pastime. Make haste slowly and you will eventually reach your goals.

Cut From the Team?

Athletes select a sport, position, or an event for a number of reasons, including personal preference, peer pressure, or parental influence. As the coach, you are guided by how the athlete can best contribute to his or her own development and to the success of the team. You can use objective measures to gain insight into an athlete's potential, and competitive performances to determine his or her current capability. Sometimes you will have to move an athlete to a different event or position, and occasionally you may have to cut an athlete from the squad.

There is no dishonor in being cut from the team. Being cut should be viewed as an indication that the athlete is not well suited for the sport and as an opportunity to seek success in another. Jerry, the runner mentioned in chapter 7, played several years of high school football in spite of the fact that his weight never exceeded 150 pounds (68 kilograms). He was bothered by minor injuries in his junior year and sustained a knee injury that finished his senior season. The injury eventually interfered with his promising collegiate career as a runner. He wished that he had been advised to bypass football to concentrate on sports he was physically better suited to play.

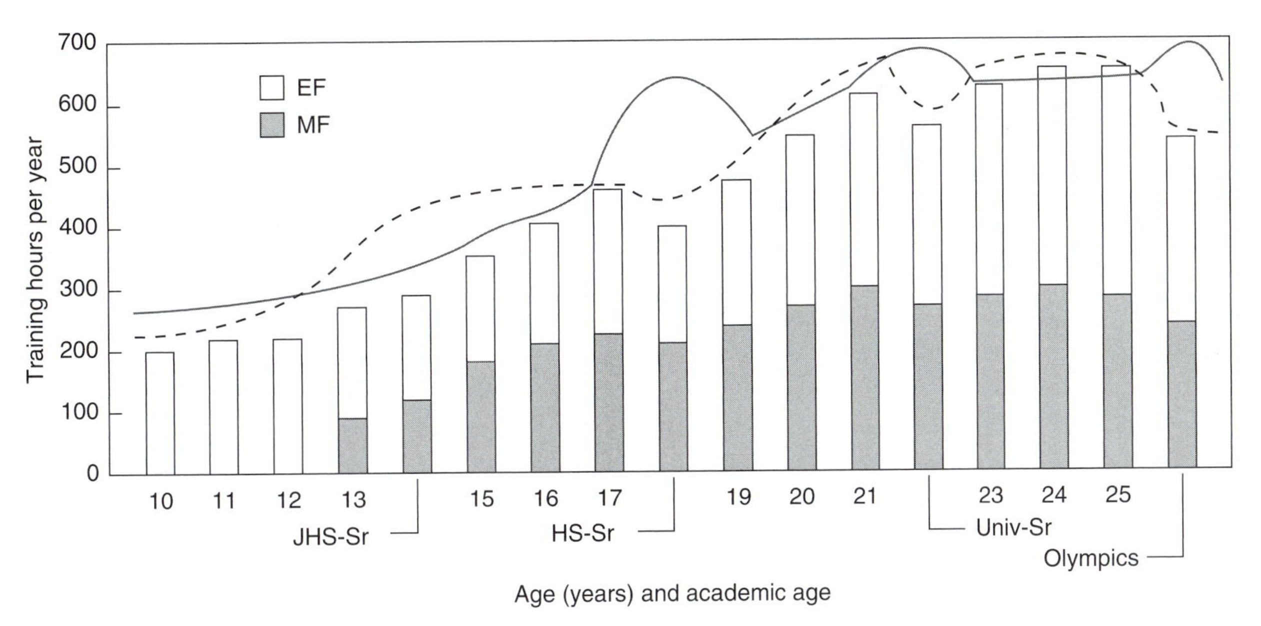

FIGURE 12.4 A career plan for a power athlete. This graph shows a theoretical model for career periodization for a power athlete. In general, there is a progression in stress each year with notable exceptions during years when higher performance is required. Note that during the senior years in high school (HS-Sr) and college (Univ-Sr), as well as a potential Olympic or World Championship year, the total training and stress (dashed line) are reduced and performance (black line) increases.

In chapter 3 we showed a model of skill development (figure 3.1, p. 32). You will notice in the figure how the major emphases change with age, progressing from general skills to specialized technique and training.

The purpose and goals of training should change as children grow and develop. Athletes in a few sports (e.g., swimming and gymnastics) seem to achieve top performances as young adults. In general, top performances are achieved by mature, well-trained, well-rested athletes. A table of age-based guidelines was presented in chapter 3 (table 3.1, p. 32).

Coaches should also consider the needs of athletes who annually compete in more than one sport. During a high school year, it is possible for talented athletes to compete in three sport seasons, and they may do more if they compete in summer or club sports. Because of the short school seasons, it is common for coaches to focus on high-intensity training. Three-season athletes may go directly from competition in one sport to interval training in the next sport. This unintended emphasis on high-intensity training may result in overtraining, poor performance, illness, or injury. When possible, the athlete and coaches of each sport should discuss a training program that maintains health and improves performance over the school year. Similar scenarios occur when an athlete participates in a club sport that overlaps with a school sport season.

When planning multiyear programs it is a common practice to reduce the training load during years of important competitions, such as the Olympics. Reducing the overall volume allows mature athletes to focus on high-intensity training and increased recovery with lower stress, resulting in peak performances (see figure 12.4). These years of reduced stress seem to improve performance for a year, but athletes then need to resume normal progression to achieve further improvements. High-intensity, low-volume training cannot be maintained successfully for many months. Knowledgeable coaches help high school and college seniors excel by varying their training appropriately.

Training volume should never be increased more than 10 percent from one year to the next. Similarly, the pressure to perform must be increased gradually. Increase training volume too fast and you risk overtraining; increase performance pressure too rapidly and you risk burnout, and possibly dropout. Don't push an athlete too hard in order to achieve short-term goals. Understand your role in the career development of your athletes, and take justifiable pride in their accomplishments.

SUMMARY

In this chapter we have presented information that will help you keep your athletes healthy. We've presented a number of health behaviors that contribute to an athlete's ability to train and achieve success. While nutrition, hydration, and adequate rest may be outside the direct control of the coach, it is important that you encourage, teach, and model good examples. Other behaviors, such as stress response, substance abuse, and disordered eating, are behaviors that too often require the attention of the coach. The chapter outlines psychological skills that help athletes play their best game more often. These skills must be taught and practiced if they are to be used in competition.

We then turned to one of the biggest problems facing coaches and athletes, overtraining. The overtraining syndrome develops when high-volume training is accompanied by insufficient time for rest and recovery. Learn the signs and symptoms of overtraining and help your athletes avoid illness and injury. The fatigue index, using the heart rate after a mild exercise test, is one way to assess accumulating fatigue.

We conclude the chapter and the book with some thoughts on sport or event selection and career development. Once an athlete has made a commitment to a sport, help the person develop a long-term plan for success—one that includes long-term goals and a plan to achieve them. Increase training volume gradually as athletes mature to avoid overtraining and the effects of burnout, a major reason why athletes drop out of competitive sport.

KEY TERMS

acclimatization (p. 233)
anorexia (p. 234)
bulimia (p. 234)
immune function (p. 232)
psychological skills (p. 234)
stress management (p. 233)

REVIEW QUESTIONS

Fill in the blank(s) with the appropriate word or words to complete the following statements.

1. _________ , _________ , and _________ are the most important health behaviors that contribute to an athlete's ability to train and achieve success.
2. It takes _________ days to make the physiological adjustments that allow one to perform in a hot environment.
3. _________ evaluate their performance in athletic contests strictly on a win–loss basis, with little regard to the quality of the performance.
4. _________ evaluate the quality of the performance, relative to a standard or ideal, attributing less importance to the final outcome.
5. The _________ develops when insufficient rest accompanies high volumes of training.

Match the stages of career development on the left with the appropriate description of that stage on the right.

6. Children (under 10 years)
7. Youth sports (11-15 years)
8. Young adults (16-19 years)
9. Adults 20 and older

a. Specialization in one sport; ready for more serious training and competition; may undertake year-round training
b. Ready for intensity and volume of training necessary to achieve the highest levels of performance
c. Learn skills and have fun; competitions should emphasize participation and races should be short; training is moderate
d. Time to refine techniques and start serious training; competitions should emphasize performance over winning

Select the correct answer or answers.

10. An athlete in training should never lose more than ___ pounds per week.
    a. 3
    b. 2
    c. 1
11. When a coach travels with the team, he or she should
    a. discuss good food choices on the menu
    b. demonstrate good choices in his or her own selections
    c. recommend at least two servings of fruits and three of vegetables
    d. all of the above

12. When you exercise, you should hydrate
    a. before the exercise
    b. during the exercise
    c. after the exercise
    d. until rehydrated
    e. all of the above
13. Naps provide a short-term solution to unavoidable sleep loss. They should be
    a. less than 20 minutes
    b. 30 to 60 minutes
    c. more than 90 minutes
    d. a or c
14. An athlete has an upper respiratory infection. She can practice if
    a. her temperature is below 100 degrees Fahrenheit
    b. her fatigue index is not elevated
    c. her symptoms are above the neck

Determine if the following statements are true or false and then circle the correct answer.

15. T/F Training must be appropriate for each stage of development.
16. T/F The overtraining syndrome develops when high-volume training is accompanied with insufficient rest.
17. T/F Competitors evaluate the quality of the experience, relative to a standard or ideal, attaching less importance to the final outcome.
18. T/F Experienced athletes don't need to practice relaxation and concentration skills.
19. T/F Coaches who practice stress management techniques improve immune function and reduce the risk of heart disease.
20. T/F Ingestion of solid or liquid carbohydrate energy during long-duration practice or competition has been shown to reduce fatigue, increase work output, and improve mood and decision making.
21. T/F Adults who get 7 to 8 hours of sleep are healthier than those who get less than 7 or more than 9 hours of sleep.
22. T/F Excessive use of tobacco products, alcohol, or recreational drugs can interfere with an athlete's ability to practice and perform.
23. T/F Eating disorders are not common in sports such as gymnastics and dance, where excess weight may have a negative effect on performance or appearance.
24. T/F Performance depends on training, skills, and tactics; therefore athletes do not need to develop psychological skills to enhance their performance.
25. T/F If you increase training volume and performance pressures too rapidly, you risk overtraining and burnout, which could lead to possible dropout from the sport.

PRACTICAL ACTIVITIES

1. What are some ways to monitor an athlete's nutrition, hydration, and rest and sleep?

2. What are some signs of substance abuse? Of disordered eating?
3. Use the fatigue index (p. 237) to gauge your fatigue and become familiar with the procedure. Do it at the same time of day on three separate days and compare results. Record results for heart rates and fatigue index.

APPENDIX A

Flexibility Exercises

The exercises presented here are a sampling of simple, easy-to-perform flexibility exercises. They stretch the major muscle groups in the body that are used for most activities. To avoid injuries, make sure that athletes have warmed up with at least 5 minutes of aerobic exercise before stretching.

Bent-Knee Stretch

This stretch should help increase flexibility in the lower back and hamstrings.

Procedure

1. The athlete should stand upright with feet facing forward and close together, legs straight but knees relaxed (not locked).
2. Have the athlete bend forward at the waist as far as possible while allowing the knees to bend slightly. The athlete should grasp ankles and pull until he feels the stretch (figure A.1).
3. The athlete should hold the stretch for 5 seconds, then relax and repeat.

FIGURE A.1 Bent-knee stretch position.

Back and Leg Stretch

This is a stretch for the lower back, hamstring, and gluteal muscles.

Procedure

1. The athlete should stand upright with feet facing forward and close together, legs straight but knees relaxed (not locked).
2. Have the athlete bend forward as far as possible while keeping legs straight. The athlete should grasp ankles and pull until she feels the stretch (figure A.2).
3. The athlete should hold the stretch for 5 seconds, then relax and repeat.
4. Athletes with more flexibility can try touching fingers or palms to the floor.

FIGURE A.2 Back and leg stretch position.

Toe Pull

Use this stretch to increase flexibility in the groin and thighs.

Procedure

1. The athlete sits on the floor with the feet together.
2. The athlete presses down on his legs with his elbows while pulling on his ankles (figure A.3*a*).
3. Athletes who already have good flexibility may try leaning forward and touching their heads to their feet or the floor (figure A.3*b*).

FIGURE A.3 *(a)* Toe pull position and *(b)* variation.

Seated Toe Touch

This three-part stretch should help increase flexibility in the back and hamstrings.

Procedure

1. The athlete should sit on the floor with legs extended out in front.
2. With toes pointed, the athlete should slide hands down legs until she feels the stretch. Have the athlete hold 5 seconds, and then relax (figure A.4*a*).
3. Have the athlete grasp ankles and pull until head approaches legs, then relax (figure A.4*b*).
4. The athlete should flex feet, drawing toes back, and lean forward, attempting to touch toes with hands. Hold 5 seconds, then relax (figure A.4*c*).
5. For variation, have the athlete try this stretch with the legs apart.

FIGURE A.4 *(a)* Part 1 of the seated toe touch stretch, *(b)* part 2, and *(c)* part 3.

Leg Pull

The purpose of the leg pull is to stretch the hamstring and gluteal muscles.

Procedure

1. The athlete should sit on the floor with legs extended out in front.
2. The athlete should pull the right leg toward and across the chest and place the foot on the floor. The left leg should remain extended in front (see figure A.5).
3. Hold 5 seconds, relax, and repeat with left leg.

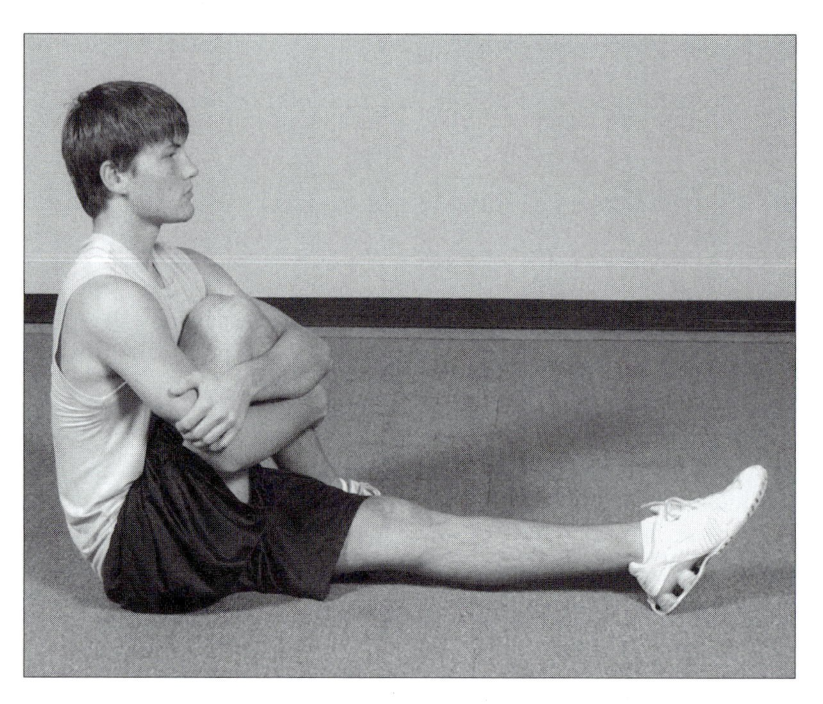

FIGURE A.5 Leg pull stretch position.

Backover

Use this stretch for the hamstrings and muscles of the lower back.

Procedure

1. The athlete should lie flat on his back with legs extended out in front.
2. With knees bent, the athlete should slowly bring his legs up and over his head.
3. The athlete should try to touch the floor with toes, feeling the stretch (figure A.6).
4. Hold for 5 seconds, then relax.

FIGURE A.6 Backover position.

Athletes with back or neck problems should not attempt this exercise.

Stride Stretch

This is a stretch for the groin (inner thigh muscles).

Procedure

1. Have the athlete stand with one foot far in front of the other with hands on a wall or chair for balance.
2. The athlete should bend legs as though taking a long stride, hold the stretch, and then relax (figure A.7*a*).
3. To accentuate the stretch, the athlete may put her arm and shoulder inside front leg (figure A.7*b*).
4. Repeat stretch on other leg.

To avoid strain on the knees, athletes should not let the knee of the front leg extend forward past the foot. The lower part of the front leg should remain perpendicular to the floor.

FIGURE A.7 *(a)* Stride stretch and *(b)* accentuated stride stretch.

Side Stretch

Use this stretch for the muscles of the arms and trunk.

Procedure

1. The athlete should begin by standing straight with good posture.
2. Have the athlete reach up and grasp hands above the head and slowly bend to one side (figure A.8).
3. The athlete should push gently, hold the stretch, then relax.
4. Repeat stretch to the other side.

FIGURE A.8 Side stretch position.

Wall Stretch

The wall stretch is designed to stretch the calf muscles and Achilles tendon.

Procedure

1. The athlete should stand about 3 feet from a wall with feet facing forward and slightly apart.
2. Have the athlete lean forward and place hands on the wall. Heels should remain on the floor. Hold, then relax (figure A.9*a*).
3. This stretch can also be performed one leg at a time by starting with one foot in front of the other (figure A.9*b*) and then repeating with the other leg in front.
4. Athletes can also try contracting their calf muscles briefly, then relaxing and feeling the stretch in the Achilles.

FIGURE A.9 *(a)*Two-legged wall stretch and *(b)* one leg wall stretch.

Shoulder Stretch

Use this to stretch the muscles of the shoulders.

Procedure

1. Athletes need to work in pairs for this stretch.
2. The athlete should stand straight with his back to his partner.
3. The athlete extends his arms behind him while the partner grasps his wrists and pulls gently upward (figure A.10).
4. Hold for 5 seconds, then relax. Partners should switch positions and repeat.

FIGURE A.10 Shoulder stretch position.

The Bow

The bow should stretch the muscles of the arms, back, and legs.

Procedure

1. The athlete should stand about 3 feet from a wall with feet facing forward and slightly apart.
2. Have the athlete bend forward at the waist, keeping back in natural position.
3. The athlete should place hands on the wall, keeping her face toward the floor and neck in line with spine, and feel the stretch from hands to heels (figure A.11).

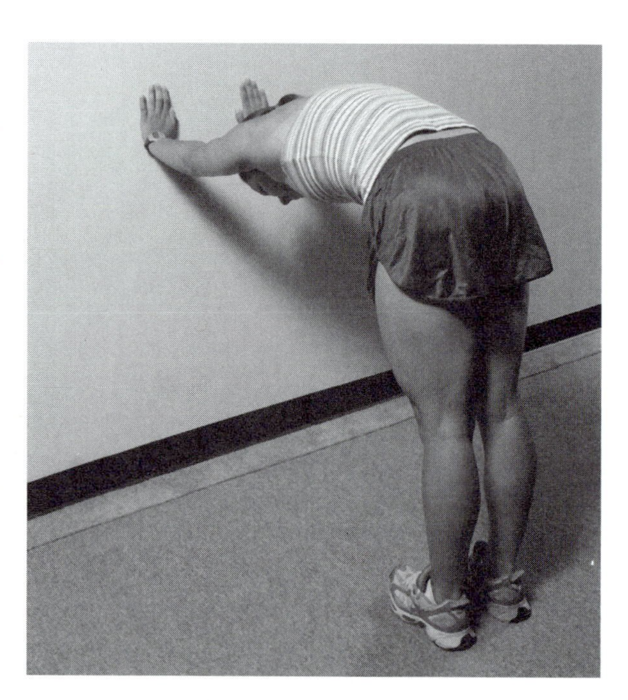

FIGURE A.11 The bow stretch position.

Side Twist

This stretch should increase flexibility in the muscles of the trunk and lower back.

Procedure

1. The athlete should begin by standing straight with good posture, arms extended to the sides at shoulder level.
2. Have the athlete slowly twist his trunk to one side as far as possible, keeping hips still (figure A.12).
3. Hold the stretch, then twist in the other direction.

FIGURE A.12 Side twist position.

Neck Circles

Use neck circles to stretch the muscles of the neck, shoulders, and upper back.

Procedure

1. The athlete should begin by standing straight with good posture.
2. Have the athlete roll head slowly in a full circle, first clockwise, then counterclockwise (figure A.13).

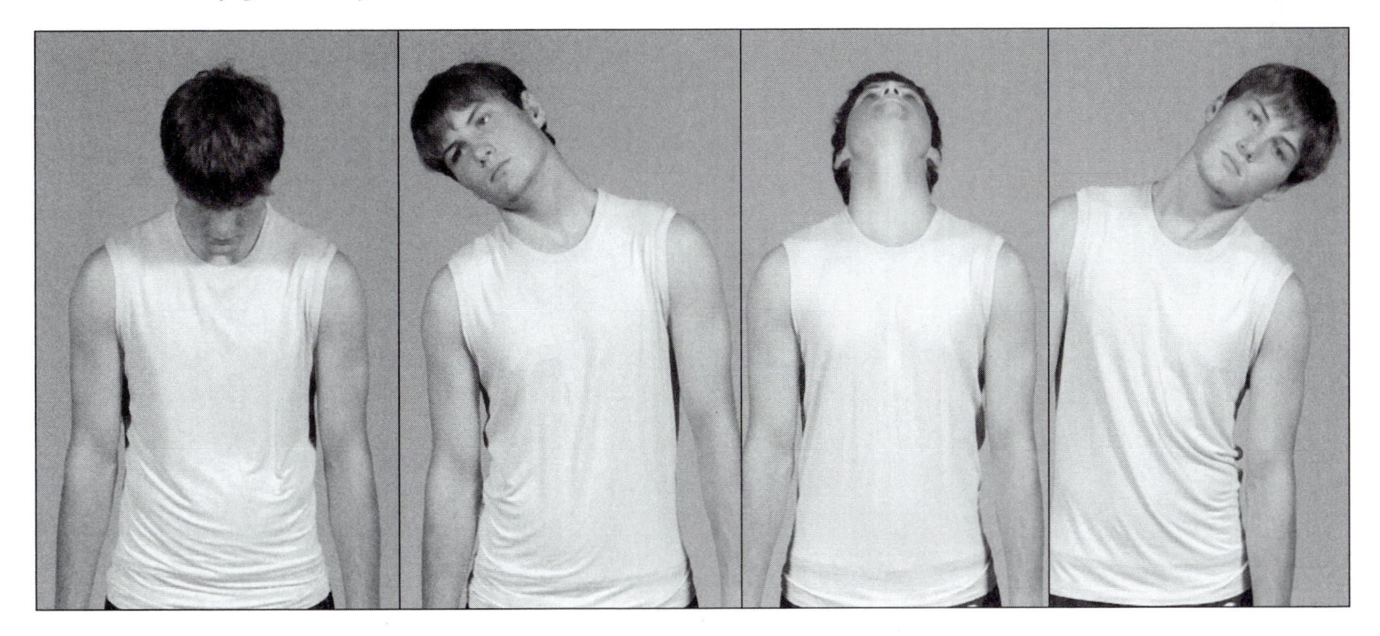

FIGURE A.13 Neck circles.

The Hug

The hug should stretch the muscles of the shoulders and upper back.

Procedure

1. The athlete should begin by standing straight with good posture.
2. Have the athlete wrap arms around herself in a hug as tightly as possible (figure A.14).
3. The athlete should hold the stretch, then relax and repeat.

FIGURE A.14 The hug position.

Quad Stretch

Use the quad stretch to stretch the quadriceps muscles (along the front of the thighs).

Procedure

1. The athlete should begin by standing straight with good posture, one hand touching a wall or the back of a chair for stability (if needed).
2. Have the athlete lift his right foot up towards his buttocks, grasping the ankle with the right hand (figure A.15).
3. The athlete should pull gently, feeling the stretch along the front of the leg. For greater stretch, the athlete should tilt his pelvis forward slightly.
4. Relax, and repeat with the left leg.

FIGURE A.15 The quad stretch position.

If there are no walls or chairs around for balance, athletes can extend their opposite arm out to the side, or try gently grasping their earlobe. Remind athletes who have trouble balancing while doing this stretch to engage their abdominal muscles.

APPENDIX B

Exercises for Core Training and Specific Muscle Groups

The exercises in this appendix start with core training exercises to stabilize the trunk and abdominal region. Chapter 6 includes an alphabetical listing of specific resistance exercises for a wide variety of sports and activities (table 6.2, p. 90). It is beyond the scope of this guide to teach coaches all of the resistance exercises. Many good books exist that can assist coaches interested in this area.

Crunch

Crunches develop strength and coordination in the abdominal region. This strength is needed for most sport activities.

Procedure

1. The athlete lies on the back with knees bent at about 90 degrees and feet flat on the floor.
2. Hands are crossed over the chest and placed on the opposite shoulders. (It is best not to put the hands behind the neck or head, as this encourages straining the neck.) See figure B.1*a*.

FIGURE B.1*a* Crunch *(a)* starting and ending position.

3. Slowly curl the upper body. Raise the head and shoulders from the floor by contracting the abdominal muscles until the shoulder blades are about 3 to 4 inches (7.6 to 10 centimeters) off the floor. See figure B.1*b*.
4. Slowly lower the back down until the shoulder blades touch the floor and repeat.

FIGURE B.1*b* Crunch *(b)* working position.

Uniwheel Exercises

Uniwheels are marketed under several brand names. They are simply a single small wheel with a long axle onto which an athlete straps the hands or feet. These simple, inexpensive devices are very useful in developing core stability because of their requirements of balance, coordination, and core strength.

Inchworm Exercises

Inchworm exercises develop flexibility, strength, and coordination in the shoulders, abdominal, back, and hip regions. This strength is needed for most sport activities.

Stationary Inchworm Procedure

1. The athlete straps both feet into the uniwheel and starts in a push-up position.
2. With the hands stationary, start by flexing at the hips and attempt to bring the feet toward the hands while balancing the wheel (figure B.2). It will take time and practice to be able to do a full range of motion.
3. Once flexed as far as possible, slowly return to the starting position and repeat.

Once athletes have mastered the stationary inchworm, they can challenge themselves with the walking inchworm.

Walking Inchworm Procedure

1. Start in the same position as for the stationary inchworm.
2. Once in the flexed position, with the feet brought up near the hands, stabilize the wheel and slowly begin to walk forward with the hands, keeping the wheel in one location, until in the starting position.
3. Repeat this exercise to "walk" across the floor.

Coaches can help athletes set increasing goals for distance and sets. Start with short distances and work up to three sets of 30 feet (9 meters).

FIGURE B.2 Stationary inchworm exercise.

Chest Press

The chest press exercise can help to develop core muscles of the chest including shoulder stabilizing muscles, pectoralis major and minor muscles, triceps, and the anterior deltoid.

Procedure

1. The athlete lies on a lifting bench with dumbbells in each hand, using an overhand grip as shown in figure B.3*a*.
2. Lift the dumbbells straight up toward the ceiling, keeping them centered over the shoulders.
3. Lift the dumbbells as high as possible by lifting the shoulders to extend the hands to maximal height (figure B.3*b*).

FIGURE B.3 Chest press *(a)* starting position and *(b)* ending position.

Seated Rows

This exercise balances the muscles developed in the chest press, including rotator cuff muscles, rhomboids, trapezius, biceps, and posterior deltoid muscles. It can be done using a seated rowing machine, a homemade pully system with ropes, or a stretch band.

Procedure

1. The athlete sits with the knees slightly flexed, holding the resistance device (figure B.4*a*).
2. Keep the upper body erect and pull the bands or bars directly toward the chest. Keep the elbows in close to the sides (figure B.4*b*).
3. Return to the starting position and repeat.

This exercise should be done slowly and with good control.

FIGURE B.4 Seated row *(a)* starting position and *(b)* working position.

Belly Blasters

Belly blasters dynamically strengthen the abdominal and hip flexor muscles and increase hamstring flexibility. Athletes of similar body weight should work in pairs.

Procedure

1. One athlete lies on his or her back and places the head between the ankles of the second athlete, who is standing. The athlete performing the belly blasters holds on to the ankles of the standing athlete.
2. The exerciser rotates slightly so that one hip is on the floor. Then, with straight legs, the exerciser flexes at the hips so that the toes approach the belly of the holder (B.5*a*).
3. At the high point, the athlete rotates to the other hip and lowers him- or herself down, stopping just before the legs touch (figure B.5*b*).
4. From this starting position on the opposite hip, repeat the exercise. Be sure to switch hips at the high point.

As dynamic abdominal strength increases, the standing athlete can catch the exerciser's feet as they come up and gently push them back toward the floor. The exerciser makes sure the legs do not touch and then rebounds quickly to bring the legs back up. It is important to alternate hip rotation with each repetition.

FIGURE B.5 Belly blaster *(a)* starting position and *(b)* ending position.

Hip Raise

The hip raise isolates the rectus abdominis muscle.

Procedure

1. The athlete lies on the back with the arms out to the side for balance and the legs pointing straight up toward the ceiling (figure B.6*a*).
2. Then raise the hips slowly upward. Keep the legs vertical and point the toes. Extend upward as far as possible.
3. Slowly lower back down until the hips touch the ground.

As athletes become stronger and develop good control, they can complete these exercises without allowing the hips to fully touch the floor each repetition. Continue to do this exercise slowly with good abdominal control.

FIGURE B.6 Hip raise *(a)* starting position and *(b)* working position.

Pilates

Pilates, a group of exercises using medicine balls, balanceballs, rollers, and other tools to strengthen the core, has recently become popular with athletic teams and some professional athletes. Little scientific research has been done to demonstrate the efficacy of Pilates, though one study evaluating elite rhythmic dancers showed improvements in leaping height and explosive power using a Pilates program. Pilates uses exercises and physical movement designed to stretch, strengthen, and balance the body with an emphasis on strength and flexibility, particularly of the abdomen and back muscles (figure B.7). Pilates is said to improve coordination, posture, balance, and core strength. Pilates classes are available at many fitness centers and may prove a useful means by which athletes can supplement their training program.

FIGURE B.7 A basic Pilates exercise.

Balance Boards

A number of low-priced balance boards, using rollers or point balance, are available on the market. These tools provide an excellent way to improve core stability of the ankles, legs, hips, and abdominal/back areas. Balance boards help develop dynamic balance and are motivating and fun for athletes (figure B.8). As with Pilates, little research has been done to evaluate improvements in athletic performance with the use of balance boards. However, these boards are widely used, and coaches and athletes continue to believe that they provide a positive benefit to training.

FIGURE B.8 Athlete using a balance board.

APPENDIX C

Plyometric Training Exercises

Lateral Hops

This exercise works the legs, hips, and ankles.

Procedure

1. The athlete stands upright with feet together or slightly apart (figure C.1*a*).
2. The athlete then jumps laterally (figure C.1*b*). These jumps can be done over a comfortable width or over a width marked with tape on the floor.
3. Upon landing, the athlete absorbs the impact and immediately jumps back to the starting point.
5. The athlete continues with repeated jumps until the desired number is achieved.

Over time, gradually increase the lateral distance.

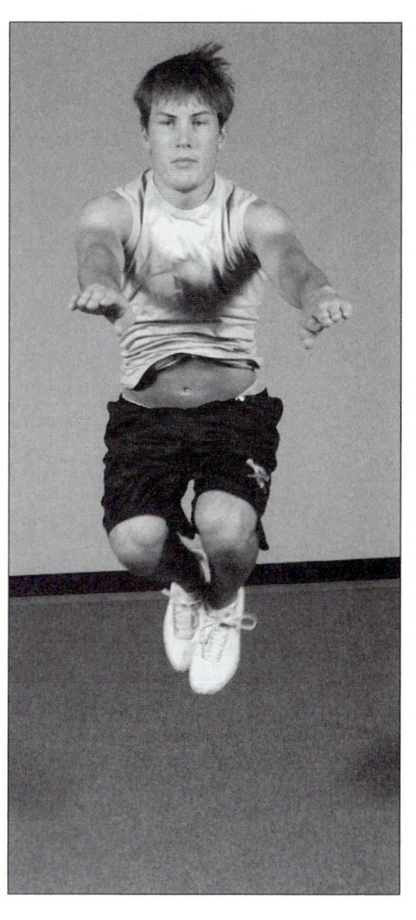

FIGURE C.1 Lateral hop *(a)* starting position and *(b)* working position.

Double-Leg Hops Over Hurdles

This exercise works the muscles of the legs and hips.

Procedure

1. Set up 4 to 12 hurdles spaced to allow athletes to jump continuously with only one touch between hurdles. The spacing will depend on the hurdle height and ability of the athletes. In general, the hurdles will be spaced about 3 to 4.5 feet (0.9 to 1.4 meters) apart.
2. The athlete stands with back straight, head up, and shoulders slightly forward. Arms should be bent to about 90 degrees (figure C.2*a*).
3. The athlete jumps upward and over the first hurdle, bringing knees up to chest and using arms for lift (figure C.2*b*).
4. On each landing, the athlete should absorb the shock and immediately explode over the next hurdle until all hurdles have been cleared.
5. Rest 2 to 3 minutes before repeating.

FIGURE C.2 Double-leg hop *(a)* starting position and *(b)* jump position.

Start with low hurdles, about 16 to 18 inches (41 to 46 centimeters) high, and gradually progress as power improves.

Note that these can be done as a moving chain of athletes. The first three athletes get down on their hands and knees, spaced as hurdles. The next and subsequent athletes hop over the athletes who are forming hurdles, then quickly kneel down to form a new hurdle. After the last athlete starts, the first "hurdle" gets up and begins to hop over all the other athletes. Thus after an athlete becomes the first hurdle, he or she jumps up over all the other "hurdles" and then becomes the last hurdle. They keep going until each athlete has completed three to six sets of hurdles.

Drop and Catch Push-Ups

This exercise works the muscles of the shoulders and arms.

Procedure

1. The athlete starts in an arms-extended push-up or modified push-up position (figure C.3*a*).
2. The athlete lowers down (figure C.3*b)* and explodes upward so that the hands come off of the floor. If possible the athlete claps the hands (figure C.3*c*).
3. Immediately, as the body comes down, lower and explode up again.
4. Repeat using fluid movements.

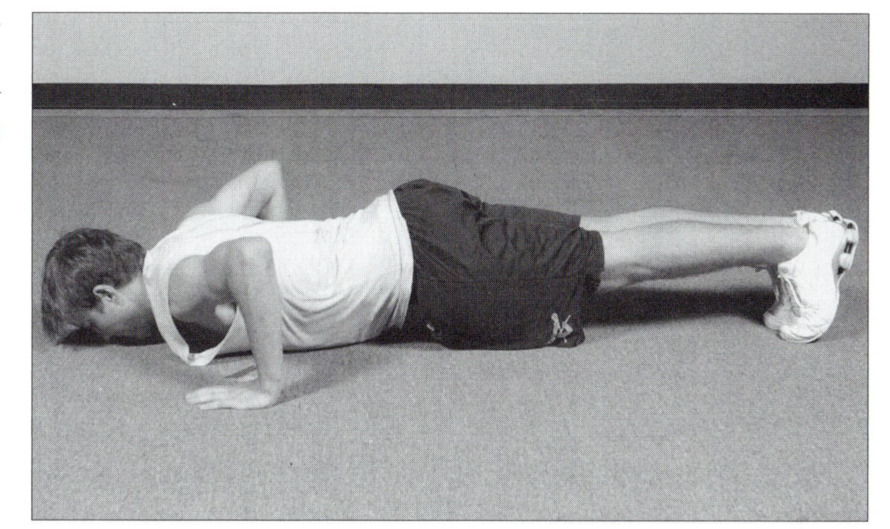

FIGURE C.3 Drop and catch push-up *(a)* starting position and *(b-c)* working positions.

Dynamic Wall Push-Offs

This exercise works the muscles of the shoulders and arms, primarily the triceps. This exercise is similar to the drop and catch push-up, but is done standing with a partner. Each athlete should have a partner of reasonably similar size.

Procedure

1. The athlete stands facing a solid wall, about arm's length away, with fingers stretched out to the wall. The assistant stands behind the athlete (figure C.4*a*).
2. When the athlete is ready, the assistant (another athlete or the coach) gently pushes the athlete's shoulders toward the wall.
3. The athlete catches him- or herself, allows the arms to bend to about 90 degrees, and then explodes back toward the assistant, who immediately pushes him or her back toward the wall (figure C.4*b*).
4. Continue for 6 to 12 repetitions.

To vary the difficulty of this exercise, have the athlete move farther away from (more difficult) or closer to (less difficult) the wall. With a little adjustment, this exercise works to give both the athlete and the assistant a good arm plyometric workout.

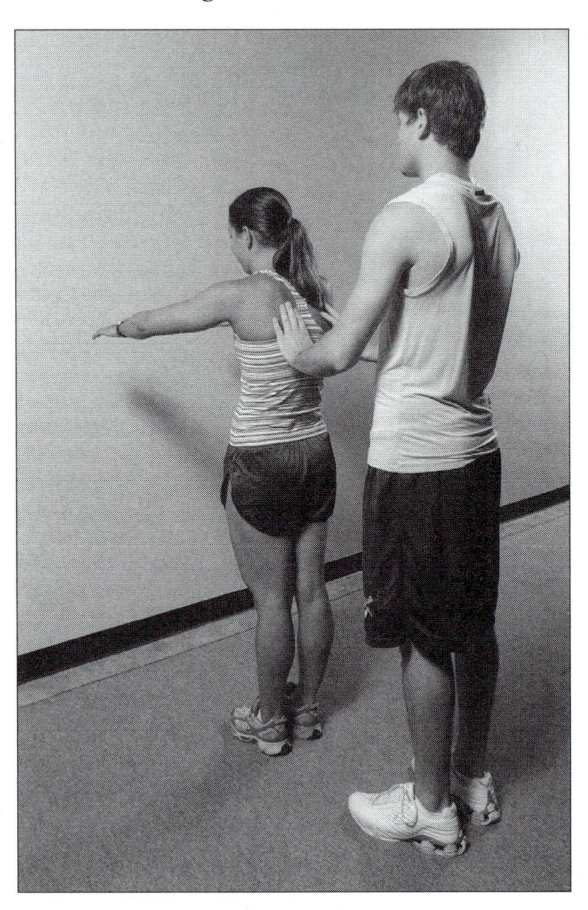

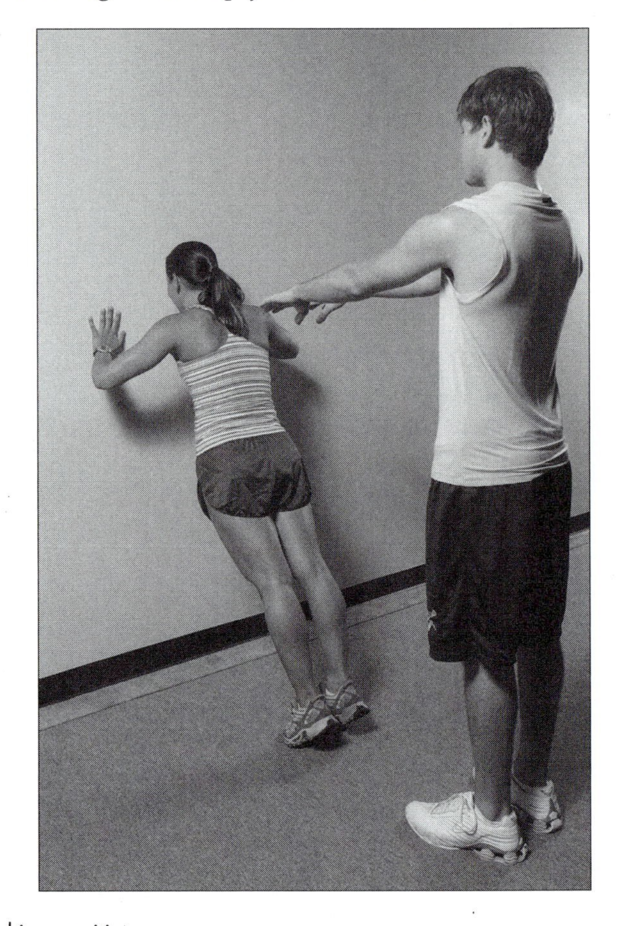

FIGURE C.4 Wall push-off *(a)* starting position and *(b)* working position.

Box Jumps

This exercise works the muscles of the hips, thighs, and lower legs. The difficulty of these jumps depends on the height of the box or bench and the speed with which they are performed. Start with a low box, 12 inches (30 centimeters) or lower, and progress to higher boxes. A standard height is the stadium bleacher, generally about 16 to 17 inches (41 to 43 centimeters). An assistant (optional) with a stopwatch may be used to time the jumps.

Procedure

1. The athlete stands facing the box or bench (figure C.5*a*).
2. On a "Ready–go" signal, the athlete jumps up and down, continuing to face the same direction. All touches should be equal on both feet (figure C.5*b*).
3. The athlete or the assistant should count the jumps.
4. The athlete continues until the "Stop" signal, generally at 15 to 30 seconds. (If a timer is not used, these jumps should be done with as little "touch" time as possible.)

FIGURE C.5 Box jump *(a)* starting position and *(b)* working position.

Depth Jumps

This exercise works the hips, thighs, knees, and lower legs. These jumps can be done using different-height boxes. They are generally the final leg plyometric added to a program. Athletes should be proficient at box jumps, squat jumps, and double-leg hops before starting depth jumps, as these jumps apply a large load to the muscles and knees. Start with low boxes of 6 to 8 inches (15 to 20 centimeters) and gradually progress to higher boxes of 14 to 18 inches (36 to 46 centimeters).

Procedure

1. The athlete stands on the box or bench, ready to jump (figure C.6*a*).
2. The athlete gently hops forward off of the bench. The legs will absorb some of the energy, and the athlete then explodes upward and jumps as high as possible (figure C.6*b* and *c*). Contact time with the floor should be minimized.
3. Athletes should rest for 30 to 45 seconds before repeating.

Alternative: A hurdle can be placed about 3.5 feet (1 meter) from the box; the athlete jumps over the hurdle after landing. As athletes become more powerful, the height of the hurdle can be increased as appropriate.

FIGURE C.6 Depth jump *(a)* starting position and *(b-c)* working positions.

APPENDIX D

Designing Appropriate Tests

Validity, reliability, objectivity, bias, test criteria, administration, and feedback for tests are discussed briefly in chapter 5. In this appendix we give you more information and some examples to show how you can develop forms for feedback and do some simple statistics to evaluate your tests and training program.

Athlete Feedback Form

Following the testing it is valuable to give each of your athletes feedback concerning his or her strengths and weaknesses. On an evaluation sheet it is helpful to have three columns for results (see figure D.1): Column 1 shows established criterion measures when available; column 2 shows the team averages, and column 3 shows the individual athlete's score. Careful examination of the results will allow you and your athletes to identify strengths and weaknesses and to individualize training programs. Figure D.1 illustrates data obtained from the fitness testing set up by the coach (Kathy) in our story at the beginning of chapter 5. The athlete, Julie, was a 16-year-old female midfield soccer player.

Posttraining Testing and Evaluating Improvements

The greatest value of field and lab testing for athletes and coaches is the immediate feedback about program effectiveness. Repeat testing will help you and the athletes determine if the training program was effective and what areas might need improvement. For the athlete, knowing that there will be follow-up testing can help provide motivation to train effectively, especially in areas of weakness. For coaches, evaluation of the team data provides feedback about the overall effectiveness of the training program.

Administering Posttraining Tests

Follow-up, or posttraining, testing should be done using the exact procedures followed for the pretraining testing. If possible, use the same assistants for each test, and perform the tests in the same order and under similar conditions. Remind athletes several weeks in advance, and check to see that they still have their pretraining

FIGURE D.1 ATHLETE EVALUATION FORM FOR PRE- AND POST-TRAINING SCORES

Name *Julie* **Age** *16* **Coach** *Kathy*

Test	Average criterion score	Team average score	Athlete score	Evaluation
Max vertical jump	12.5	13.2	10.6	Needs work
Height 4-8	12.1 (97%)	13.0 (98%)	10.6 (100%)	Excellent
Height 9-13	11.5 (92%)	12.4 (94%)	10.4 (98%)	Excellent
Height 14-18	11.25 (90%)	12.0 (91%)	10.3 (97%)	Excellent
Agility run	11.7	11.4	11.5	OK
Shuttle run	11.2	10.8	11.5	Needs work
Bench hops	47	44	59	Excellent
1.5-mile run	11:30	11:26	10:55	Excellent
Sit and reach	11.5	13.2	7.3	Needs work

Evaluation

Julie appears to have great aerobic fitness and muscular endurance. Her 1.5-mile (2.4-kilometer) test was the best on her team and ranks her in the excellent category against criterion scores. She showed great muscular power endurance on the repeated vertical jumps and had an excellent score on the bench hops. She appears to be adequate in agility compared to the criterion, but her score is lower than the team average. Julie was slow in the shuttle run, a measure of both speed and agility; had a low vertical jump (leg power), which is important in order for soccer midfielders to be able to head the ball; and did poorly on the sit and reach flexibility test. Coach Kathy makes notes to include more agility training, increased stretching, and a focus on leg power training for Julie. During individual evaluations and meetings with the athletes, Kathy talks to Julie about specifics of training that she might do in order to improve in her weak areas. She also praises Julie for her excellent aerobic conditioning, which will be a great asset for her as a midfielder.

test values. Be sure to give encouragement similar to that given during the pretraining testing.

Comparing the Pre- and Posttraining Data

It is important that all athletes receive a copy of both their pre- and posttraining test scores. Figure D.1 showed an effective method for reporting scores to athletes. Figure D.2 provides a modification for reporting both pre- and posttraining test scores. This is very effective in helping both the coach and the athlete see areas of improvement and areas that still need work. These data help guide athletes through the competitive and off-season training. In figure D.2, scores for Julie are shown for both pre- and posttraining.

Individual Evaluation

In having both the pre- and posttraining scores, coach Kathy and her athletes have great data to

FIGURE D.2 ATHLETE EVALUATION FORM FOR PRE- AND POST-TRAINING SCORES

Name *Julie* **Age** *16* **Coach** *Kathy*

Date of first testing *3/15/2004* **Date of second testing** *6/02/2004*

Test	Average criterion score	Team average score	Athlete score	Evaluation	Team Ave Test 2	Athlete Score Test 2	Athlete Eval. Test 2	Athlete Change 1→ 2
Max vertical jump	12.5	13.2	10.6	Needs work	13.7	12.5	Great job	+
Height 4-8	12.1 (97%)	13.0 (98%)	10.6 (100%)	Excellent	13.5 (98%)	12.7 (99%)		+
Height 9-13	11.5 (92%)	12.4 (94%)	10.4 (98%)	Excellent	13.0 (94%)	12.5 (98%)		+
Height 14-18	11.25 (90%)	12.0 (91%)	10.3 (97%)	Excellent	12.6 (91%)	12.3 (96%)		+
Agility run	11.7	11.4	11.5	OK	11.1	11.1	Good	+
Shuttle run	11.2	10.8	11.5	Needs work	10.3	10.6	Much improved	+
Bench hops	47	44	59	Excellent	51	58		NC
1.5-mile run	11:30	11:26	10:55	Excellent	11:16	10.57		NC
Sit and reach	11.5	13.2	7.3	Needs work	12.7	11.5	Excellent	+

use for evaluation of each athlete's training program as well as the overall training program. The final column showing the change from pre- to posttraining helps each athlete to identify whether or not training goals were met. In the case of Julie, she improved in each of the categories that had been identified as needing work, as well as in some areas of fitness in which she was already doing well. From these scores, Julie can feel good about her training. Obviously, Julie and her coach have worked well together to set up an effective soccer-specific fitness program. In the following sections, we will look at how coaches can evaluate overall team improvements using a couple of simple statistical measures.

Team Evaluation

The scores on the tests are helpful to both coaches and athletes in evaluating individual improvements. However, it is also vital that coaches evaluate overall changes in the team scores to determine if the training program for the attribute measured by the test was effective for the team as a whole. In the data for Julie, in table D.1 (p. 273), you can see that the team score for the maximum-effort vertical jump was 13.2 inches (33.5 centimeters) at pretesting and 13.7 inches (34.8 centimeters) posttraining. This appears to be an improvement of 1/2 inch (1.3 centimeters) in vertical jump. Statistically the reason for this change could be that a couple of girls improved a lot while others didn't improve—or, hopefully, that most of the girls improved about 1/2 inch.

Statistics to Evaluate Group Training

Understanding how to do some simple statistics to compare changes in your test results over time, and also to evaluate if tests are related to performance, will help you evaluate the overall effectiveness of your training program and the quality of your tests. The following sections show you how to use a Microsoft Excel spreadsheet to do these simple statistics.

Comparing Pre- to Posttraining Testing: The t-test

A simple statistical method is available that you can employ to see if results are due to some aspect of training or whether they are due to random change. This statistical method, called a t-test, is used when you want to compare team scores that have been obtained two times. With computer spreadsheet programs that are now available to most coaches, these tests are simple to do. If you don't want to do them yourself, have a math or science teacher do the analysis for you. Consider the team scores for Kathy's high school soccer team. The text below on t-tests shows how your scores would be set up in a computer spreadsheet and how the t-test analysis would be done in Microsoft Excel. Incidentally, the improvement by Kathy's soccer team was significant, meaning that the improvement was likely due to the training, not random chance.

To show how to evaluate scores when repeating a test, we will use the vertical jump test scores from Kathy's soccer team as an example. The scores for the pre- and posttesting are shown in table D.1. The team average improved 1/2 inch (1.3 centimeters). When you look at the scores for each individual you notice that most athletes, though not all, improved. For the coach to be confident that the overall team improvement was significant, a statistical analysis of the two sets of scores is necessary. The appropriate analysis is called a repeated (each athlete repeats the test), directional (you predict that they should improve) t-test (a statistic that compares the averages of two tests using the individual data).

To perform a t-test in a Microsoft Excel spreadsheet, set up your scores for the two tests in a manner similar to the example below. In a cell below the scores, label the row "t-test"; and in a cell to the right, type in

=TTEST([range1],[range2],1,1)

where

[range1] = range of cells for the first test scores

and

[range2] = range of cells for the second test score.

TABLE D.1

Sample Vertical Jump Data

		Vertical jump height (inches)	
1	A	B	C
2	Name	Test 1	Test 2
3	Megan	12	13
4	Julie	10.5	12.5
5	Heidi	14.5	14
6	Kim B.	11	12
7	Kim C.	9	10
8	Mary	13	13
9	Carlyn	15.5	16.5
10	Amie	14.5	14
11	Janette	13.5	14
12	Marin	14	14
13	Jessica	14	14
14	Jamie	17	17
15	Jolene	13.5	14.5
16	Margot	12	13
17	Whitney	14	14
18			
19	Average	13.2	13.7
20	t-test	0.01	

In this sample, [range1] is B3:B17 and [range2] is C3:C17. The formula in the t-test cell for this example would be

=TTEST(B3:B17, C3:C17,1,1).

The value that is returned for the t-test formula will tell you if the difference between the tests was significant. If the value is 0.05 or less, the change was significant, meaning that there is less than a 5 percent likelihood that the difference was due to chance.

If you want to use an Excel spreadsheet to calculate the average for each test, as shown in this example, enter the formula for an average in a row below each set of test scores. The formula for average that you type into a box is

=AVERAGE([range])

where [range] is the range of cells that you wish to average. In this example the Test 1 average formula would be

=AVERAGE(B3:B17).

Relating Test Scores to Performance: Correlations

As discussed in the previous section on designing assessment tests, validity is a major concern. Often we accept tests as having content validity and simply accept that they measure an attribute that we believe is important to success. In many cases it is valuable to test the predictive and construct validity to see if there is a relationship between the test and performance—that is, if an athlete improves in a test, are competitive results also improved? This is often difficult to do in team sports that do not involve individual scores; but for individual sports, especially those that are objectively measured, it may be possible to relate test scores to performance. The main statistical procedure that is used to do this is called a correlation.

Correlations are another statistic that has become easy for coaches to do if they are comfortable with computers and spreadsheet programs. If not, it should be easy to find a math teacher or a parent, for example, who can do this simple statistic for you. Correlations simply evaluate the relationship between two sets of numbers such as test scores and competitive results. The statistic that is returned from a correlation is called an "r" value and is a number between 0 and 1. The lower the number, the less likely it is that the two scores are related. Numbers above 0.6 are considered moderately strong correlations. The higher the "r" value, the more predictive the test is of performance. The following text on using correlations gives an example of how a correlation statistic could be used to evaluate the predictive validity of a test.

There are two basic ways in which correlations can help a coach evaluate the effectiveness of a fitness test:

1. Comparing test results with objective performance results
2. Comparing changes in test results with changes in performance results

Of course, there are many factors that affect performance, and if a test could completely predict results we could do away with competition and just test the athletes. Thus, the correlation gives a window into the validity of a test. Unfortunately, correlations show relationships but cannot be used to document cause and effect.

To demonstrate correlations we will use data from a men's college cross country running team. The coach, Dan, has his runners tested in the exercise physiology lab twice each year (preseason and three weeks before the conference championships) for oxygen utilization at the lactate threshold ($\dot{V}O_2$lt). Researchers have shown the laboratory test for $\dot{V}O_2$lt to be related to aerobic race performance. The day after each lab test, Dan conducts a 3-mile (4.8-kilometer) time trial. The scores are shown in table D.2 (p. 275).

Coach Dan has kept good records of his testing and time trial times for each athlete. From his scores, you can see that average oxygen utilization at the lactate threshold ($\dot{V}O_2$lt) increased 3.2 ml · kg^{-1} · min^{-1} and that the average 3-mile race times improved by 40.5 seconds between the two tests. These improvements were significant, as shown by the small *p* values (less than 0.05) evaluated by t-tests comparing the pre- to posttraining tests.

Dan does three correlations to see if there is a relationship between $\dot{V}O_2$lt and race times preseason and during the competitive season. The *r* values of −0.64 and −0.79 show a strong inverse relationship (as $\dot{V}O_2$lt increases, race times decrease) and tell the coach that this test is a valuable aid in evaluating athlete aerobic fitness. The third correlation that Dan does is to compare changes in $\dot{V}O_2$lt with changes in race times. Here he finds only a modest (though significant*) correlation of −0.45. This shows that changes in $\dot{V}O_2$lt are inversely related to changes in race performance, but there are probably many other factors during training that are affecting race times. Dan suspects that technique, motivation, leg power, and tactics are also important.

To perform a correlation in a Microsoft Excel spreadsheet, set up your scores for the two tests similar to the example in figure D.3. In a cell below the scores, label a row as "correlation" and also note what you are comparing. In a cell to the right, type in the formula for a correlation:

=CORREL([range1],[range2])

where

[range1] = range of cells for the first test scores

	A	B	C
1	Subject	$\dot{V}O_2$lt change	3 mile time change (seconds)
2	1	7	−39
3	2	5	−26
4	3	3	−29
5	4	−1	9
6	5	6	−76
7	6	6	−68
8	7	−3	−19
9	8	2	−15
10	9	5	−43
11	10	0	−26
12	11	5	−23
13	12	8	−101
14	13	−1	−97
15	14	1	−11
16	15	5	−49
17	Average =	3.2	−40.9
18	Correlation ($\dot{V}O_2$lt change to change in 3 mile time) =		−0.449895899

*There are tables that show when *r* values are significant based on the number of athletes being tested. A math or statistics teacher can help you determine if your *r* value is significant.

TABLE D.2

Sample Correlation Data for $\dot{V}O_2$lt Compared to Changes in Performance

	$\dot{V}O_2$lt (ml · kg^{-1} · min^{-1})			3 mile time (seconds)		
Name	**Preseason**	**Competitive season**	**Change**	**Preseason**	**Competitive season**	**Change**
Bjorn	46	53	7	946	907	−39
Carl	43	48	5	938	912	−26
Georg	52	55	3	928	905	−23
James	51	50	−1	930	939	9
John	55	61	6	919	843	−76
Karl	51	57	6	958	890	−68
Mark	49	46	−3	933	914	−19
Martin	46	48	2	955	940	−15
Per	45	50	5	953	910	−43
Tanner	47	47	0	939	913	−26
Torre	42	47	5	1009	986	−23
Tylan	53	61	8	928	827	−101
Tyrone	58	57	−1	925	828	−97
Vann	52	53	1	947	936	−11
William	50	55	5	964	915	−49
Averages	49.3	52.53	3.2	944.8	904.3	−40.5
t-test (Pre vs. Competitive)	$p = 0.001$			$p = 0.0001$		
Correlation: Pre $\dot{V}O_2$lt vs. Pre 3 mile time:				$r = -0.640$		
Correlation: Competitive $\dot{V}O_2$lt vs. Competitive 3 mile time:				$r = -0.786$		
Correlation: Change in $\dot{V}O_2$lt vs. Change in 3 mile time:				$r = -0.448$		

$\dot{V}O_2$lt = ventilation of oxygen at the lactate threshold.

and

[range2] = range of cells for the second test score.

In this example, the correlation between "Pre $\dot{V}O_2$lt" ([range1]) and "Pre 3 mile time" ([range2]), the formula would be

=CORREL(B2:B16,C2:C16).

The value that is returned for the correlation formula will be between 0 and 1. The closer it is to 1, the better the relationship. Generally, if you are testing 10 or more athletes, *r* values greater than 0.33 are meaningful and suggest tests that are related to performance. The higher the *r* value, the better the test in predicting performance. Again, remember that correlations show relationships but do not show cause and effect.

Revising Testing

Keeping good records of test scores and competitive data each year can help you to evaluate both your training program and your fitness tests. The smart coach is always looking for new ways to improve his or her training program. Good record keeping will help you figure out what has worked. Be selective about the tests you use. Be sure that they are reliable and that they fulfill at least one of the types of validity. If t-tests show that athletes are improving, then the training that you are having them do is effective. If correlations show a relationship of your test scores to performance, you can feel confident in the test validity. If you are convinced that the tests are meaningful to the athletes, and helpful to you, continue to use them. As new training tools and information become available, knowledgeable coaches should check the science behind the innovation and use only what makes sense. Good records and analysis will then allow them to later evaluate whether the tool or idea was effective.

APPENDIX E

Using Training Impulse Guidelines (TRIMPS) to Plan Training

Training impulses (TRIMPS) can be used to help periodize a training program and to judge the training stress of a program. Two tables are provided in this appendix to help you plan your training. Table E.1 is designed to work with figure 10.5, the seasonal planning sheet. Find the appropriate age-group column and use the weekly guidelines to fill in the bottom row on the form. These values are then transferred to the weekly planning form (figure 10.7*b*) in the top left box labeled "Suggested Weekly TRIMPS."

Table E.2 is then used with the weekly planning form to help you decide how much training to do on a daily basis. Write your plan, then calculate TRIMPS for each box on the weekly planning form using the duration of each training times the TRIMP value in the left-hand column. Sum the daily TRIMPS, compare to the recommendations in table E.2, and adjust as needed.

These guidelines should be adjusted based on ability level, sport goals, and individual needs. If you find that your athletes do not seem to be recovering between workouts, reduce the daily stress by reducing TRIMPS. For higher ability-level groups, try the TRIMP values for the next older age group. Experienced coaches have learned not to rush the training of junior athletes, and you are encouraged not to increase TRIMPS substantially beyond the values recommended in these tables.

TABLE E.1

Daily Age-Group TRIMP Guidelines

Training period	Week of period	Week difficulty	Weekly TRIMPS by age group					
			12	14	16	18	20	22
Basic training	1	Mod	21	30	47	51	65	70
	2	Hard	28	40	63	69	87	94
	3	Easy	14	20	31	34	43	46
	4	Mod	22	32	49	54	68	74
	5	Hard	30	42	66	72	92	99
	6	Easy	14	21	32	35	45	48
	7	Mod	23	33	52	56	72	77
	8	Hard	31	45	69	76	96	104
	9	Easy	15	22	34	37	47	51
Precompetition	1	Mod	32	45	70	77	98	105
	2	Hard	42	61	94	103	131	141
	3	Easy	21	30	46	50	64	69
	4	Mod	28	41	63	69	88	95
	5	Hard	38	55	85	93	118	127
	6	Easy	19	27	42	45	58	62
Early competition	1	Mod	25	36	56	61	78	84
	2	Hard	34	49	76	82	105	113
	3	Easy	17	24	37	40	51	55
Peak performance	1	Mod	19	28	43	47	60	65
	2	Hard	26	37	58	64	81	87
	3	Easy	15	22	34	37	48	51
	4	Mod	17	25	39	43	54	58
	5	Hard	24	34	53	57	73	78
	6	Easy	14	20	31	34	43	46

Age-group training impulse (TRIMP) guidelines are shown for 25-week programs. For each age group, read straight down the column. The corresponding training period and relative weekly stress (easy, medium, or hard) are shown. These age-based plans are designed to fit the minimum weekly recommendations for each training period and will require that the basic strength and endurance period be completed before the precompetition period. These guidelines will need to be adjusted based on ability level, sport goals, and individual needs.

TABLE E.2

Daily Age-Group TRIMP Guidelines

Training period	Easy week daily TRIMPS			Moderate week daily TRIMPS			Hard week daily TRIMPS		
	Easy day	Mod day	Hard day	Easy day	Mod day	Hard day	Easy day	Mod day	Hard day
AGE 12-15 (DEPENDING ON SPORT AND ABILITY)									
Basic training	2	2	3	2	3	5	2	4	6
Precompetition	2	3	4	2	5	7	2	6	9
Early competition	2	2	3	2	4	5	2	5	8
Peak performance	1	2	4	1	3	4	1	4	6
AGE 14-17 (DEPENDING ON SPORT AND ABILITY)									
Basic training	2	3	4	2	4	6	2	6	9
Precompetition	3	4	6	3	6	10	3	9	14
Early competition	3	3	5	3	5	8	3	7	11
Peak performance	2	2	6	2	4	6	2	5	9
AGE 16-19 (DEPENDING ON SPORT AND ABILITY)									
Basic training	3	4	6	3	7	10	3	9	14
Precompetition	5	6	9	5	10	15	5	14	21
Early competition	4	5	7	4	8	12	4	11	17
Peak performance	3	4	9	3	6	10	3	8	14
AGE 18-22 (DEPENDING ON SPORT AND ABILITY)									
Basic training	4	4	7	4	7	11	4	10	15
Precompetition	5	7	10	5	11	16	5	15	23
Early competition	4	5	8	4	9	13	4	12	18
Peak performance	3	4	9	3	7	11	3	9	15
AGE 20-24 (DEPENDING ON SPORT AND ABILITY)									
Basic training	5	6	8	5	9	14	5	13	20
Precompetition	7	8	13	7	14	21	7	20	29
Early competition	6	7	10	6	11	17	6	16	23
Peak performance	4	5	12	4	8	13	4	12	19
MATURE AND EXTREMELY WELL TRAINED ATHLETE									
Basic training	5	6	9	5	10	15	5	14	21
Precompetition	8	9	14	8	15	23	8	21	32
Early competition	6	7	11	6	12	18	6	17	25
Peak performance	5	5	13	5	9	14	5	13	20

Training impulse (TRIMP) guidelines are shown for easy, moderate, and hard days within easy, moderate, and hard weeks for each training period. These guidelines are designed to fit the minimum recommendations for each training period. The guidelines should be adjusted based on ability level, sport goals, and individual needs.

APPENDIX F

Answers to Review Questions

Chapter 1

1. Physiology
2. Lab and field research
3. status or comparison
4. experimental
5. 60%
6. c
7. e
8. a
9. d
10. b
11. b
12. c
13. a
14. T
15. F
16. Athletes have more confidence and work harder when the coach is able to provide clear reasons for various aspects of training.
17. Theories require experimental verification. Studies should be conducted on athletes, in actual sport conditions or sport-related simulations. The experimental variable should be randomly assigned to some and not to others (controls). Results are analyzed statistically to determine if the experimental variable is significantly different between experimental subjects and controls.
18. This book and others by reputable authors, and publishers such as Human Kinetics; clinics; videos. Peer-reviewed journals provide results of carefully reviewed studies concerning sport. Some journals are available online. Sport magazines are not reviewed by experts, so some information may be questionable. Web sites, such as those sponsored by reputable sport organizations (e.g., American College of Sports Medicine), provide useful information. Others are in business to sell products.

Chapter 2

1. c
2. j
3. f
4. a
5. k
6. g

7. b
8. e
9. i
10. d
11. h
12. F
13. F
14. F
15. T
16. F
17. F
18. T
19. T
20. Taper should be longer when training volume is high and the event long.
21. Gain more strength with the aid of testosterone.
22. Training leads to specific adaptations in muscle fibers. Strength training increases contractile protein.
23. a, c
24. d
25. a, b, c

Chapter 3

1. b
2. d
3. a
4. c
5. F
6. T
7. T
8. F
9. F
10. F
11. T
12. T
13. T
14. Heredity
15. Maturation
16. 7-8
17. diet or nutrition
18. e
19. b
20. e
21. c
22. b

Chapter 4

1. c
2. h
3. f
4. a
5. i
6. g
7. e
8. b
9. d
10. concentric
11. eccentric
12. actin and myosin
13. elastic recoil or stretch-shortening cycle
14. delayed-onset muscle soreness (DOMS)
15. T

Chapter 5

1. Relevant, objective, specific, appropriate, learning curve, gamelike, meaningful, individual, simple, challenging, sensitive
2. Validity, reliability, objectivity, and appropriateness
3. 30-60 seconds, 4-5 minutes, 1-3 hours
4. b
5. e
6. a
7. g
8. c
9. f
10. d
11. F
12. F

Chapter 6

1. frequency, intensity, and time
2. Agonist, antagonist
3. repetition maximum
4. Isometric or static contractions
5. Isotonic
6. isokinetic or variable
7. major, assistant, supplementary, and specialty
8. Plyometrics
9. static
10. proprioceptive neuromuscular facilitation (PNF)
11. F
12. F
13. T
14. F
15. T

Chapter 7

1. ATP and PCr
2. Anaerobic and aerobic
3. aerobic
4. anaerobic
5. lactate threshold
6. performance threshold
7. 9
8. 4
9. b
10. a
11. c
12. T
13. F
14. T
15. T
16. T
17. F
18. T
19. F
20. T
21. a
22. a and c
23. b and c
24. b
25. c

Chapter 8

1. ATP/PCr and anaerobic glycolysis
2. intensity
3. duration
4. b
5. d
6. a
7. c
8. F
9. F
10. F
11. T
12. T
13. T
14. F
15. T
16. T
17. b
18. b
19. c
20. d
21. d

Chapter 9

1. Energy fitness
2. Speed, perceived exertion, and heart rate
3. Speed-based
4. Taper
5. c
6. b
7. a
8. d
9. T

10. F
11. T
12. T
13. F
14. T
15. F
16. T
17. T
18. T
19. d
20. d
21. b
22. d
23. d

Chapter 10

1. Setting goals, analyzing needs, periodizing the training plan, and monitoring the progress and health of the athletes
2. recovery, basic training, precompetition, early competition, and peak performance
3. training impulses (TRIMPS), fatigue index
4. T
5. F
6. T
7. T
8. c
9. d
10. e
11. a
12. b

Chapter 11

1. skill
2. power
3. power endurance
4. intermittent
5. aerobic
6. T
7. F
8. F
9. F
10. T

Chapter 12

1. Good nutrition, hydration, and adequate rest
2. 5-10
3. Competitors
4. Performers
5. overtraining syndrome
6. c
7. d
8. a
9. b
10. c
11. d
12. e
13. d
14. c
15. T
16. T
17. F
18. F
19. T
20. T
21. T
22. T
23. F
24. F
25. T

APPENDIX G

Answers to Practical Activities

Chapter 1

1. Multiply 100 kg times 1.8 grams per kilogram per day to get number of grams of protein ($100 \times 1.8 = 180$ grams of protein per day).
2. Has the manufacturer conducted experimental studies that demonstrate the benefits and safety of the product? Were the subjects in the study well-fed athletes or individuals with nutritional deficiencies? Did the famous athlete use the product and, if so, for how long? Is there a less expensive way to get the benefits offered by the product, such as in food?
3. Has the training been verified in experimental studies? Will this training be too intense for young athletes? Will some athletes risk illness and injuries?

Chapter 2

1. The schedule should allow recovery time before important games. As you approach important games, shorten practice time, avoid long, hard workouts, and provide days off. Emphasize the need for good nutrition and adequate rest.
2. Look at the athlete's parents to estimate future growth. Conduct field tests that are related to performance (aerobic and muscular fitness, agility, etc.). Have athletes engage in sport-specific skill drills.
3. Follow three weeks of progression in training load (aerobic or muscular) with a week of relative rest. Periodize training within the week using the same principle, always ensuring time for recovery within the program.

Chapter 3

1a. You could illustrate extreme differences in ability; point to the effects of excess training (overtraining, illness); illustrate examples of inadequate training; and ask him to analyze reasons why athletes discontinue participation.

1b. Illustrate likely genetic differences based on knowledge or observation of the parents; help him understand differences in growth and development; and provide ways to estimate growth, as well as skeletal and sexual maturity.

1c. Review how heredity can influence potential and the response to training; how hormones associated with maturation influence growth and development; and how skill and psychological factors influence performance and the ability to sustain demanding practice sessions.

2. In energy fitness training, athletes could be grouped according to ability; heart rates or perceived exertion could be used to individualize training (see chapter 7); he could use fewer interval repetitions for inexperienced athletes. In muscular fitness training, gauge weights according to body size and strength; use fewer sets and more repetitions for younger athletes; utilize stack weight systems for newcomers and free weights for experienced athletes.

Chapter 4

1. On the continuum from endurance to power, the sports are already listed in a reasonable order. The coaches decided to break down the training focus into the need for endurance training, strength training, power training, power endurance training and speed training. Table G.1 shows what they decided for their team. These are ballpark estimates and are a good guideline, but if you were close then you would have a good idea of how you are going to start planning your training. Follow this table for percent of training in each area.

2. There are a number of ways that Mary could go about conducting a needs analysis for softball. She might start by talking to experienced coaches to find out what they do. She could read coaching and training material specific to softball, or she could begin the process herself by determining the primary needs for each position. Since all of the athletes need to hit and sprint she could develop tests, or use previously developed tests, to evaluate arm power and sprint speed. Based on athlete results, she could then develop specific training programs for each athlete to focus on weak areas and to maintain strengths. For defensive positions, most require accurate and effective throwing. She should work with a knowledgeable trainer to develop strength programs for the shoulder and throwing musculature both to improve performance and reduce injury based on each athlete's needs. Since the current season has just wrapped up, Mary has time to get each

TABLE G.1

	Endurance (%)	Strength (%)	Power (%)	Power endurance (%)	Speed (%)
10,000 m	87	2	3	3	5
5,000 m	75	3	9	6	7
1,500 m	60	5	10	8	12
800 m	55	10	10	10	15
400 m	40	20	15	10	15
200 m	30	25	15	5	25
100 m	20	30	20	2	28
Pole vault	15	35	20	2	28
High jump	15	40	25	0	20
Shot put	10	50	35	0	5
Javelin	10	45	30	0	15

athlete started on a long-term program to get ready for the entire year.

Chapter 5

1. Other tests may be appropriate: Here are some ideas:
   - *10,000 meters:* Sit and reach test; endurance tests discussed in later chapters.
   - *5,000 meters:* Sit and reach test; endurance tests discussed in later chapters.
   - *1,500 meters:* Sit and teach test; stair run to assess muscular power; bench hop to estimate power endurance; endurance tests discussed in later chapters.
   - *800 meters:* Sit and reach test; stair run to assess muscular power; bench hop to estimate power endurance; endurance tests discussed in later chapters.
   - *400 meters:* Sit and reach test; squat test using 20RMs; repeated vertical jump tests for leg power and power endurance; stair run to assess muscular power; endurance tests discussed in later chapters.
   - *200 meters:* Sit and reach test; squat test using 10RMs; single and repeated vertical jump tests for leg power and power endurance; stair run to assess muscular power; 40 yd sprint test.
   - *100 meters:* Sit and reach test; squat test using 5RMs; 40 yd sprint test; single and repeated vertical jump tests for leg power and power endurance.
   - *Pole vault:* Shoulder rotation test; sit and reach test; tests similar to 100 m sprint, but add tests for strength and power of the upper body, designed as specific to the motions of pole vaulting as possible.
   - *High jump:* Shoulder rotation test; sit and reach test; maximal vertical jump test for power.
   - *Shot put:* Sit and reach test; agility run test; 1RM strength for triceps, pectorals and squats in a motion specific to throwing the shot put.
   - *Javelin:* Sit and reach test; 1RM strength for triceps, pectorals and squats in a motion specific to throwing the javelin.
2. John has read a number of books on training, including this one. He fully understands the concept of specificity and suggests a number of changes to the tests. First, since soccer players seldom sprint a full 100 yards, but rather tend to have repeated short sprints of 15-40 yards, he proposes changing the sprint test to 30 meters and then initiating a repeated "beep test" in which the players cover 20 meters each interval. The intervals are done to a taped series of beeps so that the first 20 meter interval is slow, then the players get 10 seconds of rest, then the next 20-meter interval is slightly faster, followed by 10 seconds of rest. The "beep test" continues until players are unable to cover the 20-meter distance in the required (shortening) time.

   For shooting and dribbling, John keeps the shoots from the penalty mark, but adds a goalie to make it gamelike. For the other shots, he wants players to first dribble through a series of cones as fast as possible and then take a shot from a specific distance. He sets the distances at 5, 10, and 15 meters, as these are more reasonable distance for 12-year-olds, again with a goalie to simulate games. He times the dribbling portion. He also requires the girls to do both right- and left-footed shots. He does these tests once a month and gives each girl individual skills to work on to improve in areas of weakness. Of course, there are a number of different possible answers to this problem, but this was John's solution.

Chapter 6

1. Since John was showing clear signs of overtraining, the first step was to give him a period of 3-5 weeks with very light training allowing his body to recover. John would

then have been given a series of strength and power assessments including 1RM tests, a stair run and tackle sled tests to evaluate power. If possible, underwater weighing or skin fold measurements might have been made to assess body composition so that changes in weight could be evaluated in terms of muscle or fat. Once John was recovered and the evaluations were completed, a general program plan should be developed.

Following the recovery period, John has about 8 months before football season begins. In order to develop a good base of strength and power he will need to train for this entire time, working with a strength coach. A reasonable plan might look like the following.

TABLE G.2

Time line	Type of training
Months 1-2	Gradually build up weight for the desired exercises. Start with 20RM weights and each week add a little weight to reduce the RM by 2-3 until John is doing set-reps of 8-12RM repetitions with long rest between his three sets, three times a week. This preparatory period allows his body to adapt to the weights.

2. Athletes require flexibility slightly greater than the maximal range of motion used in their respective sports. Swimmers who must perform all events require great shoulder flexibility and good range of motion for the hips, knees, and ankles. Exercises presented in this chapter can be used for those specific areas. After an easy warm-up of either swimming or easy calisthenics, the swimmers should be encouraged to do light dynamic stretching using controlled movements, especially for the arms, shoulders, torso, and hips flexors and extensors. Three times a week after practice athletes should systematically work on flexibility with a combination of static, PNF, and controlled dynamic stretches for joints that require additional range of motion. For athletes who have adequate flexibility, maintenance should be done 1-2 times a week in addition to the easy stretching after a warm-up.
3. Explosive power to improve a vertical jump requires both increased strength and increased speed of movement. During the off-season, John should focus on gradually increasing his RM for squats and lunges until he can do 3 sets of 3-6RM two to three times a week. This may be maintained (with periodization or easy, moderate, and harder weeks) for 8-12 weeks. Towards the end of the strength period, begin to transition into slightly reduced weights (8-12RM) with more explosive movements to develop power. Vertical jump height should begin to improve during this period. As the season approaches, start plyometric training (hurdle jumps, repeated jumps for height, and finally depth jumps) about 5 weeks before the games start. To improve his heading skills, John should practice heading both in discrete drills where he is jumping as high as possible to head the ball, and in game and scrimmage situations where he is trying to win headers or score goals from corner or other kicks. Remember: add strength, then power, and finally plyometric speed. Even gains of 1-2 inches in vertical jump will help John and represent improvement.

Chapter 7

1. Conduct laboratory and field tests of anaerobic and aerobic fitness (see chapter 8). Conduct trials for each event; combine tests and trials for a look at immediate prospects and future potential.
2. The aerobic power (the maximal oxygen intake per kilogram of body weight) is the best measure. To relate to running, the test should be specific (i.e., running, on the track or treadmill). A score in the 60's (ml/kg/min) for boys (55 for girls) would be good; over 70 for boys (65 for girls) would be even better.
3. The major effects of training take place in the muscle fibers that receive training,

hopefully the same ones used in the sport. The effects of training on the heart, the ultimate endurance muscle, are subtle. Changes in heart rate are due in part (47% in one study) to increases in blood volume. Train sport-specific muscles and the heart and respiration will follow.

Chapter 8

1. Use Figure 8.1 and list the most important energy system. Indicate others that may apply (e.g., for specific events or positions).
2. Depends on sport: use sport-specific methods whenever possible (e.g., treadmill for runners, bike for cyclists, etc.). Indicate the energy system and how you will test the athletes.
3. Use your imagination and develop a field test that will be fun and informative. The field test could become a useful method of evaluation throughout the season. The best test is one that stimulates the right kind of training. For example, a 30-second stair run test as a measure of anaerobic power and short-term endurance could stimulate interest in appropriate training, such as leg strength development and stadium stair or hill sprints.

Chapter 9

1. Three methods:

- *Speed*: Refer to the sidebar "Calculating Performance Speed" on page 161. Calculate the athlete's current race pace in m/min. Calculate a speed that is 1 to 5 percent faster, then select interval length (e.g., 4 min) using table 9.3. Conduct intervals at that speed.
- *Heart Rate*: Use a heart rate monitor to determine average heart rate during an event, ignoring the first and last minute of competition (or time trial). Then determine the training interval and have the athlete complete intervals at a slightly faster heart rate (e.g., 5 beats faster than performance HR).
- *Perceived exertion*: Refer to table 9.1, page 161: begin interval training using RPE 16; as performance improves edge that up to RPE 17. Keep track of the time for each interval to indicate improvement in performance.

2. Training zones:
    - *EZ zone:* Low intensity zone, associated with 1 LT, a level that can be sustained for hours. It is used for recovery and long-term endurance training. Training improves oxidative abilities of slow twitch and fast oxidative glycolytic muscle fibers, and improves metabolic efficiency.
    - *NZ zone:* Intensity below performance speed; training here has limited effect on performance. Limit time spent in this zone.
    - *PZ zone:* Where pace or HR is slightly above current race values; associated with 2 LT. Training zone for maximum effect on performance. Improves oxidative abilities of fast twitch fibers and moves the lactate curve to the right, allowing a faster race pace.
    - *MZ zone:* The maximal speed zone is used to improve power and maximal speed. Improves the anaerobic energy system.
3. Peaking and tapering:
    - *Peaking:* Peaking involves short, intense PZ intervals, and MZ training, less EZ work and more rest. Speed training is used to sharpen and hone skills before a major competition.
    - *Taper:* An essential part of the peaking process, taper involves 3 to 14 days during which the training load is gradually reduced, allowing time for recovery before an important competition.

Chapters 10 and 11

No answers are provided here because the simplest method to check your answer is to compare your training plan with a plan for a similar sport provided in chapter 11.

Chapter 12

1. Monitoring nutrition, hydration, rest, and sleep:
   - *Nutrition*: Monitor body weight on a weekly basis; if weight drops too rapidly, discuss the need for adequate energy intake, from a variety of foods with emphasis on fruits, vegetables, and whole-grain products.
   - *Hydration*: In warm weather, monitor body weight daily; provide more fluids during practice and recommend other ways to rehydrate after practice.
   - *Rest/sleep*: Schedule rest and recovery time into the training program. Look for signs of fatigue, listlessness, moodiness and other signs of overtraining. Ask about hours of sleep, naps, etc. When problems arise, encourage more sleep—naps up to 20 min or over 90 min.
2. Signs of substance abuse and disordered eating:
   - *Substance abuse*: Unusual behavior, poor or missed practices, decline in skills, difficulty in the heat. Be on the lookout for so-called performance enhancing substances. Guide athletes to professional help for serious problems.
   - *Disordered eating*: Rapid weight loss, use of laxatives and diuretics, desire for perfection, distorted body image ("I'm too fat"). Help athletes get the psychological and medical help they desperately need.
3. *Fatigue Index*: Record heart rates and fatigue index. What was the greatest fluctuation between daily values?

APPENDIX H

Suggested Readings

Sport Physiology and Nutrition

Alter, M. (1998). *Sport-specific stretching for maximized performance.* Champaign, IL: Human Kinetics.

Armstrong, L. (2003). *Exertional heat illness.* Champaign, IL: Human Kinetics.

Bompa, T. (2003). *Serious strength training.* Champaign, IL: Human Kinetics.

Fleck, S., and Kraemer, W. (2004). *Designing resistance training programs.* Champaign, IL: Human Kinetics.

Foran, B. (2001). *High-performance sports conditioning.* Champaign, IL: Human Kinetics.

Manore, M., and Thompson, J. (2000). *Sport nutrition for health and performance.* Champaign, IL: Human Kinetics.

Radcliffe, J., and Farentinos, R. (1999). *High-powered plyometrics.* Champaign, IL: Human Kinetics.

Sharkey, B. (2006). *Fitness and health.* Champaign, IL: Human Kinetics.

Wilmore, J., and Costill, D. (2004). *Physiology of exercise and sport.* Champaign, IL: Human Kinetics.

Sport Psychology

Anderson, M. (2000). *Doing sport psychology.* Champaign, IL: Human Kinetics.

Orlick, T. (2000). *In pursuit of excellence.* Champaign, IL: Human Kinetics.

Weinberg, R., and Gould, D. (2003). *Foundations of sport and exercise psychology.* Champaign, IL: Human Kinetics.

Sport Skills

Contact Human Kinetics (www.humankinetics.com) for a complete list of sport resources.

GLOSSARY

acclimatization—Adaptation to an environmental condition such as heat or altitude.

actin—Muscle protein that works with the protein myosin to exert force and produce movement.

adaptation—The cellular and system adjustments that take place as a result of training.

adenosine triphosphate (ATP)—Primary short-term source of energy to fuel muscular contractions.

adipose tissue—Tissue in which fat is stored.

aerobic—In the presence of oxygen; aerobic metabolism utilizes oxygen.

aerobic capacity—Maximal oxygen intake in liters per minute (L/min); correlated to non-weight-bearing performance (cycling, swimming) in high-intensity effort lasting 5 to 15 minutes.

aerobic fitness—Maximal ability to take in, transport, and utilize oxygen.

aerobic power—Maximal oxygen intake in milliliters of oxygen per kilogram of body weight per minute ($ml \cdot kg^{-1} \cdot min^{-1}$).

aerobic (endurance) sports—Sports requiring exceptional aerobic power, tending to last more than 15 minutes and deriving over 80 percent of the energy from aerobic energy pathways; for example cross country running.

agility—Ability to change direction quickly while maintaining control of the body.

alveoli—Air sacs in the lungs where oxygen and carbon dioxide exchange takes place.

amino acids—Building blocks that form proteins; different arrangements of the 22 amino acids form the various body proteins (muscles, enzymes, hormones).

anaerobic—In the absence of oxygen; referring to nonoxidative metabolism.

anaerobic glycolysis—The nonoxidative breakdown of glycogen to restore levels of ATP.

anaerobic power—Maximal all-out effort for several seconds.

anaerobic threshold—The transition from mostly aerobic metabolism to increasing anaerobic metabolism; from oxidative muscle fibers to fast-glycolytic fibers.

annual program—A program that guides the year's training.

anorexia—Eating disorder characterized by inadequate energy intake, weight loss, and eventual deterioration of organs.

assistant exercises—Exercises that have been identified as having a significant training effect for a particular sport.

atrophy—Loss of size of muscle; when muscle isn't used, it doesn't turn to fat, but atrophies.

balance—Ability to maintain equilibrium while stationary (static balance) or in motion (dynamic balance).

basic training period—Off-season training for strength and endurance.

blood pressure—Force exerted against the walls of the arteries.

blood volume—The amount of blood in the vascular system; blood volume is increased with endurance training.

bronchiole—Small branch of airway; sometimes narrows, making breathing difficult, as in exercise-induced bronchospasm.

bulimia—Eating disorder characterized by binge-purge cycles, sometimes with laxatives or diuretics.

calories—Amount of heat required to raise 1 kilogram of water 1 degree centigrade (same as kilocalorie).

capillary—Smallest blood vessels (between arterioles and venules) where oxygen, food, and hormones are delivered to tissues, and carbon dioxide and wastes are picked up.

carbohydrate—Simple (sugar) and complex (corn, rice, beans, potatoes, and whole-grain bread and pasta) carbohydrates are stored in the liver and muscle as glycogen and used to produce ATP.

carbohydrate loading—A procedure that elevates muscle glycogen stores before an endurance event.

cardiac—Pertaining to the heart.

cardiac output—Volume of blood pumped by the heart each minute; product of heart rate and stroke volume.

cardiorespiratory endurance—Or cardiovascular fitness; synonymous with aerobic fitness or the maximal oxygen intake.

cardiovascular system—Heart and blood vessels.

career plan—A long-term plan to achieve success in sport.

central nervous system (CNS)—The brain and spinal cord.

circuit training—A method of putting sets of exercises together such that an athlete does one set of each exercise before repeating sets of any exercise. An example is completing one set of dips, sit-ups, and then push-ups that may be repeated two or more times.

concentric contractions—Shortening of the muscle during contraction.

contraction—Development of tension by muscle: concentric—muscle shortens; eccentric—muscle is lengthened under tension; static or isometric—contraction without noticeable change in length.

cool-down—Light exercise performed at the end of a workout to hasten recovery.

core stability—Part of the chain in power development in throwing and other sports that require transfer of power, starting with the legs and moving upward to accelerate the arms.

correlation study—A study that determines the relationship between variables like strength and performance.

dehydration—Loss of essential body fluids that leads to reduced performance; extreme dehydration can be life threatening.

delayed-onset muscle soreness (DOMS)—Muscle soreness that begins 12 to 24 hours after very vigorous or unfamiliar effort and that may last for several days to weeks.

development—Bringing out the possibilities, both physical and psychological, to achieve one's potential. Physical properties include energy and muscular fitness and skill. Psychological development comes in areas of intelligence, strategy, persistence, concentration, and the ability to defer gratification.

dynamic balance—The ability to maintain equilibrium during vigorous movements.

dynamic constant external resistance (DCER)—See isotonic.

early competition period—Training period that involves a buildup to the important part of the season, the peak performance period.

easy training zone (EZ)—Low-intensity training zone, below the first lactate threshold, for endurance and recovery.

eccentric contractions—Lengthening of the contracting muscle, as in lowering a weight.

elastic recoil—Release of elastic energy in a muscular contraction, brought about by a brief stretch or preload. The preload leads to more contractile force without an appreciable increase in the energy cost of the contraction.

electrocardiogram (ECG)—A graphic recording of the electrical activity of the heart.

electrolyte—Solution of ions (sodium, potassium) that conducts electrical current. Electrolytes lost in sweat should be replaced in hot weather.

endomysium—Connective tissue that surrounds each individual muscle fiber. The connective tissue is continuous with the tendons connecting the muscle to bone. Connective tissue limits muscle length and thus limits flexibility.

endurance—Ability to persist, to resist fatigue.

energy balance—Balance of caloric intake and expenditure.

enzyme—An organic catalyst that accelerates the rate of chemical reactions in cells.

epinephrine (adrenalin)—Hormone from the adrenal medulla and nerve endings of the sympathetic nervous system; secreted during times of stress to help mobilize energy.

epimysium—Connective tissue that surrounds entire muscles. The connective tissue is continuous with the tendons connecting the muscle to bone. Connective tissue limits muscle length and thus limits flexibility.

evaporation—Elimination of body heat when sweat vaporizes on the surface of the skin. Evaporation of 1 liter of sweat yields a heat loss of 580 calories.

exercise—Sometimes applied specifically to calisthenics; usually denotes any form of physical activity, exertion, effort, and the like.

experimental study—Study in which the experimental variable should be randomly assigned to some subjects and not to others (controls). Results are analyzed statistically to determine if the experimental variable is significantly different between experimental subjects and controls.

fartlek—Swedish term meaning "speed play"; a form of training in which participants vary speed according to mood as they run through the countryside.

fast-glycolytic fiber (FG)—Fast-twitch muscle fiber with limited oxidative capabilities; easily fatigued.

fast-oxidative-glycolytic fiber (FOG)—Muscle fiber with oxidative and glycolytic capabilities.

fat—Important energy source for moderate-intensity effort; stored for future use when caloric intake exceeds expenditure.

fatigue—Diminished work capacity, usually short of physiological limits. In short intense effort, fatigue is due to factors within the muscle (short-term energy, pH); in long-duration effort, fatigue is due to glycogen depletion, low blood sugar, central nervous system fatigue, or a combination of these.

fatigue index—Sum of heart rates taken before, during, and after a standard exercise test.

FIT—Frequency, intensity, and time (or duration); parameters used to measure training.

flexibility—Range of motion through which the limbs or body parts are able to move.

glucose—Energy used by muscles, especially during vigorous effort; the essential energy source for the brain and nervous tissue.

glycogen—Storage form of glucose, found in liver and muscles.

growth—Refers to an individual's size. Sexual maturation provides hormones that assist achievement of growth.

heart rate—Frequency of contraction, often inferred from the pulse rate (expansion of artery resulting from the beat of the heart).

heat stress—Temperature and humidity combinations that lead to heat disorders such as heat cramps, heat exhaustion, and life-threatening heatstroke.

hemoglobin—Iron-containing compound in red blood cells that transports oxygen from the lungs to muscles and other tissues.

heredity—Genetic passage of traits from parent to offspring. Genotype is the genetic constitution, while the phenotype is the observable appearance resulting from the interaction of the genotype and the environment.

hypoglycemia—Low blood sugar (glucose).

immune function—Overall performance of the immune system, characterized by low incidence of upper respiratory tract infections.

individual response—Inherited and environmental factors that influence the response to training.

inhibition—Opposite of excitation in the nervous system.

inoculated—Protected because of prior exposure.

insulin—Pancreatic hormone responsible for getting blood sugar into cells.

intensity (of resistance training)—The amount of weight, generally in terms of RM (repetitions to failure) or percent of maximum (% 1RM). Since the speed of movement also influences the amount of weight that can be lifted, the best definition of intensity is the amount of work per time (power).

intermittent (start-and-go) sports—Sports that require bursts of energy interspersed with periods of less demanding effort. Examples include soccer, basketball, field hockey, and lacrosse.

interval training—Training that alternates short bouts of intense effort with periods of active recovery.

ischemia—Lack of blood to a specific area like a muscle.

isokinetics—Contractions against resistance that is varied to maintain high tension throughout the range of motion while speed remains relatively constant.

isometrics/static contractions—Contractions against immovable object (static contraction).

isotonics—Contractions against a constant resistance.

lactate threshold (LT)—Defines the transition from aerobic to increasing levels of anaerobic metabolism.

lactic acid (lactate)—By-product of anaerobic glycolysis.

lean body weight—Body weight minus fat weight (also called fat-free weight).

major exercises—Exercises that have the greatest influence on strength development.

maturation—Progress toward the mature state, be it sexual or skeletal.

maximal heart rate—A person's maximum number of heart beats per minute.

maximal oxygen intake—Aerobic fitness or the ability to take in, transport, and utilize oxygen; synonymous with cardiorespiratory endurance and $\dot{V}O_2$max.

maximal training zone (MZ)—Very high-intensity training.

metabolism—Energy production and utilization processes often involving enzymatic pathways. Anabolism denotes the building of body tissues, while catabolism is the breakdown of tissues.

mitochondria—Tiny organelles within cells; site of all oxidative energy production.

moderation—Avoiding overtraining since too much of anything can be bad for physical and psychological health.

motor area—Area of cerebral cortex that controls movement.

motorneuron—Nerve that transmits impulses from the central nervous system to muscle fibers.

motor unit—Motor nerve and the muscle fibers that it innervates.

muscle balance—Maintaining a balance in the strength between pairs of agonist (mover) and antagonist (opposing) muscles. A classic example is the necessary balance between the hamstring and quadriceps muscles of the leg.

muscle fiber—A single contractile muscle cell. In form, they are generally long, cylindrical cells that have repeating light and dark bands, giving it the name striated muscle.

muscle fiber types—Fast-twitch fibers are fast contracting but fast to fatigue; slow-twitch fibers contract somewhat more slowly but are fatigue resistant.

muscle soreness—Muscle discomfort that follows exercise. Delayed-onset muscle soreness (DOMS) peaks about 24 hours after very vigorous or unfamiliar effort.

muscular endurance—The ability to sustain submaximal contractions. Muscular endurance is different from whole-body aerobic endurance since it usually involves small versus large muscle groups.

muscular fitness—The strength, power, endurance, and flexibility needed to succeed in sport.

myofibril—Set of contractile threads of muscle composed of the proteins actin and myosin.

myoglobin—A hemoglobin-like compound in muscle that helps bind oxygen.

myosin—Muscle protein that works with actin to produce movement.

neuromuscular recruitment—Training the brain to recruit the muscle fibers necessary for coordinated movements. Also refers to the ability of the brain to increase the recruitment of existing muscle fibers during strength training so that strength is increased without muscle hypertrophy.

neuron—Nerve cell that conducts impulses; the basic unit of the nervous system.

no-training zone (NZ)—Training that is too hard for endurance or recovery, but not hard enough to raise the lactate threshold. Athletes should not spend much time in this zone.

obesity—Excessive body fat; over 20 percent for males, over 30 percent for females.

objectivity—If a test is scored separately by different scorers or judges, the likelihood that they independently agree.

official training season—The portion of the training year when coaches work with the athletes. This may be regulated by the league, state, or national association.

overload—A greater load than normally experienced; used to coax a training effect from the body.

overtraining—Excess training volume or intensity that leads to poor performance and could compromise immune function.

oxygen debt—Recovery oxygen uptake above resting requirements to replace deficit incurred during exercise.

oxygen deficit—Lack of oxygen in early moments of exercise and when effort exceeds the oxygen intake.

oxygen intake—Oxygen used in oxidative metabolism.

peaking—The process of preparing for best possible performances. See peak performance period and taper.

peak performance period—The period designed to produce peak performances; involves a decrease in training volume, an increase in training intensity, and sufficient rest to allow complete recovery.

perceived exertion—Subjective estimate of exercise difficulty (rating of perceived exertion or RPE).

percent maximum strength (% max)—The percent, in weight, of a maximal 1RM lift that an athlete is using. Repetition maximum can be related to %

maximum: In general, a 5RM lift is about 85% max, and a 10RM lift is about 70% max.

performance speed—The average speed that an athlete can maintain during the given event.

performance threshold (PT)—Current race speed (usually measured in meters/minute).

performance training zone (PZ)—Training at or slightly above race pace to improve performance.

perimysium—Connective tissue that surrounds bundles of muscle fibers. The connective tissue is continuous with the tendons connecting the muscle to bone. Connective tissue limits muscle length and thus limits flexibility.

periodization—Scheduled alterations in training load that ensure adequate time for recovery, or the process of building systematic variation into the training plan. This variation is programmed at several levels: daily, weekly, seasonal, and career.

peripheral nervous system—Parts of the nervous system not including the brain and spinal cord.

phosphocreatine (PCr)—Another name for creatine phosphate, a high-energy compound stored in muscle.

plyometrics—A popular method to develop power utilizing both neural and metabolic systems; training the preload and elastic recoil present in many sport skills.

PNF (proprioceptive neuromuscular facilitation) stretching—Method of stretching in which the athlete first stretches a muscle or muscle group, then contracts that muscle before again stretching.

potential—One's genetic capacity to improve with training.

power—The rate of doing work, or (force × distance)/time.

power endurance—Muscular endurance developed at sport-specific movement speeds.

power endurance sport—A sport that requires the ability to do repetitions of forceful contractions as quickly as possible; for example, wrestling.

power sport—A sport requiring a high level of muscular strength and power; for example, football, particularly linemen.

power training—Type of training that entails two general loading strategies: (1) strength training and (2) use of light loads (30-60% 1RM) performed at a fast contraction velocity, with 2 to 3 minutes of rest between sets with multiple sets per exercise. It is also recommended that emphasis be placed on multiple-joint exercises, especially those involving the total body.

precompetition period—The training buildup that precedes the start of the competitive season.

preload—See elastic recoil.

progression—Planned increases in training intensity, frequency, and time (duration), as well as moving from general to specific, from part to whole, and from quantity to quality.

progressive resistance—Training in which the resistance is increased as the muscle gains in strength.

protein—Organic compound formed from amino acids; forms muscle, tissue, enzymes, hormones, and so on.

psychological skills—Relaxation, concentration, imagery, and other learned skills that help athletes play their best game more often.

puberty—The period of life when sexual maturity or the ability to reproduce begins. Pubescent means arriving at or having reached puberty, and prepubescent defines the period prior to the beginning of puberty.

readiness—Physical and psychological preparedness for training.

recovery training period—Postseason period of rest and recuperation.

regression—The loss of adaptations achieved in training as a result of a cessation of training.

reliability—The likelihood of getting similar results if a test is repeated. A perfectly reliable test should give identical scores on repeated trials.

rep (repetition)—One complete cycle of an exercise, such as one bench press or one sit-up.

repetition maximum (RM)—The maximum number of times you can lift a given weight (1RM is the most you can lift one time).

respiration—Movement of oxygen from the atmosphere to the tissues via lungs and the circulation, and transport of carbon dioxide from the tissues back out to the atmosphere.

rest period (for strength training)—The amount of time allotted for recovery between sets. For the development of strength and power, longer rest periods are often beneficial as they allow for greater recovery and better effort during each set.

sarcomere—The contractile unit of the muscle.

sarcopenia—Loss of muscle mass, or vanishing flesh.

set—The number of reps completed continuously without a rest period, such as a set of 10 bench press reps or 30 sit-ups. A set is always followed by a rest period.

set-rep training—Method of organizing training into multiple repeated sets of one exercise before the next exercise, such as three sets of 10 bench presses followed by three sets of 10 squats. The resistance and number of repetitions may be adjusted for each set.

skill sport—Sport requiring high levels of skill, with only modest demands for muscular or energy fitness; for example golf, table tennis, shooting sports (archery and rifle), and bowling.

sliding filament theory—The theory that actin and myosin slide over each other to cause muscle movement.

slow-oxidative (SO) fibers—Muscle fibers that are well suited for endurance. They resist fatigue and are able to produce energy using oxidative sources.

slow-twitch fiber—See muscle fiber types.

somatotype—Body type. Ectomorph is linear or thin; mesomorph is muscular; and endomorph is fat.

specialty exercises—Exercises selected according to each athlete's individual needs.

specificity—Principle stating that training elicits responses in the particular muscle fibers utilized and the support systems, so training must be closely related to the desired outcome.

speed-assisted training—Any training tool that a coach can safely use to help an athlete practice speeds slightly faster than current race speed and that will help to increase neuromuscular coordination, thus helping athletes become more effective at using their muscular power.

speed of movement—Composed of reaction time, or time from stimulus to start of movement, and movement time, or time to complete the movement.

sport physiology—Study of the immediate- and long-term effects of exercise on the body.

static balance—The ability to maintain equilibrium in a stationary position.

static stretching—Form of stretching in which slow movements are used to the point of a moderate stretch. The position is held for 5 to 10 seconds, and then the athlete relaxes. Static stretching is the most widely accepted form of stretching for teams.

status (or comparison) study—Compares groups, such as athlete and nonathlete.

strength—Ability of muscles to exert force.

stress management—Use of exercise, meditation, and other techniques to avoid overreacting to stress.

stroke volume—Volume of blood pumped from ventricles during each contraction of heart.

supplementary exercises—Exercises that are carefully selected but less vital than major or assistant exercises.

sustainable aerobic fitness—Ability to continue an aerobic workload for extended periods of time.

synapse—Junction between neurons.

taper—A reduction in training volume lasting from several days to two weeks prior to important competitions to allow complete recovery and best possible performances.

tendon—Tough tissue that connects muscle to bone.

testosterone—Male hormone.

threshold—The minimal level required to elicit a response.

tonus—Muscle firmness in the absence of a voluntary contraction.

training effects—Physiological changes brought on by training.

training impulses (TRIMPS)—A method of estimating or quantifying the total stress of a training session based on intensity and time of the session; used to control the weekly training stress and reduce overtraining.

training periods—Phases of training, including the recovery, basic training, precompetitive, early competitive, and peak performance periods.

training program—A plan devised to achieve training effects.

training stimulus—The activity that brings about specific training effects.

training zones—The easy (EZ), no-training (NT), performance (PZ), and maximal training (PZ) zones.

unofficial training season—The period during the year when athletes are expected to train on their own and may not be allowed to train with a coach.

validity—The degree to which a test measures what it is supposed to measure.

variable resistance—Form of resistance that changes as muscle moves through range of motion, as with devices that use cams or oval-shaped pulleys.

variation—Principle stating that training must be varied to avoid boredom and to minimize overuse injuries.

velocity—Rate of movement or speed (distance/time).

ventilation—The amount of air moved in or out of the lungs per minute, the product of respiratory frequency (f) and tidal volume (TV).

ventricle—Chamber of the heart that pumps blood to the lungs (right ventricle) or to the rest of body (left ventricle).

volume (of training)—Number of lifts × resistance.

warm-up—Light exercise performed at the beginning of a workout to prepare the body for exertion.

weight training—Progressive resistance exercise using weights to impose a load on the muscle.

work—Product of force and distance.

REFERENCES

American College of Sports Medicine. (1996). Position stand: Weight loss in wrestlers. *Medicine and Science in Sports and Exercise*, 28, ix-xii.

Balke, B. (1963). A simplified field test for the assessment of physical fitness. Rept #63-3, Oklahoma City: Federal Aviation Agency.

Bannister, E. and Calvert, J. (1980). Planning for future performance: Implications for long term training. *Canadian Journal of Applied Sport Science*, 5, 170-176.

Bompa, T. (1990). *Theory and methodology of training*. Dubuque: Kendall/Hunt Publishing Co.

Costill, D., and others. (1976). Muscle fiber composition and enzyme activities of elite distance runners. *Medicine and Science in Sports and Exercise, 8*, 96-100.

DeLorme, T. and Watkins, A. (1951). *Progressive resistance exercise*. New York: Appleton-Century-Crofts.

Fox, E. 1984. *Sport physiology*. Philadelphia: Saunders College Publishing.

Frederick, E.C. (1973). *The running body*. Mountain View, CA: World Publications.

Groppel, J.L., and Roetert, E.P. (1992). Applied physiology of tennis. *Sports Medicine, 14*(4), 160-168.

Hoffman, J.R., and Kang, J. (2003, February). Strength changes during an in-season resistance-training program for football. *Journal of Strength and Conditioning Research, 17*(1), 109-114.

Holloszy, J. (1967). Biochemical adaptations in muscle: Effects of exercise on mitochondrial oxygen uptake and respiratory enzyme activity in skeletal muscle. *Journal of Biological Chemistry*, 242, 2278-2282.

Kemi, J., and others. (2003). Soccer-specific testing of maximal oxygen intake. *Journal of Sports Medicine and Physical Fitness, 43*, 139-144.

Kenitzer, R. (1998). Optimal taper period in female swimmers. *Journal of Swimming Research, 13*, 31-36.

Kirby, R. (1971, June). A simple test of agility. *Coach and Athlete*, 30-31.

Kraemer, W.J., and others. (2002). American College of Sports Medicine position stand. Progression models in resistance training for healthy adults. *Medicine and Science in Sports and Exercise, 34*(2), 364-380.

Lemon, P. (1995). Do athletes need more protein and amino acids? *Journal of Sports Nutrition, 5*, 39-61.

Lortie, G., Simoneau, J., Jobin, J., Leblanc, J., and Bouchard, C. (1984). Responses of maximal aerobic power and capacity to aerobic training. *International Journal of Sports Medicine, 5*, 232-236.

Margaria, R., Aghemo, P. and Rovelli, E. (1966). Measurement of muscular power (anaerobic) in men. *Journal of Applied Physiology*, 21, 1662-1664.

Martens, R. (2004). *Successful coaching*, 3rd ed. Champaiagn: Human Kinetics.

Nesser, T., Chen, S., Serfass, R., and Gaskill, S. (2004). The development of upper body power in junior cross country skiers. *Journal of Strength and Conditioning, 18*(1), 63-71.

Petibois, J., and others (2003). Biochemical aspects of overtraining in endurance sports: The metabolism alteration process syndrome. *Sports Medicine, 33*, 83-94.

Ridley, M. (2003). What makes you who you are? *Time*, 161, 54-63.

Siegler, J., Ruby B., and Gaskill, S. (2003). Changes evaluated in soccer-specific power endurance either with or without a 10-week, in-season, intermittent, high-intensity training protocol. *Journal of Strength and Conditioning Research, 17*(2), 379-387.

Smith, L. (2000). Cytokine hypothesis of overtraining: A physiological adaptation to excessive stress? *Medicine and Science in Sports and Exercise, 32*, 317-331.

Tanner, J. (1962). *Growth at adolescence*. Oxford: Blackwell Scientific.

Tcheng, T. and Tipton, C. (1973). Iowa Wrestling Study: Anthropometric measurements and the prediction of a "minimum" body weight for high school wrestlers. *Medicine and Science in Sports*, 5, 1-6.

Wathen, D. (1994). Muscle balance. In T. Baechle (Ed.), *Essentials of strength training and conditioning,* National Strength and Conditioning Association. Champaign, IL: Human Kinetics.

INDEX

Note: The italicized *f* and *t* following page numbers refers to figures and tables, respectively.

G

H

I

J

L

M

U

V

W

Z

ABOUT THE AUTHORS

Steven Gaskill and Brian J. Sharkey

Brian J. Sharkey brings nearly 40 years experience as a leading fitness researcher, educator, author, and consultant to *Sport Physiology for Coaches*. Sharkey served as director of the University of Montana's Human Performance Laboratory for many years and remains associated with the university and lab as professor emeritus. Through the university, Sharkey continues to do research on ultra-endurance athletes known as "wildland firefighters."

Sharkey is past-president of the American College of Sports Medicine and served on the NCAA committee on competitive safeguards and medical aspects of sports, where he chaired the Sports Science and Safety subcommittee, which uses research and injury data to improve the safety of intercollegiate athletes.

He is also a consultant with the U.S. Forest Service in the areas of fitness, health, and work capacity. He received the U.S. Department of Agriculture's Superior Service Award in 1977 and its Distinguished Service Award in 1993 for his contributions to the health, safety, and performance of firefighters.

Sharkey has written or co-written numerous books, many published by Human Kinetics, including *Coaches Guide to Sport Physiology*, *Training for Cross Country Ski Racing*, and *Fitness and Health* (with Gaskill).

In his leisure time, Sharkey enjoys cross-country skiing, mountain biking, running, hiking, and canoeing. He and his wife, Barbara, live in Missoula, Montana.

Steven Gaskill is in the Department of Health and Human Performance at the University of Montana. He has authored or co-authored two previous books published by Human Kinetics: *Fitness Cross Country Skiing*, and *Fitness and Health* (with Sharkey).

Gaskill worked for the U.S. Ski Team for 10 years—as head coach of the Nordic Combined (ski jumping and cross-country skiing) and Cross-Country teams and as director of the coaches' educational programs. Gaskill has coached at three Olympic Games, and twenty skiers who have trained under him have competed in the Olympics. In 1992, the U.S. Ski Association named him the U.S. Cross-Country Coach of the Year.

Gaskill was the founder and first director of Team Birke Ski Education Foundation, which is dedicated to the development of excellence in cross-country ski programs for skiers of all ages. He also has published more than 200 articles about cross-country skiing and has produced six instructional videotapes. He has presented his extensive research findings on training for cross-country skiing to the American College of Sports Medicine and has written a major article for Medicine and Science in Sports and Exercise.

Gaskill lives in Burnsville, Minnesota, with his wife, Kathy. His favorite leisure activities include cross-country skiing, hiking, and mountaineering.

The authors met in 1980 when Gaskill was coach and Sharkey sport physiologist for the U.S. Ski Team. Eventually they went their separate directions, but in 1998, Gaskill applied for a position at the University of Montana, where Sharkey was retiring after 30 years. Since then they have renewed their association, conducting research and development activities in the Human Performance Laboratory and in the field.